# Statistical Curves and Parameters

# Statistical Curves and Parameters:
## Choosing an Appropriate Approach

Michael E. Tarter

A K Peters
Natick, Massachusetts

Editorial, Sales, and Customer Service Office

A K Peters, Ltd.
63 South Avenue
Natick, MA 01760

Copyright © 2000 by A K Peters, Ltd.

All rights reserved. No part of the material protected by this copyright notice may be reproduced or utilized in any form, electronic or mechanical, including photocopying, recording, or by any information storage and retrieval system, without written permission from the copyright owner.

**Library of Congress Cataloging-in-Publication Data**

Tarter, Michael E.
   Statistical curves and parameters : choosing an appropriate approach / Michael E. Tarter.
       p. cm.
   Includes bibliographical references and index.
   ISBN 1-56881-105-5 (alk. paper)
     1. Mathematical statistics. I. Title.

QA276.T28 2000
519.5--dc21                                                  99-058803

Printed in the United States of America
03 02 01 00                    10 9 8 7 6 5 4 3 2 1

# Contents

Preface....................................................... xi

## 1 Introduction ............................................ 1
1.1 Background ............................................. 1
1.2 A fictional example .................................... 4
1.3 Curves and statistical history ......................... 5

## 2 Model and Distribution Terminology ...................... 9
2.1 Modeling background .................................... 9
2.2 Representative number .................................. 10
2.3 Curve types ............................................ 13
2.4 Distribution and data terminology ...................... 16
2.5 Parameter validity and property existence .............. 18
2.6 Estimator terminology .................................. 21
2.7 Degenerate curves ...................................... 23

## 3 Variability and Related Curve Properties ................ 25
3.1 Uncertainty and variability ............................ 25
3.2 The absolute deviation curve property .................. 26
3.3 The general AD and the ADM curve properties ............ 28
3.4 Curve property selection ............................... 33
3.5 The history of variability appreciation ................ 36
3.6 Simplistic approaches and the history of probability ... 37

## 4 Moments and Curve Uncertainty ........................... 41
4.1 E and Var Geometry ..................................... 41
4.2 Higher order moments and the indicator function ........ 45
4.3 Early statistical models ............................... 47

v

|     |     |     |
| --- | --- | --- |
| 4.4 | Early statistical models and higher order moments | 50 |
| 4.5 | Curve sub-types and model choice | 52 |

# 5 Goodness of fit — 55

- 5.1 Neyman's and alternative criteria . . . . . . . . . . . . . . . 55
- 5.2 Criteria, metrics and estimators . . . . . . . . . . . . . . . 61
- 5.3 The Kolmogoroff-Smirnoff criteria . . . . . . . . . . . . . 64
- 5.4 Bernoulli variation and the Cauchy density . . . . . . . . 65
- 5.5 Comparative goodness of fit . . . . . . . . . . . . . . . . . 71

# 6 Variates, Variables and Regression — 73

- 6.1 Variates and variables . . . . . . . . . . . . . . . . . . . . 73
- 6.2 Variates and subjects . . . . . . . . . . . . . . . . . . . . 75
- 6.3 Expressions, algorithms and life tables . . . . . . . . . . . 76
- 6.4 Distinctions between curve types . . . . . . . . . . . . . . 77
- 6.5 Curve properties and symbols . . . . . . . . . . . . . . . . 79
- 6.6 Variates, variables, and $E_f(Y|x)$ regression . . . . . . . . 80
- 6.7 $\mu(x), E_f(Y|x)$ and regression alternatives . . . . . . . . 83

# 7 Mixing Parameters and Data-generation models — 89

- 7.1 An introduction to data-generation models . . . . . . . . . 89
- 7.2 Error, regression, and probit, models . . . . . . . . . . . . 91
- 7.3 Regression and data-generation models . . . . . . . . . . . 93
- 7.4 Probability, proportion, and data-generation models . . . . 94
- 7.5 The generation of contagious model and mixture model data  99

# 8 The Association Parameter $\rho$ — 103

- 8.1 Response, key, and nuisance, variates . . . . . . . . . . . . 103
- 8.2 The association parameter $\rho$ . . . . . . . . . . . . . . . . 105
- 8.3 Conditional, joint and marginal, notation . . . . . . . . . 110
- 8.4 The sample and the population correlation coefficient . . . 114
- 8.5 Correlation geometry . . . . . . . . . . . . . . . . . . . . 116

# 9 Regression and Association Parameters — 123

- 9.1 The curse of dimensionality . . . . . . . . . . . . . . . . . 123
- 9.2 Multiple variable interdependence . . . . . . . . . . . . . 128
- 9.3 Logit and linear models . . . . . . . . . . . . . . . . . . . 132
- 9.4 Dual regression functions . . . . . . . . . . . . . . . . . . 136

## 10 Parameters, Confounding, and Least Squares — 141

- 10.1 Ideal objects .............................. 141
- 10.2 Linear data-generation models and mixture models ..... 146
- 10.3 Parameter distinctiveness .................... 148
- 10.4 Representational uniqueness and model fitting ......... 150
- 10.5 Model-fitting considerations .................. 151
- 10.6 The variance curve property and bathtub functions ..... 154
- 10.7 Regression and least squares .................. 155

## 11 Nonparametric Adjustment — 159

- 11.1 Age-adjustment and logistic regression ............ 159
- 11.2 Crude and specific rates ..................... 163
- 11.3 Age-adjustment; marginal, joint, and conditional curves .. 164
- 11.4 Age-adjustment and partial correlation ............ 166
- 11.5 Direct and indirect adjustment ................. 168
- 11.6 The computation of adjusted rates ............... 172

## 12 Continuous Variate Adjustment — 175

- 12.1 Observed and expected rates .................. 175
- 12.2 Trivariate data-generation and additive regression models . 177
- 12.3 Regression and data generation ................. 179
- 12.4 Correlation, regression, and nuisance variables ........ 180
- 12.5 Trivariate Normality graphics ................. 185

## 13 Procedural Road Maps — 189

- 13.1 The organization of statistical data and statistical methods 189
- 13.2 Log and log(-log) transformations ............... 193
- 13.3 Methodological alternatives................... 196
- 13.4 Conditional and joint density models............. 198

## 14 Model-based and Generalized Representation — 203

- 14.1 Multiple properties and parameters .............. 203
- 14.2 Specification and generalized representation ......... 208
- 14.3 Identifiability of generalized versus extended model representation ........................ 210
- 14.4 The $E(X)$ curve property's relationship to location and scale 214

## 15 Parameters, Transformations, and Quantiles — 217

- 15.1 Location and scale parameter representation of continuous variates ................................. 217
- 15.2 $\rho$-focused transformations and $\sigma$-focused transformations .. 220

| | |
|---|---:|
| 15.3 Quantiles, quartiles, and box-and-whisker plots | 222 |
| 15.4 Normal ranges and box sizes | 226 |
| 15.5 Confidence bands and prediction bands | 228 |
| 15.6 Notches, stems, and leaves | 229 |
| 15.7 The log transformation and skewness | 232 |

## 16 Noncentrality Parameters and Degrees of Freedom — 237

| | |
|---|---:|
| 16.1 The $(\mathcal{C}_1\|\mathcal{A}_2)$ case and variate-variable relationships | 237 |
| 16.2 Invariance and confounding | 243 |
| 16.3 ANOVA tables and confounding | 247 |
| 16.4 Contingency tables and the parameter $\nu$ | 249 |
| 16.5 Student-t and Cauchy densities | 251 |

## 17 Parameter-Based Estimation — 255

| | |
|---|---:|
| 17.1 Likelihood and BLU estimation | 255 |
| 17.2 Censoring and incompleteness | 256 |
| 17.3 Outliers and errors | 258 |
| 17.4 Ordered variates and subscripts | 260 |
| 17.5 BLU estimators | 261 |
| 17.6 BLU estimation and censoring | 265 |
| 17.7 BLU estimators and alternatives | 269 |

## 18 Inference and Composite Variates — 273

| | |
|---|---:|
| 18.1 Curves and composite variates | 273 |
| 18.2 Specific sampling distributions | 276 |
| 18.3 The mean's variance formula and mixtures | 281 |
| 18.4 Inference and a two-valued metric | 282 |
| 18.5 The one tail z-test | 286 |

## 19 Parameters and Test Statistics — 293

| | |
|---|---:|
| 19.1 The parameter $\Delta$ | 293 |
| 19.2 Power and efficiency | 295 |
| 19.3 Power and test considerations | 298 |
| 19.4 The sample mean and the sample median | 302 |
| 19.5 Tables and the details of test construction | 304 |
| 19.6 Power, efficiency and BLU estimators | 306 |

## 20 Curve Truncation and the Curve e(x) — 311

| | |
|---|---:|
| 20.1 Expectation as a limit and the effects of truncation | 311 |
| 20.2 Truncation symmetry | 313 |
| 20.3 Truncation and bias | 317 |

|       | 20.4 Truncation and the curve e(x) . . . . . . . . . . . . . . . . . | 327 |
|---|---|---|
|       | 20.5 When are curve properties relevant and when are model parameters relevant . . . . . . . . . . . . . . . . . . . . . . . . | 326 |

## I  Models and Notation                                              329

    I.1  Notation historical background . . . . . . . . . . . . . . . . 329
    I.2  Specific models, the Normal . . . . . . . . . . . . . . . . . . 333
    I.3  Specific models, lognormals and related curves . . . . . . . 336
    I.4  Model families . . . . . . . . . . . . . . . . . . . . . . . . . 340
    I.5  Mixtures and Bayesian statistics . . . . . . . . . . . . . . . 342
    I.6  Notational conventions about moments and variates . . . . 345

## II  Variate Independence and Curve Identity                          347

    II.1  Independence and identical distribution . . . . . . . . . . . 347
    II.2  Regression notation . . . . . . . . . . . . . . . . . . . . . . . 350

## III  General Statistical and Mathematical Notation                   355

## References                                                          367

## Index                                                               375

# Preface

Although the topics discussed are elementary, this is in no way an elementary statistics textbook. Instead, it was designed as a graduate text and reference guide for researchers and students who are unfamiliar with the recent statistics literature but who have at least a one-semester background in calculus. This background is needed to consider advances made by pioneer researchers like R. A. Fisher with an eye to reconsidering their inquiries from nonparametric and computationally intensive viewpoints.

An early version of this material was first tried out at a Summer Session in Biostatistics offered in Seattle, Washington, in 1970 and later at two American Statistical Association short courses as well as a recent summer session in statistics sponsored by the Istituto di Metodi Quantitativi, Universita Commerciale, L. Bocconi, Milano. These four courses and twenty-eight years of teaching at Berkeley were based on a "full circle" approach, where exercises began with student-or-researcher-specified parameter values and ended with estimated curves together with these curves' estimated parameter values.

Calling upon data simulated in this way gets around a didactic problem. In contrast to issues addressed by many other disciplines, statistical questions about natural data seldom have clear-cut answers. Multiple interpretations and multiple data pathologies are likely to occur in what may seem initially to be straightforward data analyses. As a consequence, when learning how to deal with realistic data it is often advisable to begin with artificial samples whose complexity is subject to the instructor's and the learner's control. Besides the advantages of self-checking, the random number generation capabilities of modern statistical and mathematical program systems can be called upon to interpret difficult distinctions, such as those between probability and proportion as well as between curve properties and model parameters.

The following teaching experience occurred soon after my return to Berkeley from the 1970 Seattle Summer Session. Our *Risk and Demographic Statistics* class was then based, in part, on a computationally complicated exercise designed to illustrate the distinctions and similarities between census and registry data. One student, Jansin Lee, asked if she could construct a computer program to do the exercise in lieu of conventional calculation. There was at that time a research project underway at some other institution whose project goals included the construction of just such a program. However, when other students vouched for Jansin's computational skills she was given the go ahead to at least try. Jansin Lee's success in writing a census-registry data program convinced me that the implementation strategy phase of statistics had changed forever. Emphasis had shifted from the question, "How can a procedure be implemented?" to "Which procedure—out of a great many alternatives—should be chosen?" To answer this question, careful regard must be given to underlying assumptions.

There are of course some downsides to this change in emphasis from how to which. Successful automation of a technique does not translate necessarily to an understanding of basic concepts. In addition, multiple computer programs that do the same thing can have individual characteristics that distract users' attention away from crucial (epistemological, existential, and aesthetic) concepts.

When it comes to modern views about the philosophy of knowledge, existence or beauty, I cannot claim to have more than amateur status. Yet many years experience with statistical practice have provided the motivation to cut to the conceptual core. For example, I view a single model presented at the beginning of Chapter 6 as the trunk from which three distinct classifications of statistical uncertainty branch. The first branch is shaped by the choice of density model, the second by the choice of regression model, while lurking variables and their effects on model specification incompleteness weight down the third.

Despite the diverse branches of statistical curve uncertainty, a single idea serves to unify many seemingly diverse concepts. This is the distinction introduced in Chapter 3 between parametric model validity and curve property existence. It is argued that this seemingly esoteric issue is fundamental to choices between parametric and nonparametric approaches that have major practical consequences.

The force of this and other arguments was augmented by able editorial assistance from Alice Peters and Sarah Gillis. Comments were also forwarded by Florence Morrison and Robin Tarter. In addition, Robin was able to correspond in French with the Muse Des Augustins to obtain permission to reproduce their portrait of young René Descartes. The International Programs Center, U.S. Bureau of the Census, generously granted permission to reproduce their animated population pyramid display. Drs. Joe Brown and Andrew Salmon helped provide funding from The State of California, Interagency agreement #97-E0032, while Drs. James M. Meyers and John Miles did the same from PHS CCU912911-01 and other sources. Many students, researchers, and professors provided encouragement and specific comments. Among these, David R. Brillinger, David G. Hoel, Elizabeth H. Margosches, Thomas E. McKone and Robert C. Spear are due special thanks.

# CHAPTER 1
# Introduction

## 1.1 Background

The first sentence of I. R. Savage's book, *Statistics: Uncertainty and Behavior* (1968), is: "Probabilities are quantitative expressions of uncertainty." Thus, for Savage, the term uncertainty denotes a more general concept than the principle of quantum mechanics that earned W. K. Heisenberg a 1932 Nobel Prize. But what exactly is statistical uncertainty? One of the many possible answers to this question concerns what Savage calls "quantitative expressions."

Curves *per se* and quantitative expressions for curves can be matched well or matched poorly. For this reason, we place much emphasis on the distinctions and connections between model parameters and curve properties. Parameters are symbols within some mathematical expression that may vary between problems, but not within a given context or problem. In this sense they are representational devices that are intermediates between constants and variables.

An early recorded application of a parameter was published by the seventeenth century mathematician and philosopher Ren Descartes (Bell, 1937). Descartes realized that the concept *diameter*, when applied to a circle, resembled the *para*meter of a *para*bola, the curve property "latus rectum." Then, as now, the notational linkage of two different entities, the first a symbol and the second a curve property like the diameter or latus rectum of a particular curve, was a major advantage of the parametric approach.

The word *parameter* sounds like the word *perimeter*, yet the role played by any parameter is contrary to that played by any perimeter. Rather than defining boundaries, parameters extend flexibility. When a mathematical expression does not contain any parameters, it is completely specified to the degree that it can be tabled. On the other hand, when such an expression contains unspecified parameters, it corresponds to a whole spectrum or family of tables, one for each distinct set of parameter values.

Figure 1.1. René Descartes.

For any statistical analysis, it was once considered essential to reduce the problem under consideration to the estimation or testing of a handful of parameters or curve properties. Today, because modern computational procedures can provide thousands of times more representational scope than was available merely a few years ago, this reduction is less important than it once was. There now is a rich palette of distinct model, parameter, and curve property combinations.

As is natural when there are many strategies possible, some may be more appropriate or more effective than others. In addition, some may be more difficult than others from a computational or conceptual viewpoint. To choose among strategies, one can call upon the venerable fourteenth century approach called the *law of parsimony*: one can adopt "...the simplest assumption in the formulation of a theory or in the interpretation of data, especially in accordance with the rule of Ockham's razor" (*Compton's Interactive Encyclopedia*, 1993, 1994). Fortunately, when one wants to trim the stubble of statistical uncertainty today, some computer can serve as Ockham's electric razor.

At least in theory, today's graphical techniques can disclose how much representational scope is needed, while numerical techniques can provide the scope. Yet, in practice, if one is not to be bewildered by today's computational riches, many distinctions must be carefully made. As far as statistical uncertainty is concerned, particular care must be taken to align the parameters within the particular expression chosen to represent a curve with the true properties of the curve.

Certainty about curve and model choice is a concept that differs substantially from the idea of estimator optimality. But exactly what, in general, models and their component parameters are, is not obvious. For example, because parameters by convention are treated as individual or dis-

## 1.1. Background

crete entities, it would seem strange to encounter a model that within its symbolic representation contains three and a half parameters. Certain parameters however (such as those discussed in the chapter that deals with $\mu_{y|x} = \mu(x)$ regression) are taken to be curves, and these curves assume a continuous sequence of different values for every real number $x$ within a finite or even infinite interval. In particular, suppose a second parametric curve, such as $\sigma_{y|x} = \sigma(x)$, is defined. Then even when the values $x_1$ and $x_2$ are distinct, the quantities $\mu(x_1), \mu(x_2), \sigma(x_1)$ and $\sigma(x_2)$ are taken as partial descriptions of two parameters, $\mu$ and $\sigma$, not four parameters; $\mu(x_1), \mu(x_2), \sigma(x_1)$ and $\sigma(x_2)$. Here $\mu(x_1), \mu(x_2), \sigma(x_1)$ and $\sigma(x_2)$ are four curve evaluations, two per curve, while each of the two curves, $\mu(x)$ and $\sigma(x)$, corresponds to a distinct and unique parameter denoted by a particular Greek letter. The correspondence between parameters and symbols (such as Greek letters) explains why there cannot be half parameters. Bisecting any given symbol creates a second, distinct, symbol.

Occasionally the word *parameter* denotes what are referred to below as "variates" and "variables." However, this usage can cause confusion when, for example, researchers model risk as a product of variates, such as the amount of air taken into the lungs during one typical breath, or the amount of mercury in a gram of cooked fish. Taking air intake and mercury level as two distinct parameters differs from the statistical viewpoint that air and mercury variates have distributions and these distributions are described by many individual curves where, in turn, each of these curves is represented in terms of a particular set of parameters. That one or more entities called parameters may themselves be represented as curves, such as $\mu(x)$ or $\sigma(x)$, is complicated enough. To interchange it for one of the two terms, variate or variable, places too much of a representational burden on the single word parameter.

When it is a part of the expression for a particular curve, a particular parameter often helps to describe a variate's distribution. Thus a parameter can help specify either, or both, of two entities, a distribution or a curve. In light of this duality, it is confusing when a third or even fourth entity, a variate or a variable, is referred to as a parameter. It is best to conceptualize a variate as an entity described by a curve and, in turn, conceptualize this curve's model as an entity described, at least in part, by some parameter's value.

In early statistical texts the term "population characteristic" designated that a symbol might, depending on the context, represent either a parameter or a curve property. However, because uncertainty issues often concern the differences between model parameters on the one hand, and curve properties on the other, the term population characteristic will not appear below. Uncertainty about some curve *per se* can be quite different from uncertainty about which model provides accurate representation of this curve. The former concern addresses property existence, while the latter concerns model validity. A major thesis of this book is that existence issues are not mere theoretical curiosities. Quite the contrary, addressing these issues can avert data simulation and data analysis disasters.

The difference between the concept curve itself and the expression for a curve, is illustrated by the two equations: $sin^{-1}[sin(x)] = x$ and $(x^{-1})^{-1} = x$, which are both valid for $x$ values that satisfy the inequalities $0 < x < \pi/2$. Two -1 symbols appear within $(x^{-1})^{-1} = x$ and only one within $sin^{-1}[sin(x)] = x$. Among the symbols that form the expression $(x^{-1})^{-1}$, the exponent $^{-1}$ denotes the reciprocal operation while, on the other hand, the -1 that appears after the first sin of $sin^{-1}[sin(x)]$ helps denote a particular type of curve, the inverse sine function. A curve is often represented by an expression, but this expression is no more the curve itself than a photograph of a person is the person. It follows that curve notation and expression notation can have subtle, but important, differences.

It is easy to gloss over the distinction between (1) the curve and (2) the expression for the curve. In particular, there is little confusion concerning an expression for a curve designated by $([1 - x]^{-1})^{-1}$, since the minus sign within $1 - x$ obviously designates the operation subtraction. Yet the false simplicity of taking (1) and (2) to be the same thing is at odds with current computational and statistical thinking. In particular, if parameters as components of expressions and curve properties were the same thing, there would be no such fields of study as nonparametric statistics and model-free curve estimation.

## 1.2 A fictional example

The curve-based approach is illustrated by a somewhat far-fetched example. Both the original and the "Next Generation" series of the popular television show *Star Trek'* made frequent use of a fictional prop called a "replicator." As its name suggests, this device served as an extension of a paper copier or laser printer. Rather than merely providing paper copies, however, its fictional output consisted of solid items ranging in size from a small tea cup to a very large pumpkin.

Granted that eons from now replicator technology will eventually come of age, it is not too far-fetched to contemplate a future replicator-designer deciding between two strategies with which to process raw input. A large set of computerized plans or templates could be stored by means of which the basic features of a tea cup, a pumpkin, or other objects could be altered to suit immediate needs. This strategy is nothing more than an extrapolation of the process whereby a pair of slacks, or one of the many other items of clothing sold off-the-shelf, are altered after purchase.

From a statistical point of view, both replicator-template raw materials and an off-the-shelf item correspond to a partly specified model or expression. Alternatively, there is an approach that is radically different from the replicator template. By executing many repeated similar operations, the desired tea cup, pumpkin, or statistical equivalent, could be assembled from chemical compounds or even elements from scratch, so to speak. Because its output could conform to complex requirements, the from-scratch approach is more flexible than the template approach. For example, it could process the request for a teacup shaped like a pumpkin. Of course, from-scratch design and implementation is more demanding technically than the template approach.

Flexibility is called for in dealing with curve uncertainty, because curves targeted by data analyses are usually hybrid analogs of teacups that look like pumpkins. In practice, truly bell-shaped and smooth-contoured curves are rare. This motivates approaches based on expressions that can represent many disparate statistical structures accurately, yet a great deal can nevertheless be said in favor of conventional parametric modeling approaches, especially when they serve as an introduction to—and even when they don't serve as full length descriptions of—some scientific enterprise.

## 1.3 Curves and statistical history

Suppose a curve is constructed on a blank pad of lined paper by pressing a pen or pencil above the highest line, called $y = 0$, and with a steady rightward motion, without crossing the horizontal line $y = 0$, tracing the curve with "argument" $x$, $y = h(x)$. Then even a low-power magnifier will show that what has been drawn has much more in common with a ribbon than it has with either of the mathematical concepts of *curve* or *function*.

Six symbols, $y, =, h, (, x$ and $)$, form $y = h(x)$. These, as well as the putative curve we just tried to draw, are both grist for no less an authority than Humpty Dumpty, when he "...said in rather a scornful tone, 'it means just what I choose it to mean—neither more nor less' " (Carroll, 1872). Were we to look closely, for instance, at the $h$ symbol, we would see a ribbon-like representation that is similar to a magnified view of the hand-drawn curve. It is only because of many years of experience with the letter $h$ that we are more comfortable with this ribbon-like entity being an $h$ than we are that the hand drawn entity is a curve. In reality, both the $h$ and the curve are symbols. The difference between them is not due to one being symbolic and the other being something else. Instead the difference is that the hand-drawn curve symbolically represents something that may or may not exist while, in the case of some model, the multiple symbols by which a curve is expressed may or may not provide a valid representation or, for the selection of some lower-case Latin letter like $h$ that falls in the middle of the alphabet, a representation that is commonly recognized usage.

Many notational conventions can be traced back to games of chance studied in the seventeenth century by Pascal and Fermat. The combinatorial gambling analyses of the seventeenth and eighteenth centuries were followed in the nineteenth century by what was then called "the theory of errors." Unfortunately, these early investigations happen to have led to some of the most equivocal controversies that currently burden statistical notation and practice. This burden stems from the close connection between the probabilities intrinsic to games of chance and what today is called the "frequentist" interpretation of probability. Consequently, to introduce previously unexposed minds to seventeenth century statistical thinking, risks an overemphasis on only one out of what are today taken to be a variety of alternative perspectives.

The remainder of this introductory chapter and Chapter 2 both deal informally with basic concepts such as probability and frequency. Instead of mathematical and statistical intricacies, emphasis is placed on notation; the relationships among, and distinctions between, different types of curves; and the relationships between particular curve properties and particular model parameters.

When the word "type" precedes the word "curve," this combination has a special interpretation. For example, the bell-shaped Normal and many other models are pressed into service to represent a single type of curve, the statistical density $f$. Since the values assumed have different interpretations in terms of probability, the $f$-type curve differs from a second type of curve, represented typically by the symbol $F$.

Individual statistical models often are named for mathematicians, physicists and statisticians such as Gauss, Laplace, Weibull and Cox. On the other hand, curve types are usually distinguished from each other by particular choices from among a few letters and punctuation marks, as in the curve types $f(x), f(x,y), f(y|x), F(x), \mu(x)$ and $E(Y|x)$. A notable exception, however, is the Normal, bell-shaped, or Gaussian model. This model

$$\phi(x) = e^{-(1/2)x^2}/\sqrt{2\pi}$$

appears in the statistical literature often enough to be assigned its own Greek letter, $\phi$. In addition, with two exceptions, curve types are rarely distinguished from each other by the choice of particular Greek letters. These exceptions are the curve types $\mu(x)$ and $\sigma(x)$.

## 1.3. Curves and statistical history

Because emphasis is put on the differences and commonalities of curve types, a conscious attempt has been made to avoid presenting explicit formulas for statistical models of a single curve type. The move away from an emphasis on specific closed-form mathematical models to an emphasis on flexible and general forms of expression began during the first quarter of the nineteenth century.

Detailed "model" formulas can be found by consulting some choice of what Spanier and Oldham (1987) call an "atlas of functions," such as the Johnson and Kotz (1970) *Distributions in Statistics* series. It makes as much sense to encourage the memorization of these formulas as it would to encourage a geography student to memorize the entire content of a large world atlas. Minute details act as distractions during the search for answers to basic questions, questions that pertain to curve types and not to differences between individual models.

Two curve types differ if they play distinct roles. On the other hand, whether a curve like $\mu(x)$ is described as a parabola or as a straight line is a representational issue that affects what this curve discloses about some particular distribution. The word "location" is often identified with a curve such as $\mu(x)$, while the word "scale" is associated with the curve $\sigma(x)$. Yet both $\mu(x)$ and $\sigma(x)$ might be lines, and might even be represented by the same sort of line, say for $\mu(x) = c_1$ and $\sigma(x) = c_1$, as horizontal straight lines. Yet even if $\mu(x) \equiv \sigma(x)$, in the mathematical sense of being represented by the same curve, $\mu(x)$ and $\sigma(x)$ would still be two different types of curves.

Later chapters will consider the reasoning that underscores the distinguishing characteristics of $\mu(x)$ and $\sigma(x)$, as well as the differences between the two curves $\mu(x)$ and $E(Y|x)$ and many other regression alternatives. Four different interpretations of the term "correlation" and three interpretations of the term "bias" will also be discussed.

Thanks primarily to modern computational power, the number of alternative statistical procedures has increased vastly over recent years and presents what can amount to a bewilderment of riches. Much of this book is designed to help sort out these riches. During the era when problems presented in statistics courses often began with the statement, "Suppose one has a sample that consists of $n$ Normally distributed variates...", distinctions between conceptualizations like $\mu(x)$ and $E(Y|x)$ seemed irrelevant. Today, the identity of curves such as $\mu(x)$ and $E(Y|x)$ in the multivariate Normal case can be viewed as a unique crossroads where two well-traveled alternate routes come together.

# CHAPTER 2
# Model and Distribution Terminology

## 2.1 Modeling background

Some data analysis difficulties can be traced to curves themselves and others to the models chosen to represent the curves. Although what is meant by an incorrect model is easy to explain, problems related to curve choice per se, and not model choice, are more difficult. For instance the distinction is made below between continuous and discrete variates in terms of what are referred to as *types* of curves. The difficulties that result from this distinction are harder to study than difficulties that are due to the incorrect choice of a model, like that of the Normal curve.

Some background review can be a helpful preliminary to the examination of the subtle effects of curve-type distinctions. What Blackwell and Girshick (1954) call "the theory of games and statistical decisions" is based on earlier work by Wald (1939) that emphasized general criteria, such as the decision rules and metrics. Concurrent with these advances, Bayesian and subjective probability methodology reconsidered problems once exclusively approached by what are now called *frequentist* methods (Savage 1968).

When computational and graphical innovations enabled statisticians to view data structures comprehensively, this, in turn, motivated robust approaches that are insensitive to departures from the assumptions that underlie procedures. Statistical *robustness* refers to some right answer gotten for the wrong reasons. A robust statistical procedure is insensitive to departures from the assumptions on which it was based, such as the choice of a particular model for the curve called the *statistical density $f$*.

A robust technique can be thought of as one that is insensitive to data messiness. If some conventional technique can be called into service, then why go to the trouble to devise some new procedure? The problem is, even the most trustworthy of techniques will be robust in some respects and yet respond adversely in others. As far as procedures that predate today's intense interest in departures from simplistic assumptions are concerned, an excellent text that deals with probability and other basic concepts has been written by Cramer (1945). Many terms discussed below are concisely defined in Kendall and Buckland's (1960) and Marriott's (1960) dictionaries. The *Encyclopedia of Statistical Sciences*, edited by Kotz (1998), reviews a considerable portion of the statistical literature that has appeared since the publication of the Kendall and Buckland, and the Marriott dictionary editions.

## 2.2 Representative number

One variety of statistical uncertainty can be interpreted in terms of two curves, $f_1$ and $f_2$, where $f_2$ describes the distribution of values that are often called "outliers," while $f_1$ does the same for these outliers' non-messy counterparts. Usually when the word outlier is applied, which value goes with which curve, $f_1$ or $f_2$, is not known exactly. Now suppose that in this section the Greek letter $\mu$ temporarily designates a "representative number." A problem then arises. Does $\mu$ help to represent $f_1$ or, instead, does it help to represent the curve that will be discussed extensively below that is called the *mixture* of $f_1$ and $f_2$?

One advantage that the interpretation of $\mu$ as a location parameter has over older interpretations is that in the mixture of $f_1$ and $f_2$ context, two parameters can be assigned, as in $\mu_1$ to $f_1$ and $\mu_2$ to $f_2$. Thinking of $\mu$ as a representative number, measure of central tendency, or even the quantity called the first ordinary moment (discussed extensively below), makes it difficult to deal with mixture-type uncertainties.

The parameter $\mu$ has had a lengthy history. For example, the plots of Anthony Trollope's mid-nineteenth century Barsetshire novels hinge on the extrapolation of quantity specified in a will. Suppose this quantity was 5 English pounds (circa AD 1440). The determination of what this quantity is equivalent to today can be thought of as a distillation process whose output is a single number. This raises the question: What are conceptual, and actual, distillation process inputs?

Changes in the value of money, what a single pound is equivalent to at a sequence of subsequent dates, illustrate what statisticians call a *distribution*. The goods that money buys are distributed in a vernacular sense of this term, and their costs are *distributed* in a manner that is, technically speaking, a *function* of time.

For most practical purposes, the word "function" (as used below) can be taken to be synonymous with the word "curve." For example, a curve or a function known as a *time series* often graphs business growth or decline

## 2.2. Representative number

as a function of time. In this sense, such a curve describes how profits, as well as many other variates, are temporally distributed.

The term "index-number" is defined by Kendall and Buckland (1960) as "...a quantity which shows by its variations the changes over time or space of a magnitude which is not susceptible of direct measurement in itself or of direct observation in practice." At first glance, it seems to make sense to think of an index number such as $5\tilde{\mu}$ in terms of the price of a market basket of AD 1440 goods in today's market. The trouble is that no AD 1440 market basket could possibly contain an electric toaster and no AD 1440 English market basket could possibly contain a kiwi fruit. Like the conceptualization, outlier, incommensurable market baskets can be interpreted in terms of the two curves $f_1$ and $f_2$. Those of today's goods that were available to AD 1440 English shoppers have one distribution that may be described by the curve $f_1$, while the remainder are distributed as described by $f_2$.

The incommensurable market basket example illustrates only one out of many difficulties with the concept of a representative number $\mu$. Hence, the Greek symbol $\mu$ designates a location parameter. This particular parameter may serve as a representative number or it may be equal numerically to a third quantity called the *first ordinary moment*. However, the interpretation of $\mu$ as more than simply a location parameter, can be thought of as merely a fortunate notational coincidence.

As a preliminary to defining what a location parameter actually is, some additional clarification is needed about the word "type," applied as a hyphenated suffix to the word "curve." Curve-types, such as time series, are categorized in many different ways. For instance, in most statistical applications, allocation strategies depend on the purpose the curve serves with regard to probability. The most important example of this variety of characterization is the density type of curve represented by a lowercase letter such as $f$ that, once preliminaries are dispensed with, will in many common situations be defined below as the derivative of another type of curve represented as an uppercase letter such as $F$.

No matter what interpretation is given to the concept of probability, it is hard to envisage some change in probability as itself being a probability. The "10%" within the sentence—There is a 10% greater chance of rain two days from now than there is of rain tomorrow—is not a probability. When some $f$ designates the derivative of some $F$, it could not be the case that the values assumed both by $f$ and by $F$ are probabilities. It follows that at least one of these two curves must be associated with probability in some indirect fashion.

The dual curves $f$ and $F$ provide considerable representational latitude, in the sense that some particular property of a statistical distribution can often be represented geometrically in two different ways, the first in terms of $f$ and the second in terms of $F$. Notational devices that extend the meaning of $f$ or $F$ often can be extended in a similar manner to particular curve properties. For example, the $e_x = e(x)$ type of curve defined in Chapter 20 that is encountered often in demographic and actuarial studies, and the curve $E(Y|x)$ whose interpretation is discussed in Chapter 6, are both defined by relationships that concern the $f$-type (density) curve $f(y|x)$, called a *conditional*.

The term "property" designates some quality of, or quantitative value obtained from, a curve by a well-defined sequence of mathematical or computational operations. As $x$ changes, the curve $f$ can, potentially, assume differing properties, and the value assumed by $x$ within $f(y|x)$ serves a purpose similar to that often served by some subscript of the symbol $f$, such as the $x$ subscript within $f_x(y)$. Therefore $E(Y|x)$ denotes one particular choice of curve property—in this case where the curve balances—that as $x$ changes has the potential to change accordingly.

Were some curve like $f(y|x)$ compared to the human body taken as a whole, some curve like $E(Y|x)$ can represent one particular organ within that body. Just as different infectious agents target particular organs, different data pathologies tend to target individual curves such as $E(Y|x)$. The health of one organ may be dependent on the condition of some other organ. Similarly, procedures that target a curve like $E(Y|x)$ may respond positively or negatively to characteristics of other curves.

Prior to introducing a few more basic concepts, the curve $f$ can be thought of temporarily as taking on a value at $x$ proportionate to the chance of $x$'s occurrence. Unfortunately, since the term "proportionate" is difficult to interpret in reference to the vague word "chance," this is not a rigorous definition.

The $f$-type curve that provides a basic structure about which data analysis applications are worked and reworked, has the synonyms "density function," "probability density function" (pdf), and "frequency" curve or function. On the other hand, while there are at least three ways of designating the curve $f$ in words, the single term "time series" is universally applied to different types of curves, such as the curve that describes how the price of a prototypic item that cost one pound in AD 1440 has changed over time.

Suppose $\mu$ is thought of as a representative number that is to some particular $f$-type curve what the single quantity, 5 times $\tilde{\mu}$ pounds, might be to a particular time series. Then $\mu$ and $5\tilde{\mu}$ pounds are, respectively, to two curves what Descartes' latus rectum length of Chapter 1 was to a particular parabola. Later sections will consider the connections between the model parameter represented by the Greek letter symbol $\mu$ and curve property counterparts of some particular latus rectum length. One or more loose connections between some model parameter and some counterpart curve property often underlie what applied statisticians refer to as "data messiness."

## 2.3 Curve types

Two examples of the density type of curve are often distinguished from each other by the choice of lower case letter, as in $f$ as opposed to $g$. Below, a density curve will sometimes be called an $f$-type curve even though it may be represented by some other lower case Latin letter like $g$, or even, for one specific variety of the Normal, i.e., bell-shaped, curve, the Greek letter $\phi$.

Even before it is rigorously defined, one can see that $f$ plays many roles in statistical applications. For example, when it is evaluated at the point $y$, the curve $F$, which is called the *cumulative, cumulative function,* or *cumulative distribution function* (cdf), equals the area under $f$ over the interval between $-\infty$ and $y$, including the point $y$ itself. The proviso, including $y$ itself, is an example of the difficulty encountered when curves like $f$ and $F$ are defined without calling upon basic calculus terms and expressions. A few of these mathematical conceptualizations can help define $f$-type and $F$-type curves in the rigorous manner that their importance deserves.

Besides separating model-variety distinctiveness from curve-type distinctiveness, there is a second advantage to allocating curves among the $f(x), F(x), e_x,$ and $E(Y|x)$ types. This form of categorization helps to clarify how uncertainty about curves is related to particular assumptions. For instance, when checking whether or not two curves are identical, it is almost always the case that two examples of different $f$ curves appear to be more distinct visually than counterpart $F$ curves.

Curve $f$ and curve $F$ resolution-difference can be traced to $f$'s proportionality to chance which, in turn, implies that no $f$-type curve can assume negative values. Thus, because there is no such thing as a negative probability, an $F$-type curve records the sweeping of area under its counterpart $f$-type curve in such a way that the amount of sweepings cannot decrease. Therefore, unlike a bank account that is drawn from as well as added to, for the nondecreasing curve $F$, small changes in total accumulations are less noticeable than differences in the individual increments described by the $f$ curve.

As a consequence of $F$'s accumulating structure, when checking whether or not two distributions are the same, all other things being equal, it is preferable to compare $f$ curves in lieu of $F$ curves. Conversely, an investigator who argues on the basis of two $F$ curves' similarity that the distributions described by these curves are comparable, shows by this choice of

$F$ instead of $f$ that he or she would like it to appear that the distributions in question are, indeed, similar.

Thanks to the contributions of many great mathematicians like Sir Isaac Newton and Baron Gottfried Wilhelm von Leibnitz, it is easier to distinguish curve types than it is to define the general terms curve, parameter, and curve property exactly. Although it is difficult to specify exactly what all curves have in common, one thing about these entities can be accepted with little difficulty. Curves are not single numbers. More specifically, curve-based representational simplicity extends beyond the concept of a representative number.

As a somewhat simplistic example, consider that the final grade given in a class is a representative number in the sense that it boils down many test scores—all of which apply to a single individual—to one quantity. On the other hand, the process known as "grading on a curve" concerns an entire classroom of students. Instead of forcing his or her class's grades to conform to a specific curve an instructor can "estimate" the curve $f$ by constructing a *curve estimator* known as a *histogram*. Histograms often provide a simple way to garner information concerning the distribution of a class's grades.

The boiling down of aggregations of numbers to extract their essence is seldom a single-step process. For example, to obtain a single numerical test grade, scores given to individual questions must be combined. When a single student's multiple test results serve as the basis of the assignment of a single semester grade, this again reduces several numbers to a single number. When instructors characterize their classes, perhaps to compare one class to another, they again boil down information. Finally, for any one out of a large number of possible reasons, an entire school or university is often rated by a single score.

Commonly, at each abstracting step, only a single number is retained. Consequently, were a curve like the histogram of Figure 2.1 archived together with this number, then the histogram would provide a potential backup. Nevertheless, a histogram does not usually encapsulate all the informational content of a sample of data (for example all of a student's exact individual test scores). In order to accumulate identical values, exact numerical grades are rounded and, in this way, they determine the proportions per bar shown along the $y$-axis in Figure 2.1. Thus, something is generally lost when original data are replaced by a histogram designed to describe some distribution.

A histogram, on the other hand, like most other curve estimators, provides something that is not obtainable from unprocessed data. It is easier to see what is lost in any data analysis step than it is to discern what is gained as an overview. To assess this general perspective, a distinction

## 2.3. Curve types

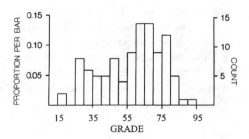

Figure 2.1. An estimated curve.

should be made between the concepts "population" and "sample." Like model parameter with curve property looseness-of-fit, the poor correspondence between some assumed, and some actual population is a concomitant of poor data quality. Unfortunately, a few more terms must be introduced and defined clearly before this basic distinction can be made in the context of parameters and curve properties.

The process "grading on a curve" implies that some curve shaped like the graph shown in Figure 2.2 plays a role in the determination of individual grades. Figure 2.2 is based on the general model for the Normal curve. This general model, in turn, is general insofar as its formula contains two distinct symbols for parameters. The location parameter $\mu$ corresponds to what was often called in the older statistical literature a *measure of central tendency*. The second, a *scale parameter*, is linked to the spread of individual grades (Appendix I.1 discusses the location and scale parameters of several common statistical models).

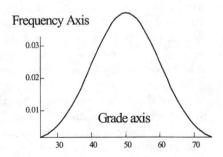

Figure 2.2. Normal curve fit using grades.

While the Figure 2.2 curve is more aesthetically pleasing than Figure 2.1, in many applications it is less informative than this histogram. For example, here the histogram has reduced data to sixteen numbers, while the two quantities, an estimated location parameter and an estimated scale parameter, determine the curve shown in Figure 2.2. Hence, were the need to resurrect original information to arise, it is more likely that this need would be satisfied by Figure 2.1 than by Figure 2.2.

## 2.4 Distribution and data terminology

Figures 2.1 and 2.3 both illustrate *estimators* of a particular $f$-type curve. A plain Latin letter that represents a curve designates an idealized entity, in the sense that this entity describes a *population*. This usage of the word "population" denotes the target of the statistical investigation in a general sense, and not necessarily any particular large group of living beings. The word population simply designates the answer to the question: "What is the scope of the study, survey, or investigation?"

While it is some aggregate of objects or *subjects* that constitutes any population, the individual entity that has a distribution is designated by the term "variate." There are many statistical applications where some parameter is taken to be distributed in a specified sense, but from a statistical perspective, objects or subjects are measured to obtain variate values—not parameter values. Just why this is the case will become clear when parameters are illustrated in the context of $f$ and other curve-type models. For now, it is best to think of variates in connection with measurements of subjects and objects, while parameters should be conceptualized in terms of one or more models that describe the structural features of one or more curves explicitly.

When two or more curves describe a single distribution it is common for these curves to be expressed in terms of a single set of parameters. For this reason, the phrase "parameters of a distribution" often refers to multiple symbols, like $\mu$ and $\sigma$, that appear in one or more curve models. Since $\mu$ generally helps disclose where a curve is located or situated, and $\sigma$ usually quantifies scale, it would be surprising to encounter two curves that each describe a single distribution in its entirety, where the model for one curve is solely expressed in terms of the parameter $\mu$ and the other curve in terms of the parameter $\sigma$. In this sense, the parameters $\mu$ and $\sigma$ can be thought of as *broad-brush parameters*. The curve that can change as a result of a change in some broad-brush parameter value is an $f$ or an $F$ type of curve. The term "identically distributed" often is applied in reference to a single sequence of broad-brush parameter values.

## 2.4. Distribution and data terminology

In this book, broad-brush parameters will be placed prior to any slash symbol that appears within a curve's symbolic representation, as is the parameter $\theta$ of $f(y;\theta|x;\omega)$. On the other hand, a parameter like $\omega$ of $f(y;\theta|x;\omega)$ is placed after the slash sign if it serves as part of a model that describes how the variable $x$'s value affects one or more broad brush parameters or one or more properties of the curve $f$ not $f$ taken in its entirety.

Suppose $G(z)$ represents a cdf designated by the term "standard." Then some second curve, with broad-brush parameters $\mu$ and $\sigma$ is often defined as $G[(x-\mu)/\sigma]$. In Section 7.3 of this book, location and scale parameters, $\mu$ and $\sigma$ respectively, will be interpreted in terms of the generation of artificial data. Among other advantages of interpreting location and scale in this way is that data generation model structure can help explain why $\mu$ is subtracted from $x$ and the result divided by $\sigma$ within $G[(x-\mu)/\sigma]$'s argument $(x-\mu)/\sigma$.

A broad-brush parameter like $\mu$ or, alternatively, some curve property that is linked to this parameter, is often modeled by an expression that includes one or more $\omega$ parameters. Generally speaking, $\omega$ parameters help individualize population subcomponents, while broad-brush parameters help describe a population taken as a whole. For example, the inclusion of $\sigma$ before the slash sign and $x$ after this sign within the notation $f(y;\sigma|x;\omega)$, designates that changes in $x$'s value do not alter $f$'s scale. In other words, $\sigma$ is constant no matter what value is given to $x$.

Although censuses sometimes enumerate an entire population, and in this way yield data, the term "data" usually refers to some proper subset of a population. In general, data are comprised of variate values. Also, while populations are rarely thought of as being composed of numbers—as opposed to being composed of entities that are measured in order to obtain numbers—data are usually spoken of as some collection of numbers, while a datum is to a single measured subject as data are to a sampled and then measured subset of population members.

Modified forms of Latin letters, such as $\hat{f}$ represent curves like the histogram shown in Figure 2.1 that are computed from data. When statistical equations and formulas are decoded, the symbolic distinction between $f$ and $\hat{f}$ provides a helpful shorthand representation. The hat symbol placed above a symbol $f$, where $f$ designates a curve, also is placed above many other types of symbols. In general, notation such as $\hat{f}$ or $\hat{g}$, as well as many statistical terms such as the word "data" itself, serve as reminders that entities like $\hat{f}$ and $\hat{g}$ are means to an end—not ends in themselves.

The word data and the symbols $\hat{f}$ and $\hat{g}$ concern a fundamental point. Unfortunately however, even an extensive mathematical education does not, in itself, provide much experience with this basic duality. Consequently, the slightest doubt concerning hat-appropriateness calls for careful checking.

## 2.5 Parameter validity and property existence

Besides curves such as $f$ and $\hat{f}$ whose differences are based on ends versus means, a key distinction between two conceptualizations applies to population-based curves like $f$. This distinction between parameter and property has had a history that can be traced back to the specific curve property, latus rectum length of Descartes' parabola, and a specific parameter within the analytic geometry representation of this parabola.

In Descartes' day, the path of an airborne cannon ball was taken to be perfectly parabolic since, among other things, no one contemplated in 1637 that some kind of star wars missile might be launched that could modify the path taken by this heavy ball. Yet, in the 1690s, long before our modern era, James Bernoulli discovered that it is the curve known as a catenary—not a parabola—which provides the arc of fixed length passing through two fixed points that has the lowest center of gravity. Picture a fine hanging chain. Because it illustrated that a curve's looks could deceive, the catenary expression's discovery exerted a major influence on our view of the complex linkage between curves as geometric entities and equations as algebraic entities.

Today many $f$ and other curves are studied without reference to some simple model or formula like those that describe some catenary or parabola. Such studies are often based on characteristics of curves and their subtended areas, for example an $f$-curve's balancing point $E$, or other properties, that can be defined without reference to a specified model. A model parameter often corresponds to some specific property of the curve that is modeled. Nevertheless, important statistical ideas, such as those that underlie the fields of model-free curve estimation and nonparametric inference, can be brought into focus only when parameters and properties are distinguished with great care.

Since credibility is basic to statistical thinking, model specification is a major sticking point. An assertion based on the assumption of a model stands or falls depending on the validity of that model. It is for this reason that the dual terms, parameter and property, underscore a basic idea. The choice of one or the other of these terms helps distinguish whether a referent is, or is not, based on a model's validity.

There is no question that some curve that could be denoted as $f$ and any particular variate's distribution fit together. For this reason, when the term "fitted curve" appears in the statistical literature, it is best to determine which of the terms "fitted model," "curve approximation," or "curve estimator," is the most suitable designation. Specifically, some true curve is often fit by something else. To specify this something else, it is good practice to modify the word curve to, among other things, place emphasis

## 2.5. Parameter validity and property existence

on one of the three alternatives: model; approximation; or estimator. This choice alerts the reader to particular distinctions between the ideal and some representation of reality.

A model is a particular symbolic representation that depicts a curve. If it is known that no exact model exists, or if it is known that such a model would be unwieldy, then an approximation is often called for. Neither the word "model" itself nor the term "approximation" indicates any connection with sampled information. Consequently when an expression for a curve is called a *curve estimator* this stresses that the expression is based on data.

Statistical modeling, approximation as well as estimation procedures depend on theorems and relationships. Those theorems and mathematical relationships that concern curve properties, as distinguished from model parameters, are based on the *existence*, and not the validity, of these properties. The existence of some curve's property, as opposed to the validity of some curve's parametric representation, is a major conceptual demarcation that will be illustrated many times in this book. How well does your shirt fit you after it has been lost by a laundry? How well does your shirt fit a tree? Both questions are strange, but their strangeness stems from the different conceptual issues, existence and validity.

The simulation of artificial data illustrates the statistical counterpart of the laundry versus tree—existence versus validity—comparison. Consider the process that yields a single random variate; say one that has the distribution described by the Cauchy density function $\sigma^{-1}\varphi([y-\mu]/\sigma)$, where

$$\varphi(z) = 1/\{\pi(1+z^2)\}$$

or, alternatively, one described by the Normal density function

$$\sigma^{-1}\phi([y-\mu]/\sigma),$$

where

$$\phi(z) = (2\pi)^{-0.5}e^{-z^2/2}.$$

The variate $G_N^{-1}(Z)$ that has a density $\phi$ is called a *standard* Normal variate and the variate $G_C^{-1}(Z)$ that has a Cauchy standard density $\varphi$ is also called a Student-$t$ variate with one degree of freedom. Here $Z$ is a random variate obtained from a standard computer package like MATHCAD, MATLAB, S+, SPSS, SYSTAT or SAS and, although much will be said about $Z$ and these two curves in later sections, for now, $G_C^{-1}$ can be thought of as the inverse standard Cauchy cdf and $G_N^{-1}$ the inverse standard Normal cdf. To simulate a Normal or a Cauchy variate, say $Y_N$ or $Y_C$, respectively, first $Z$ is obtained and then substituted into expressions such as $Y_N = \mu + \sigma G_N^{-1}(Z)$ or $Y_C = \mu + \sigma G_C^{-1}(Z)$, respectively (It is known that no model for $G_N^{-1}(Z)$

can possibly be found, but to generate a Cauchy variate, the function $G_C^{-1}(Z) = \tan(\pi[Z - 0.5])$ does nicely).

Whether a particular variate observed in nature is Normal or, alternatively, is Cauchy distributed, is a common validity-oriented concern. A less common, but nevertheless important, existence question is: Is the scale parameter $\sigma$ proportionate to one of the important curve properties—standard deviation and absolute deviation; two properties whose definition and graphical interpretation are both discussed in the next chapter. For both curve properties, the answer to this question is yes in the Normal case and no in the Cauchy case since, for Cauchy distributed data, these two curve properties do not exist.

For a physical object like a shirt, it is easy to interpret what nonexistence means. The object has vanished into the void. Similarly, a vertical line does not have a finite slope and, in the same sense of the word "have," a density curve that describes a Cauchy variate's distribution does not have either a standard deviation or an absolute deviation.

Although uncertainty about existence on the one hand, and uncertain validity on the other, have both become major statistical concerns, the literature that discusses validity is far more extensive than that which concerns existence. However, both the digital computer and improvements in statistical procedure generality have tended to shift much attention from validity to existence issues. This quest for enhanced statistical certainty has proceeded from validity to existence milestones along at least two distinct pathways. The most obvious pathway has long called upon property-based procedures, such as the histogram, to validate some choice of model. A separate pathway is paved by techniques based on curve properties, as distinct from model parameters, that are collectively said to be "distribution-free."

The term "distribution-free" can be confusing since the concept of distribution underlies all statistical thinking. Hence, parameter versus property distinctions will not be discussed here under the heading, distribution-free approaches. Rather they will be taken to merely distinguish whether what is being referred to is some model or, alternatively, some curve per se.

A basic statistical relationship is known as "Tchebycheff's Inequality." Suppose $X$ is a real valued variate and $h(X)$ is a function of $X$ that cannot assume negative values. For example, $h(X)$ might be $X$'s square, $X^2$, or its absolute value, $|X|$; both of which cannot assume negative values when $X$ is real-valued. If $C$ is any positive constant, consider the probability $P$ that the non-negative $h(X)$ is greater than $C$. Then $P$ must be less than or equal to the fraction $K/C$, where $K$ designates the place where the density curve that describes the variate $h(X)$'s distribution balances, in other words, here $K = E[h(X)]$. This inequality concerns the curve property $E[h(X)]$—not curve parameters. Consequently any mathematical or statistical argument

## 2.6 Estimator terminology

that is based on Tchebycheff's Inequality can be trusted to apply whenever $E[h(X)]$ exists—even in situations where an $f$ or any other curve that describes $h(X)$'s distribution is not, or cannot be, modeled validly.

## 2.6 Estimator terminology

Although parameters differ from properties, both concern a curve like $f$ and not some curve whose representation contains a "hat" symbol, as in $\hat{f}$ Besides this convention, the phraseology "an estimated parameter of $f$" is proper terminology, while it would be poor statistical usage to refer to "a parameter of $\hat{f}$." The *statistical properties of an estimator* are properties of some specialized density or cdf curve and the distribution they describe. The inappropriateness of *a parameter of $\hat{f}$* phraseology will become clear after certain notational preliminaries are reviewed.

In place of the parameters of its model-based representation, the properties of a curve like $f$ are often estimated. Unfortunately, when an estimator is symbolized it is rare at present for notation to disclose explicitly whether it is a curve property or model parameter that is estimated. However, when a symbol appears within a model, that unadorned would represent a particular parameter, a hat symbol is often placed above this symbol to designate some estimator of the parameter.

The important distinction between curves that describe a population, like $f$, and curves obtained from a sample of data, for example some histogram $\hat{f}$, applies to all types of curve. One example of a population-based curve of a type different from $f$ is the cdf $F$ whose estimator is illustrated in Figure 2.3.

The cdf estimator shown in Figure 2.3 corresponds to the density curve estimator shown in Figure 2.2. Both figures provide estimated descriptions

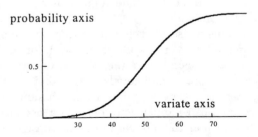

Figure 2.3. An example of an estimated cdf.

of a variate's statistical distribution. Supplementing s-shaped cdf curves and their estimators, such as the curve shown in Figure 2.3, with a special variety of cumulative distribution function helps clarify this basic idea. Besides, serving as a base for which the term "statistically distributed" can be defined, in the next chapter this special curve will help illustrate the basic concept, "curve property."

Suppose one wishes to find one single number that characterizes the curve $F_1$ whose estimator $\hat{F}_1$ is shown in Figure 2.3. Also suppose that within an interval $(x_0, x_1)$, of finite length where $x_0 < x_1, F_1$ and $\hat{F}_1$ are continuous in the mathematical sense of this term. A cdf curve, or its estimator like $\hat{F}_1$, is continuous when it is not only bounded such that, for example, it increases steadily from a value of 0 at the point $x_0$ until it reaches the value 1 at $x_1$, but it is also free of sharp jumps and wild swings. Cdf curves can contain horizontal line segments between two points like $x_0$ and $x_1$. However in this section the cumulative distribution function $F_1$ and its estimator $\hat{F}_1$ are assumed to be everywhere increasing. In other words, the paths they trace have no dips or flat sections.

A special variety of cdf referred to here as $F_2$ illustrates how the term "representative number" is associated with the concept "curve." Consider a perfectly vertical line segment stacked over a midpoint $x_r$ defined to equal $(x_0 + x_1)/2$. This type of segment is called a riser. In everyday speech the word "riser" sometimes denotes a vertical board that separates two adjacent steps of a staircase. In technical jargon, the variety of cdf that now will be defined is called a *single-step step function*.

Since Euclid, perfection has been expected of entities like lines and points in the mathematical sense. Were this not the case then, when viewed from its side, a riser would resemble a board or even a domino laid on its edge. Suppose that besides containing a single riser over the point $x_r$, the curve designated as $F_2$ assumes the value zero in some half-open interval that extends up to, but not does not include, $x_r$, and that $F_2$ evaluated at $x_r$, and beyond, equals 1. The term "half-open" helps designate an interval that extends up to, but does not include, a particular point at one end, but does contain its other endpoint.

The cdf $F_2$ contains a riser of height 1 resting on the single point $x_r$, and $F_2$ equals 0 for all $x < x_r$, while it equals 1 for all $x > x_r$. Any $F$-type curve's riser corresponds to a pile of probability. Consequently, if some cdf equals zero before $x_r$ and equals one beyond $x_r$ then it describes a *degenerate distribution* where a variate cannot assume any value other than $x_r$. Many events can happen only once to a person, and there are even some variates, such as the number of distinct designs on the backs of a deck of 52 ordinary and unmarked playing cards, that can assume only one unique value.

## 2.7 Degenerate curves

The word "degenerate" modifies a statistical or mathematical term to indicate that there is a contradiction between the vernacular and technical definition of the term. For example, when in conversation something is said to be *distributed* this designates that someone has sown, strewn, or spread out something in the way a farmer spreads seeds or a fairy tale monarch scatters coins to his or her subjects. Hence, coins cast so that they all land in this king or queen's own pockets might be taken to be a degenerate case of the concept "distribution." Similarly the exclusive piling over a single point described by the $F_2$ cumulative distribution function implies the contradiction—some entity or variate is distributed in place.

A *ray* is a line segment of infinite length that begins or ends at a specified set of finite valued coordinates. A degenerate $F_2$ curve that consists of two horizontal rays separated by a single riser at $x_r$ of height one over the $x$-axis is the curve-based way of specifying that there is no *variability* about the value $x_r$. In largess-tossing terms, were a fairy-tale monarch to distribute coins in an $F_2$ manner, he or she would find that somehow magically these coins have ended up back in his or her own purse.

Largess distribution is sometimes motivated by charity. In modern politics, sad to say, largess redistribution is motivated typically by other considerations, including the desire to avoid, or forestall, being deposed or otherwise removed from office. From the vantage point of the potential deposer this implies that a potential "deposee" should be rated on the basis of the beneficence variability properties of largess distribution. For example, suppose the dispersed gold, silver, and base metal coins return to the ruler's pockets. This would be the zero beneficence variability counterpart of the $F_2$ curve.

The wealth dispersal example, although far-fetched, illustrates a point central to many forms of data analysis. Many curve properties and parameters are, in actuality, based on combinations of curves. For instance, as will be discussed in detail in the next chapter, some particular $F$ curve will often be compared to some $F_2$ degenerate, all-or-nothing curve. A second example, called a *shift alternative* is, for models, based on the distance between the location parameters within the representations of two curves that would be, but for their location parameter values, identical.

Suppose the parameter $\mu_1$ appears within a model for density $f_1$, and the parameter $\mu_2$ appears within a model for density $f_2$. Then $\mu_1 - \mu_2$, the difference between two parameters, is itself a parameter. However, suppose there are two populations, the first comprised of subjects who have been given a medical treatment under a strict protocol and the second comprised of controls. Then instead of the validity of the models chosen to represent

$f_1$ and $f_2$, the following existence issue is often formulated: Does there exist a constant $K$ that, when added to the values that prior to addition have density $f_2$, now—after $K$ is added to their values—have density $f_1$? Much of the subject matter subsumed under the heading "nonparametric inference," is based on existence, not validity assumptions. For example, the existence of $K$ does not hinge on the capacity to model some curve that can be translated to match some second curve perfectly.

# CHAPTER 3
# Variability and Related Curve Properties

## 3.1  Uncertainty and variability

Because its value is obtained when a sequence of well-defined numerical operations is executed, a property is to a curve as a one-word abstract is to a sentence. Examples are $E$, whose interpretation in terms of curve geometry is discussed in the next chapter, and the property that in this chapter is represented by the six capital letters M-E-D-I-A-N, and "a mode," a point over which a density assumes a relative maximum value. (The possibility that a density has multiple modes is discussed at length in Chapter 6.)

It will be shown that $E$ can be interpreted as the area between two curves. Thus, because area is involved, the abstracting process that defines $E$ is a statistical counterpart of the mathematical concept, integration. Unlike $E$'s connection to an area, in the many situations where it exists unambiguously, the MEDIAN curve property is defined as a value assumed by the inverse of a distribution's cdf.

Despite their importance, neither $E$ nor some MEDIAN is usually paired with the concept "uncertainty." Instead, for a variety of reasons, the term "variability" is paired with uncertainty. For example, within the environmental risk analysis literature it has been found convenient to separate doubts concerning the model for a density curve from the scale parameter $\sigma$ that appears within this same model's symbolic representation. Uncertainty refers to the model per se and variability to $\sigma$'s magnitude.

Like the multiple interpretations of the word "uncertainty," as in uncertainty about some curve property's existence and uncertainty about some model's validity, there are many ways that variability can be conceptualized. In other words, there are many ways some model parameter like $\sigma$ can be interpreted in terms of the geometric properties of any one of the many curves that describe any distribution. However, both in terms of the extensive literature that concerns them, and in terms of the ease with which they can be interpreted, two interpretations stand out.

Before considering the absolute deviation, a curve property that has a particularly simple geometric interpretation, as well as the standard deviation together with its near identical twin, the population variance, a third concept should be mentioned. This is the *range*. In several later chapters much use will be made of the concept, "support region" which, because its definition requires notation that is peripheral to the concept *variability*, will be defined in a later chapter. While the *range of a sample* is defined unambiguously as the numerical difference between the largest and smallest observed value, because of its infinite value the *range* curve property does not exist for Normal or Cauchy distributions. This is one reason why Chapter 20 is devoted exclusively to the related concepts of support regions and truncated curves.

## 3.2 The absolute deviation curve property

Two properties, like $E$ and MEDIAN, are taken to be distinct if the specific operations that determine their values can yield differing results. This leaves the question: How should all the information contained in a curve, whose shape is one out of an infinite number of possibilities, be abstracted? Among many alternatives there is one graphically elegant answer to this question in terms of the curve $F_2$ and the curve $F_1$ that are shown in Figure 3.1. This curve property, "the absolute deviation" (AD), will be discussed in this chapter extensively because, unlike the alternative—the standard deviation—the AD can be displayed geometrically in a straightforward way.

Unless it is defined in one—and only one—particular way, the AD curve property has a blind spot. The standard deviation curve property has a closely related weakness that has motivated a large number of remedial approaches. That neither the AD nor standard deviation is flawless is not surprising since, as the multiplicity of terms like: (1) variability; (2) dispersion; (3) deviation; (4) error; (5) spread; and (6) distribution implies, one single number cannot be expected to disclose all nuances of these six terms.

The weakness shared by both the AD and the standard deviation is sensitivity to data value extremes at the expense of mid-portion insensitivity. When a researcher is uncertain whether or not outliers are mixed with relevant data, this weakness is particularly troublesome. It would be unfortunate if a curve property that discloses how much variation is present within a population is swayed most appreciably by that same portion of a sample whose validity is particularly doubtful.

Suppose the single-step step function $y = F_2(x)$, which has a riser at $x_r$, intersects a continuous cdf $y = F_1(x)$ at the point where the line $y = 1/2$

## 3.2. The absolute deviation curve property

also passes through $y = F_1(x)$. The $x$ coordinate of this intersection is the *population median*, $x_r =$ MEDIAN. Another way to interpret the population median is that there are exactly 50/50 odds of $X$'s being greater than $x_r =$ MEDIAN; and equal, 50/50, odds that $X$ is less than $x_r =$ MEDIAN.

Although pristine in most ways, the median is not completely trouble-free. Some unusual, yet perfectly well-defined, density curve could consist of two halves, each subtending an area equal to 0.5, and these areas could be separated by a finite interval, say $[a, b]$. Were this the case then the median would not exist as a unique value but, instead, could be interpreted as being the midpoint, $0.5(a + b)$, or defined equal to any other point within $[a, b]$.

Although the median can, at least in theory, have $[a, b]$ bursitis, the concept of uniqueness will only be considered further below in reference to an additional location-associated curve property, "population mode." It is worth noting that much has been written concerning the uniqueness of estimation equation solutions. However, because of the close association between the three topics: (1) degenerate curves; (2) infinite curve properties and; (3) statistical uncertainty, any curve property that exists will here be taken to be both finite valued and single valued.

Figure 3.1 shows the two curves $F_2$ and $F_1$. The area shown above the line $y = 1/2$ that is bordered by $F_2$ on the left, $F_1$ on the right, and the line $y = 1$ from above, plus the counterpart area below $y = 1/2$, is the absolute deviation from the median, (ADM) curve property.

The ADM curve property increases as the $F_1$ curve's variability—relative to the $F_2$ curve—increases. For example to help determine whether or not a level of dispersion, variability, or the spread of a distribution is sufficiently high or low to merit action, the ADM's magnitude can be compared to a value referred to as a discriminant point.

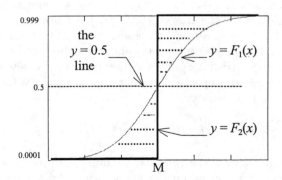

Figure 3.1. The ADM Curve Property.

While the curve property, MEDIAN, serves as a central value or representative number upon which the solutions of many statistical problems are based, the ADM property serves a different purpose. The ADM indexes dispersion, as in the largest distribution example of the previous chapter. It is one of several alternative ways to interpret the algebraic symbol $\sigma$ geometrically.

Suppose prior information of two—and only two—sorts is available. The nonzero numerical value of the parameter $\sigma$ is known, and it is also known that the particular variate, one whose density model contains $\sigma$ has a Normal distribution. Given only this information, the variate's value will not be known. For instance in those applications where a density could be either Normal or Cauchy, uncertainty extends beyond variability. Exactly how much further it extends or whether uncertainty can be quantified clearly as a single number, remains a subject for debate.

Because much is already known about scale parameters and the curve properties that represent them geometrically, emphasis will be placed in the next two chapters on the concept of variability, and the crucial issues, when (and whether) variability can (and should) be taken to be a single number.

## 3.3 The general AD and the ADM curve properties

Much attention has been paid in the two previous chapters, and will continue to be paid below, to many statistical distinctions, such as that between parameters, and curve properties. Given such a large cast it is not surprising that there are also many interconnections and relationships between cast members. Two such relationships will be described in this section. The first has considerable bearing on curve uncertainty, but is little known; the second is better known, but of primarily theoretical interest.

Figure 3.2 shows an $f$-type curve known as a triangular density. This density $f$ is zero outside the interval $(0, 2)$, $f(x) = x$ over the closed interval $[0, 1]$, and $f(x) = -x + 2$ over the half closed interval $(1, 2]$. (A half closed or half open interval like $(a, b]$ contains the endpoint $b$ but not the endpoint $a$.)

Besides the geometric interpretation illustrated previously by Figure 3.1, the absolute deviation property of the curve $f$ about the value $m'$ can also be interpreted as the quantity

$$\int_{-\infty}^{\infty} |(x - m')| f(x) dx.$$

Here the absolute value of the quantity, $x$ minus $m'$, designates the distance between the points $x$ and $m'$ on the $x$-axis, while $f$ discloses how likely it is for $x$ to be a distance $|x - m'|$ from $m'$.

## 3.3. The general AD and the ADM curve properties

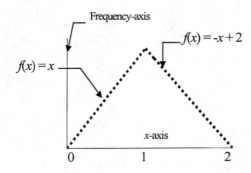

Figure 3.2. A triangular density.

Now suppose $m'$ is set equal to 0.5. Then, because the integral

$$\int_{-\infty}^{\infty} |(x - m')| f(x) dx$$

equals the integral

$$\int_{0.5}^{1} (x - 0.5)x \, dx + \int_{1}^{1.5} (x - 0.5)(2 - x) dx$$

which equals 0.375, plus

$$\int_{0}^{0.5} (x - 0.5)x \, dx + \int_{1.5}^{2} (x - 0.5)(2 - x) dx$$

which equals 0.166...; it follows that for this triangular density curve, $ADm'$, the absolute deviation about $m'$, equals $0.375 + 0.166... = 0.541667$.

The integral that defines $ADm'$ was broken into four parts to facilitate a comparison between the $ADm'$ property of the curve shown in Figure 3.2 with the $ADm'$ property of the curve shown in Figure 3.3. The apex reversed density shown in Figure 3.3 was formed from its Figure 3.2 counterpart by reversing triangle apex halves within the interval (0.5,1.5). As far as the $ADm'$ curve property of this apex reversed curve is concerned, the sum of definite integrals

$$\int_{0.5}^{1} (x - 0.5)x \, dx + \int_{1}^{1.5} (x - 0.5)(2 - x) dx = 0.375$$

that involve the curve shown in Figure 3.2 are here replaced by the sum of definite integrals

$$\int_{0.5}^{1} (x - 0.5)(1.5 - x) dx + \int_{1}^{1.5} (x - 0.5)^2 dx = 0.375,$$

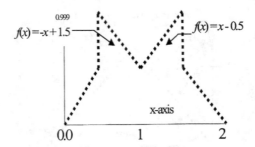

Figure 3.3. An apex reversed density.

which together, numerically equal the integral of the product formed by the triangular density over the region (0.5, 1.5) times the absolute deviation $|x - m'|$ from $m'$.

Despite the apex spreading of the triangular density shown in Figure 3.3, the density curves shown in Figures 3.2 and 3.3 both have the same AD$m'$. This example illustrates that the AD$m'$ curve property can often be insensitive to dispersion change; here induced by apex reversal but due, in other instances, to other distribution features. Here, with regard to variability changes that occur in some neighborhood of x = 1, the AD$m'$ curve property has a blind spot.

The size of this blind spot will vanish when the absolute deviation curve property is defined with an $m'$ equal to the population MEDIAN. However, in practice, a researcher is as unlikely to know the position of the true population MEDIAN as he or she is in the position to know the true value of the population absolute deviation. Consequently, although it is possible that blind spot size is inappreciable, a researcher who proceeds with this assumption without appropriate checking does so at his or her own risk.

Suppose that for both the triangular and apex-reversed densities, instead of the point $m' = 0.5$, the absolute deviation is measured about the median, $m$, which in this example equals 1. Then because

$$\int_{0.5}^{1} |(x-1)|x\, dx + \int_{1}^{1.5} (x-1)(2-x)dx = 0.1666...,$$

$$\int_{0.5}^{1} |(x-1)|(1.5-x)dx + \int_{1}^{1.5} (x-1)(x-0.5)dx = 0.208333...$$

and

$$\int_{0}^{0.5} |(x-1)|x\ + \int_{1.5}^{2} (x-1)(2-x)dx = 0.1666...$$

## 3.3. The general AD and the ADM curve properties

it follows that since 0.1666... is smaller than 0.208333..., the absolute deviation of the triangular density is smaller than the AD of the density shown in Figure 3.3—a finding that, in light of the apex reversal of the Figure 3.3 curve, is reasonable.

A mechanical model based on the AD illustrates how most curve properties that disclose variability are interpreted. Let $m$ = MEDIAN and assume that the distribution function $y = F(x)$ designates the outline of a flexible steel band that pivots at point $(m, 1/2)$ within the $(x, y)$ plane while a compressible substance is confined as shown by the shaded area of Figure 3.1 in such a way that the substance is held in place by the steel band. Consequently, an increase in the absolute deviation corresponds to the relaxing of the steel band which, in turn, permits the substance to disperse about the median.

The association of model parameter $\sigma$ with a curve property like AD$m'$, or the curve property standard deviation that is defined at the beginning of the next chapter, is no more a clear-cut choice than is the issue of association of the location parameter $\mu$ with either $E$ or the MEDIAN curve property. Like its AD$m'$ first cousin, the standard deviation is sensitive to curve extremes at the expense of the curve center. Moreover, the standard deviation has an additional weakness that is traceable to the substitution of the squaring operation for what, in the case of the AD, is the absolute value operation $|\ |$.

The blind spot of the AD$m'$ curve property can be seen more clearly than the standard deviation's weakness. The absolute deviation (AD$m'$) about some point $m' \neq$ MEDIAN can be described geometrically in terms of the cumulative and inverse cumulative distribution functions as follows: Let $m'$ equal an arbitrary point and m equal the median of density function $f(x)$ that has a finite absolute value. Then the AD$m'$ curve property is equal to

$$\text{ADM} - 2\left[\int_m^{m'}(x-m)f(x)dx + (m-m')([1/2]-F(m'))\right]$$

which, in turn, if the cdf $F$ has a well-defined inverse $F^{-1}$, equals

$$\text{ADM} + 2\left[m'\{F(m')-1/2\} - \int_{1/2}^{F(m')} F^{-1}(x)dx\right].$$

Over intervals within which $f$ is identically equal to zero, the associated cdf does not have a well-defined inverse. For a cdf function $F$ that has a

32  3. Variability and Related Curve Properties

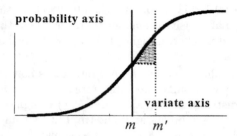

Figure 3.4. Absolute value increment.

well-defined inverse $F^{-1}$, the shaded area shown in Figure 3.4 is equal to

$$\left[ m'\{F(m') - 1/2\} - \int_{1/2}^{F(m')} F^{-1}(x)dx \right].$$

Consequently, since the shaded area will be zero if $m = m'$, this figure illustrates a relationship that the absolute deviation variation disclosing curve property has with the location disclosing curve property MEDIAN.

The absolute deviation about the median's value cannot exceed the absolute deviation calculated about any alternative value of $m'$. The variability indexing curve property that is called the *standard deviation* has the same sort of relationship with the $E(X)$ curve property. Specifically, the standard deviation determined by assigning $E(X)$ to be the central value about which deviations are taken, must be less than or equal to any other modified standard deviation where deviations are taken about any value other than $E(X)$.

Figure 3.5 illustrates why the equal absolute deviations about 0.5 of the apex reversed density and the usual triangular density is an equality

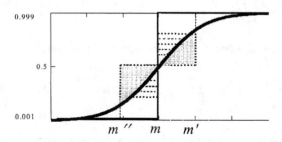

Figure 3.5. The AD$m'$ curve property of a symmetric density.

that is shared by many other $f$-type curves. Suppose, for example, that $m$ represents the median of a symmetric density with corresponding cdf shown in Figure 3.5. Then as long the equality $m - m'' = m' - m$ is satisfied for $m'$ and $m$, it makes no difference to this AD$m'$ value where the cdf passes through the shaded area shown in Figure 3.5. As far as this value is concerned, the cdf can allow area to disperse to the edges of the shaded area or it could confine it within this area in any other way.

Whenever deviation is determined by taking distances to some value different from the exact value of the MEDIAN curve property, property AD$m'$ determined in this way has a zone of indifference within which it is insensitive to variation. Dispersion, spread and variability are not associated as clearly with any one particular curve property as any one of Descartes' parabolas is to the geometric construct, "latus rectum."

## 3.4 Curve property selection

Throughout the last section no mention was made of curve representation. In particular, the AD and its blind spot are concepts that are defined without reference to any model, such as that for the Normal curve. The symbol $\sigma$ appears within the usual formula for the general Normal density. The AD about the Normal's median equals 0.79788 times the parameter $\sigma$. But this does not mean there cannot exist some other density, also defined in terms of some parameter $\sigma$, whose AD curve property does not equal $0.79788\sigma$.

The AD is just one way out of a large number of alternative ways to abstract information about the variability of a distribution. The Normal is just one, out of many, alternative ways a particular type of curve can be represented. Yet the spectrum of alternative curve properties differs substantially from the collection of curve representations. It differs because only one model-based representation is deemed to be correct while, on the other hand, for almost all densities, curve property correctness is a moot point as long as the curve property in question exists. It is important, however, that curve properties be matched together properly, as is illustrated by the absolute deviation's match with the median and the standard deviation's match with the curve property $E$.

The AD$m$'s blind spot shows that one curve property may be deemed to be inferior to some alternative choice of property, just as one brand of wine may be superior to another brand. Yet while a curve property connoisseur may fit the brand to the occasion, he or she does not second-guess nature, as would be the case were a Normal taken to provide accurate representation. Indeed, whether or not a distribution is represented by a Normal curve is neither relevant to the ADm's blind spot nor to the choice of the property, MEDIAN, to be the point about which absolute deviations are minimized.

In general, when the information content of an entire curve is boiled down to a single distillate (in other words, when a curve property is selected) effectiveness depends on

1. skewness and other characteristics of a curve that are unrelated to curve variability and curve location, and yet are often relevant to decision making;

2. the particular choice of variability-indexing property (even assuming that it is only variability that is important to a scientific investigation, there are several available alternative curve properties that quantify variability, in the sense that each alternative has generated some, albeit in some cases small, portion of the statistical literature).

3. data usage (in practice, decision-making must be based on "randomly collected" data by means of some curve like $\hat{F}_1$ and not on some *known* $F$ curve.)

The hypothetical situation or thought experiment in which data quantity is unlimited, in the sense that the sample size $n$ is infinite, illustrates the importance of considerations (1) and (2). The AD and the population standard deviation are merely two of a great many alternative choices that reduce the general conceptualization, variability, to a single number. Hence, even were $n$ infinite, some particular property weak points and strong points are germane and some are irrelevant. For example, some curve properties may be hypersensitive and some may be minimally sensitive to outlying distributional characteristics over what are called the *tail regions*.

Suppose the true value of the AD about the median, and the true value of the population standard deviation about the curve property $E(X)$, are both known exactly. However, suppose also that for the indexing, risk assessment, and discriminant procedures known to apply to these curve properties, the procedure based on the standard deviation and that based on the AD suggest contradictory actions or decisions. This circumstance is far from impossible. In addition, it has as little to do with data availability as does consideration (1).

In some contexts, papers that concern (1) and (2) are grouped together collectively under the heading, "descriptive statistics." This is an unfortunate term since the word "descriptive" suggests the word "graphic," a word that denotes procedures that apply to all branches of data and risk analysis, not just to investigations where considerations (1) and (2) are important. For this reason analyses that are concerned more with clues

## 3.4. Curve property selection

than with evidence are characterized by the term "exploratory," while the search for statistical evidence is referred to as a *confirmatory* process.

The clue/evidence, exploratory/confirmatory dichotomy can be useful. Unfortunately, like the division of statistics into descriptive and "inferential" categories, the correspondence between these two dichotomies, on the one hand, and the trichotomy (1), (2) and (3), is unclear. For this reason, the terms "descriptive," "inferential," "exploratory," and "confirmatory" will not appear below. Instead of these concepts, attention will be focused on curve-based considerations. For example, the word "population" within the term, "population standard deviation," designates a hatless curve $F$ or $f$'s property.

In almost all applications, the curve $\hat{f}$ satisfies the few requirements for a curve to be characterized as an $f$-type curve. When it does satisfy these requirements it is sometimes considered to have the same type of properties that are possessed by its population based counterpart. However, even when terms like "sample absolute deviation" or "sample median" are employed, it is confusing to call these quantities curve properties. Below, when the term "curve property" is applied in a strictly statistical, as opposed to the vernacular sense, it refers to a population-based curve. For instance, suppose some curve $\hat{F}$ or $\hat{f}$ is substituted for its $F$ or $f$ counterpart in a formula that defines a curve property. When this substitution is performed and, more generally, when data values are combined, the resulting quantities are called *composite variates, estimators,* or *statistics,* not *curve properties.*

When the term "population" is applied in reference to curves $f$ or $F$, a distribution, or a curve property, this term emphasizes that $f$ and $F$ are idealized entities whose outlines are actually viewed only in simulation experiments, not in natural data-based experiments or observational studies. In nature, a curve like $f$ or $F$ serves the role of a hidden target's bull's eye towards which some corresponding statistical procedure, test, or estimator is aimed.

Statistical notation has conventions that can only be taken to be historically, rather than logically, justified. For instance, the symbolic representation of the population standard deviation counterpart of the AD curve property differs substantially from the notational forms that have been defined so far. The particular parameter of the Normal model that is usually represented by $\sigma$ is numerically equal to the curve property, standard deviation. On the other hand, what is actually represented by the letter $\sigma$ is not identical conceptually to the curve property, standard deviation. This can be confusing. For now, suffice to say that as interest in density representation other than by the Normal has increased, the importance of this distinction has also tended to increase.

Bad habits are hard to break once they are learned. For this reason, the definition of the population standard deviation will be postponed until enough preliminaries are dispensed with to clearly distinguish the scale parameter $\sigma$ from the curve property standard deviation. These specifics are best appreciated after basic questions are considered such as: In the process of boiling down an entire curve's information content, what general purpose do curve properties, like MEDIAN, $E(X)$, AD, and population standard deviation actually serve?

## 3.5 The history of variability appreciation

Studies that concern the conceptualization "variability," can be traced back to seventeenth century investigations into the laws of probability by Pascal and Fermat. From that time on, models for generalizations of $F_2$ curves were applied to study games of chance. Unfortunately, as described by C. E. Sarndal (1971), these seventeenth and eighteenth century models are now known to represent many $f$ and $F$ curves encountered in applications (other than games of chance) poorly.

Because imperfections in modeling processes are somewhat like holes in dikes, during the later years of the nineteenth century, Lexis (1877) and other pioneer statisticians came to appreciate the dimensions of the patching problem. Why they became disillusioned with simple models that had been represented typically by means of fewer than a dozen individual symbols, is illustrated by the variate, the number, $x$, of females born out of $n$ live births to a given mother.

Prior to Lexis's investigations, the model known as the *binomial*,

$$f(x) = [n!/(x!(n-x)!)]P^x(1-P)^{n-x},$$

was taken to provide an adequate description of this variate's distribution at any nonnegative integer $x$. A parameter $P$ of this model can be thought of in terms of the following curve property: chance of a daughter being born in any single live birth. The trouble is, as illustrated by photographs in the *Life Science Library* (Bergamini, 1963), there have been instances where, after giving birth to five daughters, a couple has proceeded to have six consecutive male births.

This sort of occurrence led Lexis to doubt that the parameter $P$ should be taken to be a single value that applies to all women during all stages of their childbearing years. It could be the case, for example, that the chance of a female birth, $P$, changes as a function of age or other factors. In several later chapters, different forms of curve uncertainty will be classified in terms of the way that variates and parameters are linked to quantities that will be distinguished from the variates and parameters themselves, by applying the term "variables." For example, many models that describe the distribution of a single variate within which at least one parameter is not constant-valued are called *mixtures*. A model for what is called a *finite mixture* is associated with a lurking variable.

Besides mixtures, A. C. Cohen (1951) asserts that "the logarithmic normal distribution provides a useful theoretical model for studying a number of biological populations, certain economic populations involving income distributions, and others in which the standard deviation of individual observations is approximately proportional to the magnitude of the observations."

Here, unlike the case of a finite mixture, rather than a parameter taking on a finite number of distinct values, what in 1951 Cohen called a *standard deviation*, and is called a *scale parameter* here, varies over a continuous range of values.

## 3.6   Simplistic approaches and the history of probability

The logarithmic Normal on the one hand, and a mixture of Normal or binomial curves on the other, are only two out of many other instances where nature—in the sense that some parameter is constant valued—cannot be taken to repeat herself exactly. Nondegenerate mixtures, where rules change in midstream, also stand out in stark contrast to their probability-based science counterparts, many of which can be traced back to Pascal's and Fermat's investigations of gaming practices, where rules of play are constrained to be constant unless, of course, the game is broken up prematurely. Consequently, the question arises: Why has it taken so long for the limitations of simplistic statistical approaches to be brought to light?

The games studied by Pascal and Fermat were played by gentlemen of leisure. Unlike Pascal, and most probably Fermat, these individuals or their friends had the wherewithal to not only gamble for high stakes, but to have elaborate game rooms built for their pleasure. Later, along this elite pathway, Cullen (1975) quotes Adam Sedgwick, who in the mid-nineteenth century served as a president of the British Association. In his concluding address to that scientific society, Sedgwick dealt with the issue of forming a new statistical section by stating that "...if we transgress our proper boundaries, go into provinces not belonging to us, and open a door of communication to the dreary world of politics, that instant will the foul Daemon of discord find his way into our Eden of philosophy."

In the mid seventeenth century, less than a decade after Pascal and Fermat's correspondence concerning probability theory, the Englishman John Graunt published his *Observations upon Bills of Mortality*. This 1662 monograph described investigations of plague-deaths and led, albeit in a roundabout fashion, to many of today's applications of statistical methodology to public health, epidemiological, and environmental problems. However, beside their nationalities, Pascal and Fermat differed in another way from their contemporary, John Graunt.

The studies that Pascal and Fermat conducted on the European Continent dealt with games that resemble their modern card, dice, and roulette counterparts. These games were played in artificial, controlled environments that had little to do with the outside—plague-ridden—world. It was not until the later part of the nineteenth century, in particular when Florence Nightingale corresponded with the pioneer statistician Francis Galton, that the first partially successful effort was ever seriously attempted that had the potential to link real-world problems, such as those studied by Graunt, to the methodology devised originally by Pascal and Fermat.

Had it taken less than two centuries for problems and methodology to span the English Channel, styles of statistical notation, practice, and instruction would differ from what they are today. For example, in 1865, I. Todhunter published a 624-page volume in which a description of Graunt's work begins with the sentence, "The history of the investigations on the laws of mortality and of the calculations of life insurance is sufficiently important and extensive to demand a separate work; these subjects were originally connected with the Theory of Probability but may now be considered to form an independent kingdom in mathematical science..."

One of the reasons why the origins of probability can be traced far back to games of chance while, in 1865, the solution of life or death problems was taken to be an independent kingdom, is that games of chance are easier to deal with than life and death problems. In this sense, games of chance were statistical training wheels.

The distribution of identical event outcomes, such as the roulette wheel spins, are described adequately by simple formulas, like that of the binomial. On the other hand, when environmental risk factors or the survival time variates of bubonic plague or AIDS victims are studied, these variates have complex $f$ and $F$ curves. For example, an important demographic and actuarial tool is known as the *population life table*. This tool is designed to deal with particular highly complex curves which as their name *bathtub functions* suggests, are far from bell-shaped.

To study $f$ and $F$ curve-complexities, computer power is required. Until the recent blossoming of computer graphics, curve-based methods were difficult. It was tedious and time consuming to graph extensively. Final findings have long been presented with the aid of graphical methods, but before the advent of the digital computer, the *initial* screening of raw data, as well as subsequent intermediate steps of statistical investigations, rarely involved so simple a graph as the histogram shown as Figure 2.1.

As the term "life table" suggests, until the early 1950s, tables and formulas rather than graphs were the primary means by which aggregations of single numbers were abstracted. Thus, like the lengthy gambling strategy phase of statistical development, redirection was long overdue. For example, the histogram shown in Figure 2.1 is based on class interval midpoints that represent multiple distinct values. This is because, not too long ago histograms were based on the then-popular devices known as *frequency tables*. Like simple histograms, these tables utilized class interval midpoints that remained after variation within these class intervals was spread out evenly in the interval. Thanks to the power of the modern computer, except as one of the steps required in order to construct some histogram, this tedious approach is seldom encountered today.

# CHAPTER 4
# Moments and Curve Uncertainty

## 4.1 E and Var Geometry

The general formula for the Normal density is expressed in terms of the two parameters $\mu$ and $\sigma$. Consequently, it is not at all surprising that two curve properties, the mean and the standard deviation, have long been paired with these two parameters.

In the third edition of their classic text, Dixon and Massey (1983) indicate that: "We shall denote the arithmetic mean of the population by the Greek letter $\mu$ (mu) and the variance of the population by $\sigma^2$ (sigma squared)." Based on this 1983 definition, both mu and sigma squared are curve properties that are also often called the *first* and *second ordinary moment about the mean*, respectively.

After providing geometric interpretations of the first ordinary moment, which here will be denoted $E(X)$ to distinguish this quantity from a location parameter, and then repeating the process for the second ordinary moment, here designated by $\text{Var}(X)$, the remainder of this chapter will deal with curve properties known as *higher-order moments*. For an integer $k > 1$, the $k^{th}$ ordinary moment about the population mean is defined as

$$\int_{-\infty}^{\infty} [z - E(X)]^k f(z) dz.$$

In addition, for $k > 0$, the $k^{th}$ absolute value moment about the population median, MEDIAN, is defined as

$$\int_{\text{MEDIAN}}^{\infty} [z - \text{MEDIAN}]^k f(z) dz - \int_{-\infty}^{\text{MEDIAN}} [z - \text{MEDIAN}]^k f(z) dz.$$

Higher-order ordinary and absolute value moments are sometimes applied to deal with curve uncertainty issues. When the first two lower-order moments, $E$ and $\text{Var}$, are called upon to match the parameters $\mu$ and $\sigma^2$ of a model such as that of the Normal density, this leaves the higher-order

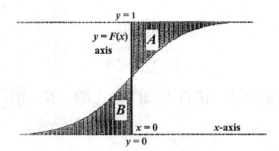

Figure 4.1. $E(X)$ in terms of cdf $F$.

moments free to help decide whether or not the actual density in question, say $f$, and this model can be matched validly.

It is noteworthy that as long ago as 1983, even though the two moments, $E(X)$ and $\text{Var}(X)$, appear in almost every chapter, Dixon and Massey chose to omit the two quantities that are based on the third and fourth moments, "skewness" and "kurtosis," from their book. Similar doubt was cast on the usefulness of $E$'s and Var's higher-order counterparts when the 1960 Kendall and Buckland dictionary defined skewness to be "An older and less preferable term for asymmetry..."; and when, in reference to K. Pearson's "moment ratio" (the fourth ordinary moment about the mean of the variate $X$, divided by $[\text{Var}(X)]^2$), they state that "It is doubtful, however, whether any single ratio can adequately measure the quality of 'peakedness.'"

Today's silence about higher-order moments resembles that famous Sherlock Holmes character, "the dog that didn't bark." Hence, following preliminary first- and second-order moment geometry, the remainder of this chapter will explain why it is that $E$ and Var's once noisy litter-mates are currently drowned out.

Much is learned from the graphical presentation of curve properties that cannot be gleaned from algebraic expressions. Like the absolute deviation $\text{AD}(X)$, the $E(X)$ curve property has a simple geometric interpretation in terms of the curve $F(x)$. While $\text{AD}(X)$ is the sum of two areas, such as those labeled with the letters $A$ and $B$ in Figure 4.1, the property $E(X)$ equals $A - B$, the difference between these two same areas. $E(X) = A - B$ geometry is valid for any finite area $A$, above the value $F(0)$, which is bounded by the cdf $F$ on the right and the $y$-axis on the left, and also lies below the horizontal line with height one, specifically, $y = 1$. Correspondingly, area $B$ of Figure 4.1 is below the cdf $F$ and has both the $y$-axis and the horizontal line $y = 0$ as borders.

## 4.1. E and Var Geometry

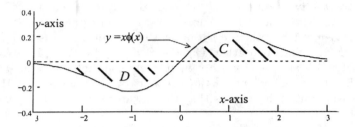

**Figure 4.2.** Interpretation of $E(X)$ in terms of $xf(x)$.

Figure 4.2 illustrates an alternative geometric interpretation of the curve property $E(X)$. For $f$ chosen to be the standard Normal density $\phi$, Figure 4.2 shows the composite curve $xf(x)$, a function formed by multiplying the density curve $f$ by the 45° slanted line that passes through the origin. Here, as was the case for Figure 4.1, the curve property $E(X)$ is the difference between two areas (denoted in Figure 4.2 by the letters $C$ and $D$). If $C$ represents the area over the positive part of the $x$-axis that is subtended by $xf(x)$ (in other words, $C$ is the area that faces and is under the curve $xf(x)$, while $D$ is the area over the curve $xf(x)$ and under the negative part of the $x$-axis) then $E(X) = C - D$.

For certain $f(x)$ curves that will be discussed below extensively, such as the Cauchy density $\sigma^{-1}\varphi([y-\mu]/\sigma)$, where

$$\varphi(z) = 1/\{\pi(1+z^2)\},$$

it will not be the case that the areas $C$ and $D$ are finite. One motivation for statistical techniques that are based on curve properties other than $E$ is that, for some simple and often-encountered curves such as

$$\varphi(z) = 1/\{\pi(1+z^2)\},$$

areas $C$ and $D$ are infinite.

Consider the generalizations of $C$ and $D$ for higher order moments

$$\int_{\text{MEDIAN}}^{\infty} [z - \text{MEDIAN}]^k f(z)dz - \int_{-\infty}^{\text{MEDIAN}} [z - \text{MEDIAN}]^k f(z)dz$$

or

$$\int_{-\infty}^{\infty} [z - E(X)]^k f(z)dz.$$

For $k > 1$ the two assumptions, (1) that no $f$-type curve assumes negative values, and (2) that all $f$-type curves must have integrals over the entire $x$-axis that equal 1, together do not imply that any of the areas over some infinite region defined beneath $xf(x)$, $[x - E(x)]^k f(x)$ or $[x - \text{MEDIAN}]f(x)$, are finite. Because no variate can, in reality, be infinite-valued, infinite areas like $C$ and $D$ are not major concerns by themselves. Yet their unlimited sizes are danger signs that the researcher should not—and often *cannot*—disregard.

If the curve property $\text{Var}(X : u, x)$ is defined by the relationship

$$\text{Var}(X : u, x) = \int_u^x [z - E(X)]^2 f(z)dz,$$

then the "variance," $\text{Var}(X)$, equals $\text{Var}(X : -\infty, \infty)$ or, alternatively,

$$E\left([Y - E(X)]^2\right).$$

Because $C$ and/or $D$ are not necessarily finite,

$$E\left([Y - E(X)]^2\right),$$

as well as the population standard deviation $\sqrt{\text{Var}(X)}$, are not always well-defined. In the sense that their variances are not being well-defined, there are important density curves whose variances do not exist. (In the next chapter it will be shown that one such density curve is encountered commonly when ratios of counts, such as the number of disease cases over the number of individuals at risk, are calculated.)

For a variate whose distribution is described validly by the standard Normal curve $\phi$, or any other density for which Var exists, the Var curve property

$$E\left([X - E(X)]^2\right)$$

equals the area under a counterpart of the curve $x^2 \phi(x)$ shown in Figure 4.3. This curve assumes the value zero over the point on the abscissa with coordinate $E(X)$, here $E(X) = 0$. Any curve

$$[x - E(X)]^2 f(x)$$

is everywhere nonnegative. Consequently, unlike the geometrical interpretations of $E(X)$ shown in Figures 4.1 or 4.2, the two areas $A$ and $B$ or, alternatively, the two areas $C$ and $D$, need not be determined separately

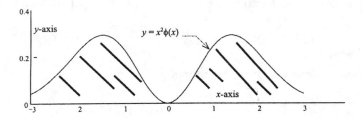

**Figure 4.3.** Geometric interpretation of Var($X$).

and then subtracted. Instead, the curve property Var($X$) is merely the area under the single curve

$$[x - E(X)]^2 f(x)$$

given, of course, that this area is finite.

## 4.2 Higher order moments and the indicator function

There have been many attempts to define skewness as a curve property (distinct from a curve parameter). Except for attempts at skewness quantification based on differences between the three properties, $E$, MEDIAN, and an assumed unique mode, all measures of skewness have been based on higher-order moments. Taken as a whole, and even proceeding one step further to include the curve property "kurtosis," little research is conducted today that concerns higher-order moment curve properties. One reason for this loss of interest stems from the same problem that is illustrated in Chapter 3 for the AD. The key issue, sensitivity to the extremes at the expense of the middle, applies with even greater force to conventional measures of skewness and kurtosis than it does for the AD and population standard deviation.

Continuing the convention where the AD is thought of as an order one absolute moment, and the variance is taken to be an order two ordinary moment, it follows that skewness can be defined either in terms of an order two absolute moment, or an order three ordinary moment (Here and below, for $r > 1$, the variate $X$'s order $r$ ordinary moment will be assumed to refer to the $r^{th}$ power deviation from $E(X)$, while the order $r$ absolute value moment refers to deviations from the MEDIAN curve property.)

Kurtosis is either expressed in terms of the order three absolute moment

$$\int_{\text{MEDIAN}}^{\infty} [z - \text{MEDIAN}]^3 f(z)dz - \int_{-\infty}^{\text{MEDIAN}} [z - \text{MEDIAN}]^3 f(z)dz$$

or, more commonly, the curve property

$$[\text{Var}(X)]^{-2} \int_{-\infty}^{\infty} [z - E(X)]^4 f(z)dz - 3.$$

Here the number three is subtracted from the ratio of the fourth ordinary moment to the squared variance in order to assure that this kurtosis-related curve property equals zero when $f$ is some Normal density.

Before illustrating what kurtosis is supposed to show about curve-model-alternatives, it should be noted that *kurtosis* is a composite property formed from the two simpler properties, Var itself and Var's fourth-order-moment counterpart. Therefore, the difficulties with kurtosis that are described in this chapter are related solely to the population under study, not to the data that are applied for the purpose of studying this population. The same sorts of problems with kurtosis that are discussed in this chapter all have their outlier-sample-value counterparts.

Because curve uncertainty issues often concern either some piece of a function or, alternatively, some subinterval of the entire $x$-axis, the indicator function, $I_{[a,b]}(x)$ is a valuable notational tool. The curve $I_{[a,b]}(x)$ is defined to equal one, both over any $x$ within the closed interval $[a, b]$, and over this interval's boundaries, while it is defined to equal 0 elsewhere. Its counterpart, $I_{(a,b]}(x)$, equals $I_{[a,b]}(x)$ everywhere except for the argument $x = a$, over which $I_{(a,b]}(x) = 0$. Similarly

$$I_{[a,b)}(x) = I_{[a,b]}(x)$$

when $x \neq b$, while $I_{[a,b)}(b) = 0$ and; finally,

$$I_{(a,b)}(x) = I_{[a,b]}(x),$$

except when $x = b$ or $x = a$, while

$$I_{(a,b)}(b) = I_{(a,b)}(a) = 0.$$

Technically speaking, since it cannot be represented by some finite assemblage of the log, exp, power, root, sin, cos and tan functions whose properties constitute the subject matter of high school intermediate algebra and trigonometry classes, the curve denoted by the upper case letter $I$ is not an elementary function. Hence, strictly speaking, the function $I_{[a,b)}(x)$, like the inverse standard Normal cdf $\Phi^{-1}(x)$, is a curve that cannot be represented exactly by any model. Nevertheless, like the representation $\Phi^{-1}(x)$, the symbolic representation $I_{[a,b)}(x)$ is a useful sequence of symbols. It is useful because any curve that $I_{[a,b)}(x)$ multiplies is, in effect, blanked or zeroed out beyond $[a,b)$. For instance, with the aid of the special function $I_{[a,b)}(x)$, one example of the important curve known as an "exponential density" is defined by the single expression $e^{-x}I_{[0,\infty)}(x)$; in notation that avoids the complex representations, $f(x) = e^{-x}$ for $x \geq 0$, and $f(x) = 0$ elsewhere.

## 4.3 Early statistical models

Johnson and Kotz (1970) summarize the early history of the Normal curve as follows: "The earliest published derivative of the Normal distribution (as an approximation to the binomial distribution) seems to be that in a pamphlet of de Moivre, dated 12 November 1733.... In 1774, Laplace obtained the Normal distribution as an approximation to [the] hypergeometric distribution, and four years later, he advocated tabulation of the probability integral ($\Phi(x)$ in our notation). The work of Gauss in 1809 and 1816 respectively established techniques based on the Normal distribution, which became standard methods used during the nineteenth century."

The Johnson and Kotz quotation raises two issues: Why, if the French mathematician Laplace played such an important role in the early history of Normal curve investigations, is the Normal referenced by Kendall and Buckland under the heading Gauss Distribution? In addition, why is the density model $f(x) = (2\sigma)^{-1}\exp(-|x - \mu|/\sigma)$ known as Laplace's first law of error?

A clue comes from what James and James (1959) call the "Laplace transform,"

$$h(x) = \int e^{-xt} f(t) dt.$$

No matter what interpretation is given here to the integral sign that defines $h$, it is obvious that an $f$-type curve proportionate to $\exp(-|x-\mu|/\sigma)$, when substituted into the definition of the Laplace transform, yields a simple expression. While the Normal density, like its Laplace first law of error counterpart, also has a simple Laplace transform, it seems, at least at first glance, more natural to blend the $-xt$ exponent of $e^{-xt}$ within the Laplace transform integrand together with the first power of $x$, than it is to blend the $-xt$ exponent and some squared quantity, such as the exponent of the Normal density model.

Yet, despite Laplace's clear preference for $\exp(-|x-\mu|/\sigma)$, this curve is not often chosen today to be the Normal's close competitor. Why this is the case is illustrated by a three-step process that can be called upon to generate artificial, Laplace-distributed variates. As a first step, a quantity called a standard *uniform* or *rectangularly* distributed variate is generated, let's call it $Z$, in such a way that $Z$ has a density $I_{[0,1)}(z)$. Once $Z$ is obtained, its value $Z = z$ serves as the argument of a function $F^{-1}$, where $F$ is the cdf that corresponds to the Laplace, or other density, of the variate $F^{-1}(Z)$. Finally, in order to obtain the value of any one of the many types of variate that are discussed in later chapters, desired parameter values are affixed to $F^{-1}(Z)$.

The way the standard Laplace variate

$$F^{-1}(Z) = I_{(0,0.5]}(Z)ln(2Z) - I_{(0.5,1)}(Z)ln(2(1-Z))$$

is defined, suggests that nature has changed her mind within some neighborhood of 0.5. She has done so in the sense that $-ln(2(1-Z))$ is substituted for $ln(2Z)$ when the running variable value z assumed by variate Z passes beyond 0.5 and then proceeds through the 0.5 intersection. Because of this change from ln(2Z) to -ln(2(1-Z)), the slope of the Laplace density has a mathematical discontinuity at this curve's mode.

Does nature change her mind exactly where a density assumes its maximum value? A negative answer to this question suggests that the Laplace variate

$$I_{(0,0.5]}(Z)ln(2Z) - I_{(0.5,1)}(Z)ln(2(1-Z))$$

should be replaced by the variate $ln(Z) - ln(1-Z)$. Unlike any Laplace variate's density's slope, any $ln(Z) - ln(1-Z)$ variate's density's slope is continuous at zero. (When simplified, $ln(z) - ln(1-z)$ equals the "logit function," $ln[z/(1-z)]$.)

The density of the variate $ln[Z/(1-Z)]$ does not have a sharp peak over its mode at zero. On the other hand, for both large and small extreme values, in other words, for values that occur within tail regions, the logistic variate $ln[Z/(1-Z)]$ has statistical properties that resemble those of its Laplace variate counterpart.

As illustrated by Figure 4.4, logistic and Normal distributions often resemble each other. Nevertheless, the conventional model chosen to represent the standard Normal density has defined away a problem that must be attended to in the logistic case. Consider, for example, the number 2 that appears within

$$I_{(0,0.5]}(Z)ln(2Z) - I_{(0.5,1)}(Z)ln(2(1-Z))$$

## 4.3. Early statistical models

as a coefficient of both the $Z$ of $ln(2Z)$, and the $(1-Z)$ term of $ln(2(1-Z))$. Both the logistic variate $ln[Z/(1-Z)]$ and the logit function $ln[z/(1-z)]$ are defined without the coefficient 2 and, as a consequence, a slight modification is required before logistic and Normal curves can be compared meaningfully.

As will be commented upon in several later chapters, the reason why the parameter $\sigma$ within the general Normal density model

$$[\sqrt{2\pi}\sigma]^{-1}\exp[-(1/2)[(x-\mu)/\sigma]^2]$$

can be taken to equal the curve property standard deviation $\sqrt{\mathrm{Var}(X)}$, is that the standard Normal model upon which

$$[\sqrt{2\pi}\sigma]^{-1}\exp[-(1/2)[(x-\mu)/\sigma]^2]$$

is based is defined with a quantity $(1/2)$ within the exponent of

$$\exp[-(1/2)[(x-\mu)/\sigma]^2].$$

Unlike the standard Normal variate $\Phi^{-1}(Z)$, the variate $ln[Z/(1-Z)]$ does not have either unit variance or standard deviation equal to 1. For this reason the author (1968) determined a constant 0.55133 that can help match a logistic variate to its Normal counterpart. The value, C = 0.55133, minimizes

$$E([Cln[Z/(1-Z)] - \Phi^{-1}(Z)]^2),$$

a quantity called the *mean squared error* (MSE). Here the logistic density that corresponds best to the standard Normal density is

$$g(x) = 1/[0.55133(2 + \exp\{x/0.55133\} + \exp\{-x/0.55133\})].$$

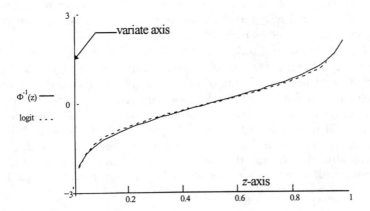

Figure 4.4. Logit curve fit to the inverse normal cdf.

The actual functions that begin with a value for $Z$ and end with either a standard Normal or a $g(x)$ distributed variate are shown in Figure 4.4. Figure 4.4 illustrates how similar the logit and Normal inverse cdf appear. However, like a comparison based on cdf curve types, inverse cdfs can appear similar even though there are important distinctions between distributions, distinctions that are emphasized by other types of curves. For this reason, we will now consider how well the curve property kurtosis can serve as a litmus test for distinguishing between Normal, logistic and Cauchy distributions. It will be shown that although kurtosis has considerable curve-uncertainty-resolution potential, this potential can be harnessed to solve only a few real-world problems. Before considering kurtosis further, some background concerning statistical models will be introduced.

## 4.4 Early statistical models and higher order moments

Describing $f$ as a symmetric curve implies that there exists some value $K$ for which $f(x)$ is identically equal to $f(K - x)$. The curve known as the lognormal density was the first suggested example of an $f$-type curve that, unlike the Normal, Laplace, Cauchy and logistic densities, is not symmetric. Although many, if not most, densities of variates observed in nature have been found to be asymmetric, sixty-three years separated Gauss's 1816 publication that linked the Normal density to the sum of independent variates from Galton's seminal study of the lognormal model (Johnson and Kotz 1970). In 1879 Galton associated the product of independent and positive variates to what is now called a *lognormal limiting distribution*. Following Galton's work, in 1903 Kapteyn wrote his definitive book *Skew Frequency Curves in Biology and Statistics* within which the asymmetric lognormal density occupies a place of honor.

The Normal, Laplace, Cauchy and logistic densities are symmetric. In general, prior to 1879 the skewness of an $f$-type curve was, in effect, deemed irrelevant. This historical ordering of priority will be maintained throughout this chapter, in the sense that the ideas that underlie kurtosis will be presented here, while characteristics of asymmetric curves like the lognormal density will be considered in later chapters. As will be seen below, curve symmetry does, indeed, raise new and difficult conceptual issues.

Traditionally, the Normal density has been taken to provide a zero-valued benchmark about which alternative curves have been indexed. For example, suppose

$$\mathrm{AD}_3 = \int_{\mathrm{MEDIAN}}^{\infty} [z - \mathrm{MEDIAN}]^3 f(z)dz - \int_{-\infty}^{\mathrm{MEDIAN}} [z - \mathrm{MEDIAN}]^3 f(z)dz$$

## 4.4. Early statistical models and higher order moments

is calculated for some Normal $f$ that is thought to describe the density of the variate $\mu + \sigma\Phi^{-1}(Z)$. The MEDIAN here equals $\mu$ and, consequently,

$$\text{AD}_3 = \sigma^3 \left[ \int_{0.5}^{1} \left[\Phi^{-1}(y)\right]^3 dy - \int_{0}^{0.5} \left[\Phi^{-1}(y)\right]^3 dy \right].$$

When data are actually normally distributed with a standard deviation curve property equal to $\sigma$, since

$$\left[ \int_{0.5}^{1} \left[\Phi^{-1}(y)\right]^3 dy - \int_{0}^{0.5} \left[\Phi^{-1}(y)\right]^3 dy \right] = 1.5957691,$$

the curve property AD3, when divided by $\sigma^3$ and then reduced in value by the constant 1.5957691, provides a zero-valued baseline toward which kurtosis estimators can be targeted.

Continuing this line of reasoning regarding kurtosis, if the fourth ordinary moment is applied in place of the third absolute value moment, the integer 3 of

$$[\text{Var}(X)]^{-2} \int_{-\infty}^{\infty} [z - E(X)]^4 f(z)dz - 3$$

plays the role that is played by the constant 1.5957691 in the previous paragraph. A comparison between the Normal and logistic densities will now illustrate that serious practical difficulties arise in many applications that greatly limit the value of fourth ordinary moment-based kurtosis.

Unlike simulated Normal, Laplace, logistic or Cauchy data obtained by calling upon the uniformly distributed variate $Z$ with density $I_{[0,1)}(z)$, natural data assumes values that are bounded. Hence the symmetrically truncated Normal is an important statistical density. However, truncation that occurs symmetrically about the curve's unique mode is an even less realistic assumption than Laplace's assumption that nature changes her mind at this same mode. Nevertheless, until asymmetric truncation is considered in depth in later chapters, for our purposes it will be assumed here that the density curve in question assumes the value zero between the two points at which the curve is truncated, the first $b$ units to the left and the second $b$ units to the right, of the point $x = 0$.

For example, when $g$ represents either the standard Normal $\phi$, or the standard logistic density

$$1/[0.55133(2 + \exp\{x/0.55133\} + \exp\{-x/0.55133\})],$$

then it follows that the truncated density in both cases can be represented as

$$I_{[-b,b]}(x)g(x)/[G(b) - G(-b)].$$

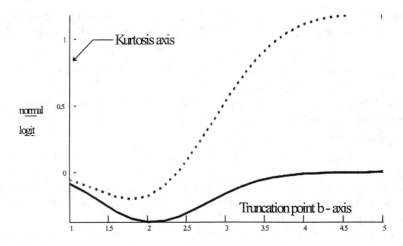

**Figure 4.5.** The kurtosis of normal and logistic densities as a function of the amount of truncation.

Here, when $G$ designates the cdf that corresponds to pdf $g$, the divisor $[G(b) - G(-b)]$ assures that the area beneath the truncated curve

$$I_{[-b,b]}(x)g(x)/[G(b) - G(-b)]$$

integrates to one.

Figure 4.5 was calculated in order to illustrate the effect that truncation has on kurtosis. (It is helpful to keep the well-known value $z = 1.96$, in mind, since for a variate $Z$ that has a standard Normal density, $\Phi$ almost exactly 95% of the area under $\phi$ is located over the interval [-1.96, 1.96].

The degree of Normal density truncation was matched to its logistic counterpart, the truncated modification of

$$g(x) = 1/[0.55133(2 + \exp\{x/0.55133\} + \exp\{-x/0.55133\})],$$

by setting the total area under $g$ snipped off symmetrically over each tail equal to the amount snipped off the Normal's tails. (Chapter 20 considers the implications of truncated model asymmetry from the point of view of applied research.)

Due to the constant 3 that is subtracted from the fourth-ordinary-moment-over-the-squared-variance ratio, as predicted, beyond $z = 4$ the solid curve shown in Figure 4.5 has a height that is almost exactly equal to zero. For values of $z$ that occur between 1.96 and 2.5, it is the truncated logistic, not the truncated Normal, whose kurtosis is nearly zero. This

## 4.5  Curve sub-types and model choice

example illustrates that even in the rare situation where tails are symmetrically snipped off in exactly the same amounts at each side of a density, kurtosis can be trusted to distinguish between Normal and logistic models solely when a very light trim has merely snipped off each tail's tip.

## 4.5  Curve sub-types and model choice

A truncated density can be thought of as a specialized subtype of curve. Consider some finite value of either the constant $a$ or the constant $b$ or, alternatively, finite values of both constants. A truncated density can be formed by first multiplying some curve like a Normal density

$$[\sqrt{2\pi}\sigma]^{-1}\exp[-(1/2)[(x-\mu)/\sigma]^2],$$

a logistic density $g((x-\mu)/\sigma)/\sigma$, where

$$g(x) = 1/[0.55133(2 + \exp\{x/0.55133\} + \exp\{-x/0.55133\})],$$

or a Cauchy density

$$\sigma^{-1}\varphi\left([x-\mu]/\sigma\right),$$

where $\varphi(z) = 1/\{\pi(1+z^2)\}$, by the indicator function $I_{[a,b)}(x)$. Then, as a second step, this product must be divided by some constant like $G(b) - G(-b)$ to obtain a truncated curve representation like

$$f(x) = I_{[-b,b]}(x)g(x)/[G(b) - G(-b)].$$

Hence Figure 4.5 illustrates how uncertainty concerning curve subtype, here truncated versus non truncated, can be related to model-choice uncertainty, for this figure, the Normal model versus the logistic model.

There is an even more general issue than that of curve-subtype-with-model-choice uncertainties. Questions regarding truncation are subsumed by the interrogative *where*, while many alternative subclassifications of curve uncertainty are subsumed by the interrogative *why*. For example, if data cannot be trusted to provide accurate estimates over tail regions because a curve under study is likely to be a truncated subtype, then higher-order moments cannot be trusted to help choose between elementary function model alternatives and if data cannot be trusted to provide accurate estimates over tail regions because there may be one or more lurking variates then, until checking procedures assure that some lurking variable's value does not index an outlier data subset, higher-order moments cannot be trusted. (In general, because of the advantages of birds in the hand, it is easier to deal with the where variety of uncertainty, as in

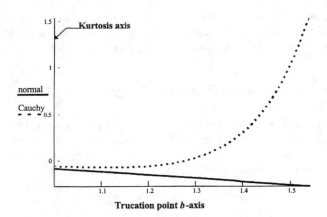

**Figure 4.6.** The kurtosis of normal and Cauchy densities as a function of the amount of truncation.

truncated curves, than it is to deal with the why variety of uncertainty, as in lurking variates.)

As far as truncation-induced uncertainty is concerned, an $f$-type curve can have tails that are so heavy that even their stumps are identifiable clearly. For example, Figure 4.6 shows the Normal-with-Cauchy equivalent of the Normal-with-logistic comparison shown by Figure 4.5. Here a standard Cauchy model was chosen so that both its median, $F^{-1}(0.5)$, and its interquartile range, $F^{-1}(0.75) - F^{-1}(0.25)$, equal that of the standard Normal $\phi$. (No constant $C$ exists that equates Normal and Cauchy standard deviations in the same sense that

$$E([Cln[Z/(1-Z)] - \Phi^{-1}(Z)]^2)$$

is minimized. Later chapters discuss the effect Cauchy constant $C$'s nonexistence has on uncertainty about curve symmetry.)

The Cauchy model has such pronounced kurtosis that, as long as truncation occurs beyond the value 1.6, the truncated standard Normal is unlikely to be mistaken for its truncated Cauchy counterpart. Yet, this example notwithstanding, it would be handy to have some way to distinguish curves that, unlike the kurtosis function shown in Figure 4.6, draws attention to central portion differences, not tail differences. We will see in several later sections that it pays to attend to those regions within which empirical evidence abounds and de-emphasize regions where information is hard to come by.

# CHAPTER 5
# Goodness of fit

## 5.1 Neyman's and alternative criteria

Like goodness of fit for suits, dresses or trousers, one curve may fit another well in some places and fit badly in others. For this reason, Neyman (1949, p. 259) discusses a modified version of the Poisson density; a curve that in its untruncated form equals $e^{-\lambda}\lambda^x/x!$ for $x = 0, 1, 2, \ldots$ and equals 0 elsewhere. Consider, for example, a fictional population comprised of sultans who father 50 or more children. Suppose the variate of interest here is the number of sons within such a family who at some time in their lives are more than 6.3 feet tall. Then, because its value is determined by counting an exact number of rare events, variates like this are often taken to be "Poisson distributed."

Here the Poisson is based on the assumption that the chance of giving birth to a son who will grow to be 6.3 feet or taller is a constant. For instance, odds do not vary from sultan to sultan according to the values assumed by some lurking variate. (Mothers' heights might be family-specific, in the sense that there might be two types of sultans, those who have short wives and those who have tall wives—hence the lurking variate, wife height, that casts doubt on the unmixed Poisson's validity.)

*Hypothesis testing* (HT) has an immense literature that only intersects curve uncertainty reduction issues in a few places. This intersection is skimpy primarily because the early development of HT took place during the era when computer methodology was primitive. Consequently, a considerable gap separated the pioneering work of K. Pearson, R. A. Fisher, and J. Neyman from the careful robustness studies conducted since the 1960s.

Details of one topic germane to both curve uncertainty and HT are discussed and illustrated in Chapters 18 and 19. These chapters have, as their goal, uncertainty reduction based on the robust procedure known as "best linear unbiased estimation." Unfortunately, it is premature to discuss this topic here because there are so many issues that are central to curve uncertainty reduction that must be covered first. What if a reader is not already familiar with basic HT ideas such as those that concern null and alternative hypotheses? To answer this question, enough introductory material is placed in the early sections of Chapters 18 and 19, so that a reader who would like to review HT basics before proceeding further can turn to these sections.

In his 1946 paper, the Poisson model was chosen by J. Neyman to illustrate his seminal "chi-square" goodness of fit testing procedure. Neyman's test compares the null hypothesis, that over a finite number of integer values the variate $X$ is Poisson-distributed with density $e^{-\lambda}\lambda^x/x!$, to the alternative that, in reality, unlike the specific model $e^{-\lambda}\lambda^x/x!$, no prior information concerns the probability that variate $X$ assumes the value $x$.

Neyman considered the modification

$$f(x) = I_{[0,\nu-2]}(x) e^{-\lambda}\lambda^x/x!$$

of the usual Poisson model where, for him, $\nu$ was taken to be a known, but possibly large-valued integer such that $\nu > 3$, while $f(\nu-1)$ was set equal to the quantity

$$1 - \sum_{j=0}^{\nu-2} f(j).$$

By defining $f(\nu-1)$ in this way,

$$\sum_{j=0}^{\nu-1} f(j)$$

equals one, an assumption that is for the Poisson what the assumption, that the area under $f$ equals one, is for a continuous density $f$. Thus here the interval $[0,\nu-1]$ is assumed to be of finite length, an assumption that is tantamount to assuming first that the Poisson density is truncated and then defining $f(\nu-1)$ accordingly.

While his null hypothesis tentatively assumes truncated Poisson model validity, Neyman's alternative hypothesis asserts the much more general assumption that for any integer $x$ less than the value $\nu-1$, each density curve evaluation $f(x)$ is an arbitrary-valued curve property. Since there are $\nu-1$ of these distinct curve properties, the alternative amounts to a sequence that contains $\nu-1$ distinct parameters.

Neyman's restriction that the value assumed by $\nu$ be finite and hence, that a density being studied is truncated, was characteristic of this perceptive thinker. Many of the topics considered later in this monograph are

## 5.1. Neyman's and alternative criteria

motivated by the dilemma that an infinite number of possibilities cannot be stacked up easily against a finite number of possibilities. For Neyman, one model that is based on one parameter is stacked up against a second, more general model, one that is based on $\nu - 1$ parameters. (Note, the Poisson is defined over the value zero and hence, under the alternative hypothesis, the curve properties taken as parameters are $f(0), f(1), ..., f(\nu - 2)$, because $f(\nu - 1)$ must equal

$$1 - \sum_{j=0}^{\nu-2} f(j)$$

in both the null and the alternative cases.

Unlike Neyman's choice of truncated Poisson, the untruncated Poisson allows for the possibility that $\nu$ is actually infinite. For Neyman, an infinite value of $\nu$ would have been an unfortunate choice, since when he wrote his classic paper, values within a set of infinite size could not be stacked up against the single parameter $\lambda$. The finite value of $\nu$ permits the numerical difference between the number of parameters specified by the null and the alternative hypotheses to be determined. This difference is crucial both to Neyman's, and to popular goodness of fit procedures based on what is called *likelihood ratio* (LR). As mentioned previously, curves can resemble trousers, in the sense that they rarely fit equally well everywhere. For this reason, the smaller $\nu$'s value, the easier it will be to guard against local loose fit, and conversely, an infinite $\nu$ causes big trouble.

For a candidate curve estimator $\hat{f}$ chosen to estimate some density curve $f$, Neyman's goodness of fit criterion is a modification of a much earlier criterion defined by K. Pearson which, in modern notation, can be expressed as

$$\int_{-\infty}^{\infty} \left\{ \left[\hat{f}(x) - f(x)\right]^2 / f(x) \right\} dx.$$

Neyman chose the estimator, rather than the estimated, curve as the denominator of his criterion

$$\int_{-\infty}^{\infty} \left\{ \left[\hat{f}(x) - f(x)\right]^2 / \hat{f}(x) \right\} dx.$$

Much of today's curve estimation literature is based on a simplified version of both Pearson's and Neyman's criteria that is called the "integrated square error," (ISE),

$$\int_{-\infty}^{\infty} \left[\hat{f}(x) - f(x)\right]^2 dx.$$

Besides its computational simplicity, the ISE criterion often places wanted emphasis on the middle of a targeted curve. Tarter and Lock (1993) discuss in Chapters 3 and 4 the details of the ISE's center-weighting characteristics.

The ISE is easy to work with when both $\hat{f}$ and $f$ are expressed as Fourier series and, also, when $\hat{f}$ is one of today's many generalizations of the conventional histogram based on what are called *kernels*. A fourth criterion, $\max|\hat{F}(x) - F(x)|$, underlies the well-known Kolmogoroff-Smirnoff test, and assesses goodness of fit in terms of the maximum difference between estimated and hypothesized cdfs.

Besides the four general goodness of fit criteria above, there are many specialized techniques for curve uncertainty reduction. Most of these approaches depend on particular relationships between model parameters and curve properties. For example, suppose $\bar{x}$ represents the arithmetic average, in other words, sample mean

$$n^{-1} \sum_{j=1}^{n} x_j$$

of the $n$ values assumed by variates from a random sample. Then the specialized goodness of fit measure referred to as the "Poisson index of dispersion,"

$$\sum_{j=1}^{n} (x_i - \bar{x})^2 / \bar{x},$$

is called upon in circumstances where parameter $P$ of the binomial $f(x) = [n!/(x!(n-x)!)]P^x(1-P)^{n-x}$ discussed in Section 3.4 is known to be very small and $n$ is known to be very large.

The Poisson index of dispersion can be thought of as the limiting case of a quantity known as the *binomial index of dispersion*,

$$\sum_{j=1}^{n} (x_i - \bar{x})^2 / [\bar{x}(n - \bar{x})/n]$$

which is, as its name implies, a check of binomial $f$-type curve validity. At the end of this chapter it will be shown geometrically why the statistic $\bar{x}(n - \bar{x})/n$ is called upon in the binomial case to play the role assumed by the Poisson index of dispersion's denominator $\bar{x}$. The remainder of this section will consider commonalties and differences between the binomial and Poisson indices of dispersion on the one hand, and estimators of the Pearson and Neyman Chi-square, ISE and Kolmogoroff-Smirnoff criteria on the other.

An obvious difference between

$$\int_{-\infty}^{\infty} \left[\hat{f}(x) - f(x)\right]^2 dx$$

## 5.1. Neyman's and alternative criteria

and
$$\sum_{j=1}^{n}(x_i - \bar{x})^2/[\bar{x}(n-\bar{x})/n]$$

is that, in the former expression, what the two curves that $\hat{f}$ and $f$ actually represent are not spelled out in detail while, in the latter, an actual formula specifies exactly how the binomial index of dispersion is calculated. Yet, despite the specificity of the binomial and Poisson indices, the rationale for these indices is not obvious. On the other hand, it is easy to see what the ISE as well as Pearson's and Neyman's goodness of fit criteria are driving at, even though, as will be illustrated in Section 5.5, the role played by the denominators of the latter two criteria is not perfectly obvious.

The rationale for binomial and Poisson indices of dispersion depends upon the simplicity of both the $e^{-\lambda}\lambda^x/x!$ and the $[m!/(x!(m-x)!)]P^x(1-P)^{m-x}$ density curve models. The value of $m$ is taken to be a known constant in most binomial applications. Consequently, unlike the general Normal density model, whose assembled symbols include the two parameters $\mu$ and $\sigma$; in the Poisson case the parameter $\lambda$, and in the binomial case the parameter $P$, are the sole means by which the model can be altered in order to achieve a good fit.

There is only a single parameter within the expression $e^{-\lambda}\lambda^x/x!$. Similarly, because $m$ is taken to be a known constant, $P$ is taken to be the sole parameter within the expression $[m!/(x!(m-x)!)]P^x(1-P)^{m-x}$. Thus for both the conventional Poisson and binomial models there is only a single parameter and, consequently, unlike the Normal case in which a location-related property like $E(X)$ or MEDIAN is related to lone parameter $\mu$, while any scale-related property like AD or $\sqrt{\text{Var}(X)}$ has the leeway to be related to a second sole scale-related parameter $\sigma$, in both the Poisson and binomial cases all four curve properties; $E(X)$, MEDIAN, AD and $\sqrt{\text{Var}(X)}$; are, technically speaking, functionally related to a single parameter.

Different Poisson models can have both different scale-related properties like $\sqrt{\text{Var}(X)}$ and different location-related properties like $E(X)$. Therefore if the value nature assigns to the parameter $\lambda$ is thought of as a control-setting; such as that achieved by turning a dial, then the turning of this single dial must, in turn, alter $\sqrt{\text{Var}(X)}$ and $E(X)$. Hence given that the two properties $\sqrt{\text{Var}(X)}$ and $E(X)$ can both change as $\lambda$ changes, a necessary condition for some density $f$ to be represented by the Poisson model is that there exists some relationship between the two properties $\sqrt{\text{Var}(X)}$ and $E(X)$.

In the Poisson case $\text{Var}(X)$ must equal $E(X)$. In the binomial case where, for example, the variate in question is the number of sons $X$ in a family of small to moderate size $m$, if

$$f(x) = [m!/(x!(m-x)!)]P^x(1-P)^{m-x},$$

it follows that $E(X) = mP$ and $\text{Var}(X) = mP(1-P)$. Hence, the two indices

$$\sum_{i=1}^{n}(x_i - \bar{x})^2/\bar{x}$$

and

$$\sum_{j=1}^{n}(x_i - \bar{x})^2/[\bar{x}(n-\bar{x})/n]$$

call upon the correspondences between dual curve properties and a single model parameter in order to help make decisions concerning Poisson and binomial goodness of fit.

Hoel's (1954) pioneering text illustrates how these two indices are applied in conjunction with the important chi-square density curve. However, as far as the general ideas that underlie curve uncertainty resolution are concerned, two points are worth mentioning. The first is that the above two indices have much more in common with estimators of kurtosis than they have with ISE, as well as both Pearson's and Neyman's criteria. Low kurtosis when Normality is of interest, as well as small values (relative to some tabled chi-square inverse cdf evaluation) assumed by Poisson and binomial indices, are necessary—but not sufficient—conditions for good fit. As elaborated upon in Chapters 18 and 19, tests are designed with an eye to rejecting—not to accepting—hypotheses, such as that kurtosis equals zero and, as a consequence, an accepted null hypothesis has lulled more than one researcher into a false sense of security.

The second point that is confusing is far less vital than the point that concerns necessary, but not sufficient, conditions. The term "chi-square" has a variety of referents. For instance, when a Poisson and/or binomial index is stacked up against some tabled chi-square inverse cdf evaluation, it may turn out that both they, and the chi-squared goodness of fit criterion, suggest the same conclusion. Yet it is three distinct entities called, respectively: *contingency tables;* *hypothesis tests;* and *densities,* that often are prefixed by the term chi-square. Hence below, for the purpose of clarity, the criteria,

$$\int_{-\infty}^{\infty}\left\{\left[\hat{f}(x) - f(x)\right]^2/f(x)\right\}dx$$

and

$$\int_{-\infty}^{\infty}\left\{\left[\hat{f}(x) - f(x)\right]^2/\hat{f}(x)\right\}dx,$$

will be designated as Pearson's *criterion* and Neyman's *criterion*, respectively, even though in many books and papers, they are referred to as forms of *chi-square* goodness of fit.

## 5.2  Criteria, metrics and estimators

The two criteria, Pearson's

$$\int_{-\infty}^{\infty} \left\{ \left[\hat{f}(x) - f(x)\right]^2 / f(x) \right\} dx$$

and Neyman's

$$\int_{-\infty}^{\infty} \left\{ \left[\hat{f}(x) - f(x)\right]^2 / \hat{f}(x) \right\} dx,$$

are both defined in terms of both an estimator $\hat{f}(x)$, and the curve $f(x)$ that is targeted by the estimation process. Yet both

$$\sum_{j=1}^{n} (x_i - \bar{x})^2 / \bar{x}$$

as well as

$$\sum_{j=1}^{n} (x_i - \bar{x})^2 / [\bar{x}(n - \bar{x})/n]$$

are themselves estimators and, consequently, can be calculated from sample values without the need to specify some specific choice of $f$. Instead of a specified $f$, these two indices rely on known relationships between curve properties like $\sqrt{\text{Var}(X)}$ and $E(X)$ for some specific density model like that of the Poisson density.

Despite the difference between the two criteria and the two indices with respect to blanks to be filled in, all four of these statistics share a common feature. They vary in value from sample to sample. Each in its own individual way helps blend the information provided by some single sample with some tentative model or other assumed curve representation. In other words, they all are validity checks.

An almost universal method is often called upon to augment some goodness of fit criteria's negotiability. The symbol $E$ is placed prior to the symbols

$$\int_{-\infty}^{\infty} \left\{ \left[\hat{f}(x) - f(x)\right]^2 / f(x) \right\} dx \quad \text{or} \quad \int_{-\infty}^{\infty} \left[\hat{f}(x) - f(x)\right]^2 dx$$

to define the two metrics, the Pearson metric,

$$E \int_{-\infty}^{\infty} \left\{ \left[ \hat{f}(x) - f(x) \right]^2 / f(x) \right\} dx,$$

and the Pearson metric's mean integrated square error (MISE) counterpart. Yet, despite the seeming similarity between criteria and metrics, in practice they are applied to deal with distinct statistical problems.

Chapter 13 will deal with curve estimation details that are related to distribution uncertainty. However here it should be mentioned that the symbol $\hat{f}$ within the formula for a metric is specified often by one of three alternative representational methods: (1) generalized histograms based on averaged kernels; (2) Fourier series and wavelets; (3) spline functions. This monograph is not written with an eye to elaborating the advantages of, and trade-offs between, these methods. All three approaches can represent all but the most far fetched of conjectured curves and any realistic curve associated with actual data.

The change of emphasis from criterion to metric is based on the interpretation of the curve $f$. Within a criterion, this symbol represents some guess about some density's structure. Within a metric, this symbol represents the true, but of course unknown, density. This suggests the crucial question, how can the MISE or some other metric whose argument is $f$ actually be applied when the curve $f$ is the unknown target that is aimed at when a population is sampled?

This question is answered better by example than it is by any intrinsic explanation. A basic statistical tool is considered in detail in Chapter 18, where this tool is referred to as the *standard error of the mean*. This quantity is often represented by the subscripted Greek letter $\sigma_{\bar{x}}$. The estimation of $\sigma_{\bar{x}}$ is a vital step in the process that often leads to decisions concerning one or more location parameters, for example, the single parameter $\mu$. It is not necessary to actually know exact value of $\mu$ in order to gainfully employ the concept of standard error. Neither is it necessary to know the exact shape or other structural feature of $f$ in order to successfully employ the concept "MISE."

Conceding the point that some model-free or nonparametric curve estimator $\hat{f}$ can be found that has desirable MISE, brings us to the conjecture that is central to this monograph. An accurate, in the MISE sense of this word, curve estimator $\hat{f}$ can be inserted into a criteria such as

$$\int_{-\infty}^{\infty} \left[ \hat{f}(x) - f(x) \right]^2 dx,$$

and then a candidate model or another choice of curve representation can be inserted in place of $f$.

## 5.2. Criteria, metrics and estimators

Before dismissing the above metric-then-goodness-of-fit approach as being roundabout, it is worth reconsidering the many instances where the term chi-square goodness of fit appears in the statistical literature, as well as in the many applications of the Kolmogoroff-Smirnoff (KS) criterion. Both classes of technique do informally the same thing that the metric-followed-by-goodness-of-fit approaches do formally. In the case of chi-square, it is the conventional histogram, and for KS, it is the cdf estimator known as the sample cdf, that play the same role that a MISE-determined $\hat{f}$ plays within

$$\int_{-\infty}^{\infty} \left[\hat{f}(x) - f(x)\right]^2 dx$$

(Tarter and Lock (1993, Section 8.3) considers the relationship between certain curve types like $f$ and $F$ and estimator optimality. For example, under what conditions will a MISE-determined density estimator $\hat{f}$ yield a reasonable cdf estimator when estimator $\hat{f}$ is integrated over the appropriate interval?)

Prior to the extensive attention paid in the last thirty years to nonparametric estimators of $f$-type and $F$-type curves, the exclusive sway of the conventional histogram and the sample cdf seemed reasonable. Today, for two distinct reasons, the histogram-sample-cdf cartel is breaking up. The most important reason for estimator divestiture has been increased interest in curves more complex than the marginal types of curve, $f(x)$ and $F(x)$. For example, many chapters below will be devoted individually to either uncertainty about conditional density, $f(y|x)$, or alternatively, joint density, $f(y,x)$. In addition, substantial consideration will be given to the question: Which of the two types of curve representation, $f(y|x)$ or $f(y,x)$, best describes the distribution that is the target of a statistical investigation?

A second area of application for histogram and sample cdf generalizations calls upon individual metrics to tune these curves so that they can better serve individual research needs. For example, both the Pearson and MISE metrics are special cases of the generalized MISE metric

$$E \int_{-\infty}^{\infty} \left[\hat{f}(x) - f(x)\right]^2 w(x)\, dx.$$

If low-dose contaminant exposure or, alternatively, long-term survival, is of particular interest, then the choice of the weight function w can be made accordingly.

With regard to the choice of weight function w, much can be gained by those interested in curve uncertainty issues from the designers of modern photographic equipment. For example, in the same way that the weight choice, $w(x) \equiv 1$, assures that the squared difference between an estimated curve and its estimator is the sole form of trade-off, a macro lens provides edge-to-edge image sharpness.

Even those who prefer unretouched photos often call upon lens choice, as well as the choice of some polarizing or other filter, to implement representational tradeoffs. In the sense that it originated from or, alternatively, was reflected from a three-dimensional object during the short time a shutter was open, the light that ends up on some two-dimensional piece of film has been distorted by some non uniform counterpart of $w$. The capacity to select $w(x)$ within a particular metric, or to choose between several alternative metrics like those of Pearson, Neyman, Kolmogoroff, and Smirnoff, can provide the data analyst's equivalent to lens or filter choice. Unfortunately, tunable-metrics do have a substantial downside. As curve type complexity increases, the potential to harness and control the effects produced by these metrics diminishes substantially.

## 5.3 The Kolmogoroff-Smirnoff criteria

Much more is now known about generalized extensions of $f(y|x)$ and $f(y,x)$ curves than is known presently about the interplay of any metric's alternatives and generalizations. For example, consider the following little-known interpretation of the Kolmogoroff-Smirnoff metric or criterion.

Suppose both $F^*$ and $F$ of max $|F^*(x) - F(x)|$ are differentiable cdf's with associated pdf's, $f^*$ and $f$ respectively and, as shown in Figure 5.1, suppose that $f^*$ and $f$ intersect at a single point. Then the maximum

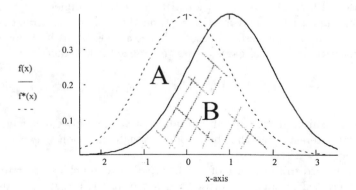

Figure 5.1. Alternative Kolmogoroff-Smirnoff interpretation.

value of the difference between the cdf's $F^*$ and $F$ must occur at the point where their derivatives are equal. In other words, when this point is designated by the subscripted symbol $x_{\min}$, then it must be the case that $f^*(x_{\min}) = f(x_{\min})$. The value $|F^*(x_{\min}) - F(x_{\min})|$ then must equal the area labeled A that lies between the two density curves shown in Figure 5.1. Consequently, since the crosshatched area labeled $B$, plus area $A$, equals 1; the area $B$ under the curve formed by the minimum of the two curves, $f^*(x)$ and $f(x)$, when subtracted from 1, here equals the Kolmogoroff-Smirnoff criterion max $|F^*(x) - F(x)|$. For instance, as the overlap of these two curves decreases to zero, it follows that the area $B$ will also decrease to zero. Correspondingly, $\max|F^*(x) - F(x)|$ will approach one.

It will be illustrated in many sections below that joint densities, as well as conditional densities such as $f(y_1, y_2, ...|x_1, x_2, ...)$, are much easier to work with than their cdf counterparts. Thus the $f$-type curve formulation of the Kolmogoroff-Smirnoff criterion has potential that cannot be realized by its cdf counterpart formulation.

What the symbols $x_1, x_2, ...$ that follow the bar symbol of

$$f(y_1, y_2, ...|x_1, x_2, ...)\text{'s}$$

argument actually represent is of vital importance to many questions that concern curve uncertainty. Appendix II illustrates what the term "identically distributed" (id) refers to in reference to slices of a distribution whose position and shape are not changed when one or more variables among $x_1, x_2, ...$ varies. Statistical analyses in which a researcher is uncertain about a *single* curve concern variates that are identically distributed. For this reason, it is easier to quantify uncertainties about identically distributed variates than it is to deal with their more complex counterparts that are not identically distributed and, as a consequence, involve multiple curves.

## 5.4 Bernoulli variation and the Cauchy density

Before considering extensions of simple marginal curve types like $f$ and $F$, there remains an additional example of these types of curve that can shed light on curve uncertainty issues. In common with models like those of the Normal, Laplace, triangular, and logistic densities; the Cauchy model describes a symmetric curve. Yet the Cauchy's tails are so heavy that "Cauchyness-detection" is one of, if not the most important of the many candidate pathological data alarm system components.

Events, such as the number of disease occurrences, are often counted. As a second step it is far from unusual to divide a count's value by the value assumed by some second variate, such as the number of individuals at risk. For fractions and ratios formed in this way uncertainty about Cauchyness must be attended to prior to many other data analysis considerations.

Suppose that in any individual trial, $q$ designates the chance of the value 1's occurrence, while $(1-q)$ designates the probability of 0's occurrence. Statistical texts sometimes refer to these occurrences of the values 0 and 1 in samples of fixed size $m$ as "Bernoulli variation" within a sequence of "Bernoulli trials." For instance, when, for the constant $q$, the binomial variate $X$ enumerates the number of occurrences $x$ of the value 1 within $m$ independent Bernoulli trials, then $X$'s distribution is described by the binomial density curve $f(x) = [m!/\{x!(m-x)!\}]q^x(1-q)^{m-x}$. (The statistical concept, *independently distributed* (id), is discussed in the first section of Appendix II, where its interpretation in terms of the concept, *randomly chosen subject*, is illustrated by a simple geometric example. Much about this topic is found in Cramer (1945, Section 14.4). Variates that are distributed both independently and identically are often referred to as iid variates.)

Besides disease occurrences, two supposed binomial variates, say two particular types of white blood cell count, often constitute the numerator and denominator of a ratio. For this reason it is a pity that, at first glance, what seems to be a simple way of comparing two statistics, the construction of their ratio, is subject to Cauchy variate vagaries. (There is, however, one application of Cauchy variates where these difficulties are, at least in part, counterbalanced by a simple connection between parameters.) In the remainder of this chapter, the term, "binary-valued variate" designates a measurement that takes on either the value 0 or 1 in the Bernoulli trial fashion, and does so in such a way that it is meaningful to take the parameters $m$ and $q$ to be constant-valued.

Before considering what is called *ratio variate Cauchyness*, first consider the proportion of those members that take on the value 1 within an id sample containing m Bernoulli trial binary-valued, 0/1, variates. The value 1 often is the code which designates that the subject has a specified disease, and the value 0 that the subject has not, at least as yet, been diagnosed as having the disease. This proportion variate can be conceptualized as the sample mean of binary variates, say $\bar{x}$, because the sum of $n$ ones or zeros, when divided by $n$, equals the *proportion* of 1's. For example, $(1+0+0+1+0+1+1)/7 = 4/7$, which is both the sample mean of the sequence of sample elements 1, 0, 0, 1, 0, 1, 1 and the proportion of ones in this sample.

The two-valued binary-valued Bernoulli trial variate has a variance that equals the product $q(1-q)$. A review of AV geometry illustrated by Figure 3.1 will help explain why the variance equals $q(1-q)$. Figure 5.2 is the

## 5.4. Bernoulli variation and the Cauchy density

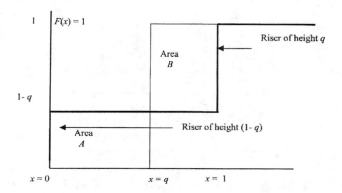

**Figure 5.2.** Binary AD curve property.

modified version of Figure 3.1 that pertains to a single Bernoulli binary-valued variate where the probability of the value 0's occurrence is $1-q$, and probability of 1's occurrence equals $q$. The cdf that describes this variate's distribution is shown as the heavy curve with two risers; one of height $1-q$ at 0, and the other of height $q$ at 1. (Unlike the degenerate $F_2$-distributed variate of Chapter 2, the binary-valued variate has a two step, rather than a one step, cdf.)

The AD-about-$E$ property equals the sum of areas $A$ and $B$, specifically, $2q(1-q)$. (The rectangles that have these areas are both bordered by one of the step function's risers.) Variance curve property geometry can be based on a modification of AD geometry. In place of the absolute deviations of the risers from the $E$ curve property that is equal to $q$, instead, for the variance, the squares, not the absolute values, of these deviations contribute to Var.

This is shown by Figure 5.3, which is identical to Figure 5.2, with the exception that here new vertical line segments replace what had been, for the AD, step function risers.

What had been rectangle $B$'s left border is now located over the squared deviation, $(1-q)^2$ units away from the $E = q$ curve property. Similarly, rectangle $A$'s right border is a riser $q^2$ units from $q$. What had previously been Area $B$ in Figure 5.2, for the variance illustrated in Figure 5.3 now equals $q(1-q)^2$; and what had been Area $A$ is now $(1-q)q^2$. The sum,

$$(1-q)q^2 + q(1-q)^2 = q(1-q),$$

a quantity equal to half the AD-about-$q$ curve property of this binary-valued variate.

Since $0 \leq q \leq 1$, the variance $q(1-q)$ of a binary-valued variate is

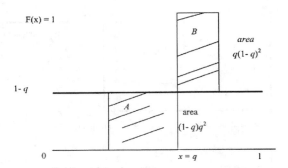

Figure 5.3. Binary variance curve property.

finite and thus the Central Limit Theorem (CLT) discussed at length by Cramer (1945) and considered in terms of curve uncertainty issues in Chapter 20 implies that, as $m$ increases, any single binomial proportion variate's density approaches the density of a Normal variate. To see what binary-valued variates have to do with the Cauchy density, first consider the variate constructed as the total number or proportion of disease cases that occur among exactly $m_j$ susceptible individuals. Such a variate is often calculated in each of c different census tracts or other regions within which, like the proportion or count numerators, individual members of the denominator sequence $m_j, j = 1, 2, ..., c$ vary in value.

In count or proportion ratio cases, it is not only numerators but also variate denominators that often are accurately modeled as Normal variates. Because they are ratios of Normals, these ratios often have a density that can be described by some Cauchy model more accurately than it can be described by some Normal model. However, for the following several reasons, it is not the standard form of the Cauchy that is modeled.

The binomial variates whose values supply both the numerators and denominators of rates and ratios do approach Normally distributed variates. Yet it is unlikely that a given ratio's denominator is a standard Normal variate that happens to be independent of this same ratio's numerator. Both the numerator and denominator seldom have Normal densities where parameter $\mu$ equals 0 and $\sigma$ equals 1. Consequently, rather than being Cauchy distributed exactly, ratios are likely to have a distribution described by a generalization of the Cauchy density called the *non-central-t*. (An estimated pdf of this curve will be illustrated and discussed in Chapter 19. For now, even though standard Cauchy distributed variates are one degree of freedom Student-t distributed, any fraction formed by dividing

## 5.4. Bernoulli variation and the Cauchy density

a standard Normal variate by a second standard Normal variate is standard Cauchy distributed only when numerator and denominator variates are independent.)

For instance, if the chance that a binary-valued Bernoulli variate $\tilde{X}$ takes on the value 1 equals $q$, while $1 - q$ is the chance that this variate takes on the value 0, then the sum, $X$, of the $m$ iid binary variates;

$$\{\tilde{X}_j\}, j = 1, ..., m;$$

is a binomial variate. Here $E(X) = mq, \mathrm{Var}(X) = mq(1-q)$ and, consequently, the CLT implies that as $m \to \infty$ the statistic $X/m$ has a density which approaches a Normal density, while the "standardized fraction,"

$$X' = ((X/m) - q)/\sqrt{q(1-q)/m},$$

approaches a variate that is $\phi$-distributed.

Now suppose that for denominator variate

$$Y, E(Y) = MQ, \mathrm{Var}(Y) = Q(1-Q), Y' = ((Y/M) - Q)/\sqrt{Q(1-Q)/M}$$

and, in addition, the values of $m$ and $M$ are moderate, such as 20, to large in size. Then for practical purposes, not only $X'$, but also $Y'$, can be taken to be a standard Normal variate. However it is unlikely that these two variates are mutually independent. Consequently, suppose $\rho$ is the additional parameter which, alongside the variate location and scale parameters, characterizes the bivariate Normal model of $X'$ and $Y'$'s joint density. (Later chapters will discuss the interpretations and applications of association or correlation parameter $\rho$ in detail. As we will now see, $\rho$ need not be restricted to its usual guises as either an association or a correlation parameter, but can also be thought of as a specialized location parameter.)

For ratios whose numerators and denominators are possibly dependent, when both $X'$ and $Y'$ have a standard Normal density $\phi$, a new variate, called a *residual*, $Y''$, can be determined by calling upon the relationship, $Y'' = Y' - \rho X'$. Values of variate $Y''$ defined in this way are parts of some sample's values; $y_i, i = 1, ..., n$; that are not explained or predicted collectively by some curve $y = \alpha + \beta x$. In the ratio numerator and denominator example based on $Y'' = Y' - \rho X'$, within $y = \alpha + \beta x$ the parameter $\alpha$ equals 0 and the parameter $\beta$ equals $\rho$.

Bivariate joint Normality implies that the residual $Y' - \rho X'$ left unexplained by $X'$ about $Y'$ is independent of $X'$; in other words, $Y''$ and $X'$ are independent variates. Thus when $m$ and $M$ are large enough to assure the bivariate Normality of $(X', Y')$, the residual $Y''$ is purified, or "partialled," of all influence of $X'$. For $Y'' = Y' - \rho X'$ the variate $Y'$ equals $(Y' - \rho X') + \rho X' = Y'' + \rho X'$. It follows that $Y'/X' = Y''/X' + \rho$.

The ratio

$$Y''/\sqrt{1-\rho^2}X')$$

has a distribution described by an $f$-type curve that here approaches a standard Cauchy density. Consequently, the ratio $Y'/X'$ approaches a variate whose distribution is described by a nonstandard Cauchy density with location parameter $\rho$ and scale parameter $\sqrt{1-\rho^2}$. This density is called a *noncentral-t curve* with degree of freedom $\nu = 1$. Within

$$Y'/(\sqrt{1-\rho^2}X')$$

the $\sqrt{1-\rho^2}$ multiplier of the variate $X'$ is the standard deviation of $Y''$, a curve property that will be discussed further in Section 10.2. For the variate $Y''/X' + \rho$ the location parameter $\rho$ plays the role that the noncentrality parameter $\delta$ (which will be applied in both Chapters 16 and 19) plays for the important quantities that are often called F-statistic ratios.

Variates like $Y'/X'$ often are assembled from numerator and denominator variates like white blood cell counts, or counts of members within geographical regions such as contiguous census tracts. For example, suppose two counts are taken, the first of the number of I-V drug users and the second of the number of HIV-positive individuals. If count values are then standardized by subtraction of the average count $\hat{q}$ among all regions, and then by dividing each of these differences by the statistic

$$\sqrt{m\hat{q}(1-\hat{q})}$$

then, as $m \to \infty$, the fraction with numerator equal to the standardized count of I-V drug users and denominator equal to the standardized count of HIV-diagnosed cases, has a density that approaches that of the nonstandard Cauchy distribution with location-noncentrality parameter $\rho$ and scale parameter 1. Specifically, the non-standard Cauchy or equivalently non-central-t-density is $\varphi(x-\rho)$, where

$$\varphi(z) = 1/\{\pi(1+z^2)\},$$

and the associated cdf is

$$\pi^{-1}\tan^{-1}[x-\rho] + 1/2.$$

We will now see that the differences between Cauchy and Normal variates can be so substantial that even trace amounts of Cauchyness can have disastrous goodness of fit implications.

## 5.5 Comparative goodness of fit

The cdf form of the indefinite integral of density

$$\varphi(z) = 1/\{\pi(1+z^2)\} \quad \text{is} \quad \pi^{-1}\tan^{-1}(z) + 1/2$$

and it is known that no indefinite integral of the standard Normal curve $\phi(z)$ exists. This is only one of many substantial differences between the standard Cauchy and standard Normal density curve.

To explore these differences, rather than considering Pearson's, Neyman's, and the ISE criterion from a sampled data perspective, suppose

$$\int_{-\infty}^{\infty} \left\{ \left[\hat{f}(x) - f(x)\right]^2 / f(x) \right\} dx, \quad \int_{-\infty}^{\infty} \left\{ \left[\hat{f}(x) - f(x)\right]^2 / \hat{f}(x) \right\} dx$$

and

$$\int_{-\infty}^{\infty} \left[\hat{f}(x) - f(x)\right]^2 dx$$

are different ways to compare two densities, denoted by $f_1$ and $f_2$. For example, consider

$$\int_{-\infty}^{\infty} \left\{ [f_2(x) - f_1(x)]^2 / f_1(x) \right\} dx = \int_{-\infty}^{\infty} \left\{ [f_2(x)]^2 / f_1(x) \right\} dx - 1$$

where here $f_1$ is standard Normal and $f_2$ is standard Cauchy. Then in accord with the difference between standard Normal and Cauchy curves, here

$$\int_{-\infty}^{\infty} \left\{ [f_2(x)]^2 / f_1(x) \right\} dx$$

is infinite.

To see why for a null standard Normal density $f_1 = \phi$, and an alternative standard Cauchy density $f_2 = \varphi$, the quantity that is estimated by Pearson's goodness of fit criterion is infinite, consider that here

$$1/f_1(x) = \sqrt{2\pi}\, e^{x^2/2} = \sqrt{2\pi} \sum_{k=0}^{\infty} (x^2/2)^k / k!,$$

while $[f_2(x)]^2 = 1/\{\pi^2(1+x^2)^2\}$. Thus for any $x$, the ratio

$$[f_2(x)]^2 / f_1(x) = \sqrt{2}\,\pi^{-1.5} \sum_{k=0}^{\infty} \frac{x^{2k}/2^k k!}{(1+x^2)^2},$$

all terms of which are positive valued. Hence for any value of $x$, the function

$$\sqrt{2}\pi^{-1.5}\frac{x^6/48}{(1+x^2)^2}$$

must assume a value that is less than the value of $[f_2(x)]^2/f_1(x)$. Since the integral of

$$\sqrt{2}\pi^{-1.5}\frac{x^6/48}{(1+x^2)^2}$$

is infinite over the entire $x$-axis, the integral of $[f_2(x)]^2/f_1(x)$ must, likewise, be infinite.

An alternative way to interpret the infinite value of what the Pearson goodness of fit criterion estimates in the null-Normal-alternative-Cauchy case, is to pose the question: How well does the Normal fit the Cauchy? In terms of Peason's criterion, the answer to this question is, infinitely badly. While this answer seems to be reasonable for these two curves, there is a simple way to generalize the Cauchy that raises many interesting issues.

Suppose, for $0 < P < 1$, that $f_2$ within

$$\int_{-\infty}^{\infty}\left\{[f_2(x)]^2/f_1(x)\right\}dx$$

is the density called the *standard Cauchy-Normal mixture*, $Pf_3 + (1-P)f_1$. This choice of density illustrates an important area of concern about situations where $P$ is small and $Pf_3 + (1-P)f_1$ is the density of a variate that can be taken to be standard Normal $f_1(z) = \phi(z)$ usually, but, in rare instances due to the change in value of a lurking variable, has a distribution described by the standard Cauchy $f_3(z) = 1/\{\pi(1+z_2)\}$.

Consider

$$\int_{-\infty}^{\infty}\left\{[f_2(x)]^2/f_1(x)\right\}dx,$$

where $f_2 = Pf_3 + (1-P)f_1$. It follows that

$$\int_{-\infty}^{\infty}\left\{[f_2(x)]^2/f_1(x)\right\}dx = P^2\int_{-\infty}^{\infty}\left\{f_3(x)^2/f_1(x)\right\}dx + (1-P)(1+P).$$

Consequently, no matter how small the finite value of the mixing parameter $P$ happens to be, the quantity estimated by the Pearson goodness of fit criterion is infinite. There is something infinitely bad about even trace amounts of Cauchy data mixed with Normal data. Thus the term "lurking variable" definitely does live up to its sinister name.

# CHAPTER 6
# Variates, Variables and Regression

## 6.1 Variates and variables

It is handy to call upon the endings of the two words, "variate" and "variable," to denote the difference between the entity that takes on the value $x$ within $f(y, x)$ and its $f(y|x)$ counterpart. In particular, the words "lurking variable" help denote an important component of many curve uncertainty analyses. Consider, for example, the standard Laplace variate,

$$F^{-1}(Z) = I_{(0,0.5]}(Z)ln(2Z) - I_{(0.5,1)}(Z)ln(2(1-Z)).$$

Within this definition of an inverse cdf, the standard uniform random $Z$ plays dual roles. It not only appears within the arguments of the two functions, $ln(2Z)$ and $ln(2(1-Z))$, but it also provides the arguments of the indicator functions $I_{(0,0.5]}(Z)$ and $I_{(0.5,1)}(Z)$, that determine the choice of either $ln(2Z)$ or $ln(2(1-Z))$. In these two ways, $Z$ not only determines what should happen once the decision to change gears has been made, but it also determines when gears should be changed. Surely this is a lot to ask of the single simple random quantity $Z$.

Suppose instead of the dual usage of a single quantity to determine both when nature should change her mind and also exactly how her mind should be changed, two quantities, $Z_1$ and $Z_2$ are selected. Then a particular inverse cdf can be selected by the first of these quantities, while the second can serve as this inverse cdf's argument. For instance, suppose a variate $Y$ is modeled by the simple expression $Y = \alpha + \beta x + \sigma \Phi^{-1}(Z)$. Then $Z_1$ can be the quantity that determines the value $x$, while $Z_2$ provides the random number $Z$ that, in turn, helps provide the standard Normal variate $\Phi^{-1}(Z)$'s value.

The term "lurking variable" reflects that unfortunate but realistic circumstance where a model like $Y = \alpha + \beta x + \sigma \Phi^{-1}(Z)$ is the valid representation of a data generation process but, for any number of possible reasons, $Y$ is thought incorrectly to result from data generation described by a model like $Y = \mu + \sigma\Phi^{-1}(Z)$. When $x$ assumes more than a single value, and $\beta \neq 0$, the simple exchange of the symbol $\mu$ for the symbol $\alpha$ or any other symbol that represents a constant, does not fix the lurking variable problem. When $x$ takes on two distinct values, $x_1$ and $x_2$, the density of $Y$ is a mixture of Normals, the first with location parameter $\alpha + \beta x_1$ and the second with location parameter $\alpha + \beta x_2$.

As discussed in the previous section, some lurking variable may determine the choice between the standard Normal inverse cdf $\Phi^{-1}(Z)$ and its inverse standard Cauchy counterpart. Alternatively, this variable may supply the subscript of a location, scale, or combination of location, scale, and other parameter types. When one or more of such variables help choose between some finite number of alternatives, the density that describes $Y$'s distribution is represented by a finite *mixture* model.

Finite-mixture-model data analyses are often designed with an eye to the possibility that some variable is lurking. On the other hand, many statistical variables supply the value for $x$ within $f(y|x)$ because they have actually been assigned as part of a controlled experiment's protocol. Obviously, since its value is known, the exact dosage given to an experimental subject, or the time or date an observation is made, is simply a variable, not a lurking variable or lurking variate.

There can be three distinct reasons for the poor fit provided by a model like $Y = \alpha + \beta x + \sigma\Phi^{-1}(Z)$. Often the inverse cdf $\Phi^{-1}(Z)$ should, in reality, have been some alternative such as the inverse Cauchy cdf. The difference between $Y$ and $\sigma\Phi^{-1}(Z)$ may, in reality, not be a linear function of $x$. There may be one or more lurking variables that supplement $x$'s contribution to $Y$. Many of the topics discussed in later chapters will address each of these three issues.

Before proceeding to discuss the general distinction between variates and variables in detail, it is worth introducing a general theme. Some of the most challenging problems faced by data analysts today occur when two or all three of the above issues must be dealt with simultaneously. For example, the two basic findings, Cochran's Theorem and the Gauss-Markoff Theorem, allow two general types of statistical procedures to be applied in many cases where it is not actually known that an inverse cdf is actually $\sigma\Phi^{-1}(Z)$.

Although it will not be discussed extensively here, a procedure called *stepwise regression* often is used to choose from among candidate variables. This method requires not only that variables be specified, but it requires the formal specification of the model like $Y = \alpha + \beta x + \sigma\Phi^{-1}(Z)$. Hence emphasis will be placed below on procedures that facilitate the choice of a term like $\Phi^{-1}(Z)$, without requiring that a term like $\alpha + \beta x$ be specified fully.

## 6.2 Variates and subjects

A variate (or random variate) assumes values associated in some well-defined way with probabilities. Here, with the exception of the type of quantity referred to as an *attribute, nominal* or *qualitative* variate, the term "well-defined" designates the existence of some particular $F(x)$ curve whose *argument* $x$ is a variate value associated with the probability $F$ of the value's occurrence at, prior to, or at a value less than, $x$.

Height, weight, survival time, gestation time, age and education level, production level, meteorite, and tumor size are examples of variates studied by applied statisticians and demographers. It is important to distinguish these and other variates themselves from the numerical values they assume. Usually lowercase letters like $x$, $y$, or $z$ represent the latter, while capital letters like $X$, $Y$ or $Z$ represent variates themselves.

A variate, like height, is no more the same as a quantity, like 6 feet, than an experimental unit or observed subject whose measurement results in the number 6, is the number 6. This is not a moot point, since variates, which are themselves not taken to be numbers, are described by curves that relate pairs of numbers. Although whether it is the variate itself that is described by a curve like $f$ or $F$, or it is the distribution of the variate that is thus described, is subject to debate, here $f$ or $F$ describe the distribution of a variate, such as the variate $X$.

While $F(x)$ designates a curve, $F(X)$ designates a variate formed by modifying $X$ to yield a new variate, $Y = F(X)$, where $X$ happens to be transformed by the same function that serves as $X$'s cdf. For example, suppose that when $x$ is positive, the curve $F(x)$ equals $1 - e^{-x}$. Then the notation $Y = F(X)$ denotes that $Y$ is a variate whose values could have been obtained by starting with a sequence consisting of values of the variate $X$, then repeatedly looking up the appropriate values of $e^x$ in a table of exponentials or, less directly, logarithms, and then calculating $1 - 1/e^x$ repeatedly.

For most statistical problems: (1) the term "variate;" (2) the real number that serves as this variate's value; (3) the "marginal" density that describes the distribution of this single variate; and (4) the joint distribution that characterizes this variate and some "covariate's" joint distribution, are every bit as important, as the word "snow" is to the Inuits and Aleuts. The many words that Eskimos have for snow is an example called upon commonly by linguists to illustrate the impact of culture upon language. Similarly, the distinctions between variates, variate values, and many curves and quantities associated with variates, is an example of the richness, not the prolixity, of statistical thinking. The remaining sections of this chapter will introduce several additional distinctions with an eye to illustrating how the linkage between curve properties and model parameters on the one hand, and the values assumed by one or more variables on the other, helps to simplify statistical procedures.

## 6.3 Expressions, algorithms and life tables

It is convenient to reserve the term "model" to designate expressions comprised of a finite number of explicitly represented symbols. In this way, the terms "model-free" and "nonparametric" can be treated as near synonyms. A model thus describes the sort of curve studied in high school and college analytical geometry courses. As anyone who has gone on, mathematically speaking, beyond these courses knows, analytic geometry does not provide the end-all of alternative curve representations.

Statistical expressions, such as those discussed below that describe the distribution of bivariate and multivariate Normal variates, can be written much more concisely using vector - matrix notation than they can using conventional notation. Yet, to discuss key aspects of this material it is not necessary to present multivariate models in explicit form.

In this era of economical digital computing, the actual details of some statistical formula are less important than an understanding of how the curve that the formula describes is related to other curves, computational subroutines, and algorithms. Even in cases where so-called *exact formulas* exist, subroutines are often based on approximations. In addition, such basic statistical tools as the *life table* are not actually expressions in the usual sense of this term. Instead, they are *algorithms*, in other words, recipes by which the evaluations of one curve's estimator are applied to estimate evaluations of some other curve.

A curve whose evaluations constitute the data entered as computer input, is often easy to construct, yet hard to interpret. On the other hand, in order to apply information for scientific purposes, a second derived or estimated curve is often informative. For this reason, an algorithm like a life table facilitates the construction of the second curve based on evaluations of some other curve.

While for computational reasons one known curve is often treated as an *approximation* of some other known curve, the *estimation* of a curve using natural, as opposed to simulated data, concerns some true curve, such as some $f$ or $F$, that is never actually known. If the true curve were known, there would be no reason to estimate it. On the other hand, there are many circumstances where, especially for computational reasons, an explicit expression, such as a model, is replaced by an approximation. Yet, as discussed in Chapter 2, the words "approximated" and "estimated" are not synonyms. The distinction between these two terms stems from the difference between deductive, strictly mathematical, or computational procedures that often call upon approximations, and inductive statistical procedures based on natural data.

The difference between the terms "estimation" and "approximation" underscores a characteristic that is vital to many branches of modern statistics. Statistical topics have often been discussed in the context of a particular curve, such as the Normal, or equivalently, the Gaussian, bell-shaped model for the curve $\phi$. Starting with an explicit model, it was natural for the issues of representation to concern approximations of such a model. For example, truncated continued fractions, power series, and Fourier series were often applied in this way.

The number $e$, raised to any real-valued power, is never zero. Hence, over the entire real-axis at no point does any Normal curve equal zero. Consequently, since it is impossible for them to assume zero or negative values, height, weight, body temperature, intelligence, stock values, and many other variates of importance in the psychological, social, and economic sciences, and even most variates encountered in the physical sciences, cannot, in reality, be Normally distributed. Thus in most contexts both the explicit formula for a Normal curve and this curve's power or Fourier series expansion are approximations and neither is, in reality, an approximation of the other.

## 6.4 Distinctions between curve types

Without familiarity with statistical subject matter, it is easy to overlook a distinction between curves that are estimated and other curves and functions. Because functions encountered in statistical investigations are not dealt with in the way that functions are measured or evaluated in other scientific studies, like the term "approximation," the term "curve fitting" can be misleading. For instance, when an astronomer considers the path of a comet, he or she measures its position through observations, in other words, evaluations of the values attained along this path. To better compare this path to statistical processes, only the simple two-dimensional number-pair case will be discussed in this section, even though realistic problems often concern four or more time and space coordinates.

While one member of the astronomer's number pair is typically the counterpart of many forms of statistical data, the second member is rarely the counterpart of the value attained by curves of statistical importance. Specifically, with the notable exceptions of time series and functions lumped under the collective heading, regression curves, most curves dealt with by data analysts assume ordinate values that, in some specific way, equal or are proportionate to probabilities.

In the same sense that the path taken by a comet is seen, probability is not seen. Yet it is the specific way that chance and probability are brought into the picture that distinguishes one curve type from another curve type. For example, since the curve $F(y)$ enumerates the probability that a variate will assume a value less than or equal to $y$ when $F(y)$ describes the distribution of a continuous variate, the curve $f(y)$ has a different, and more complex connection with probability. Yet while the cdf $F$ cannot be applied in connection with variates that are said to be nominal, qualitative, or attribute-valued, for these same variates there always exists some $f$-type curve that does, in this case, assume values that are actual probabilities. (When applied below, the term "nominal" is defined in terms of these $f$-type and $F$-type curve characteristics.)

Unlike the spatial position of a comet, probability isn't measured. Rather, it is (1) estimated through the inductive analysis of sampled data or else; (2) deduced by strictly mathematical means; or, as is almost always the case, (3) some combination of inductive and deductive approaches are applied.

For example, since it is not simply due to measurement instrument limitations that $f$ is known imprecisely, the symbol $f$ that designates the curve estimated by the histogram shown in Figure 2.1 is not like the position of a comet in the sky. Unlike the comet's position, the curve $f$ itself cannot actually be measured. Rather, it is determined by one or more of a variety of statistical procedures, such as the grouping of data within class intervals, or what in the computational literature are called *bins*, to estimate each proportion per bar shown on Figure 2.1's ordinate-axis. Allocating data to bins by what are called *frequency table* methods has had such a long history that it is easy today to lose track of the important fact that bin heights are actually counts or proportions, not true probabilities.

The distinction between observed curves like time series, and probability-related curves like $f$ and $F$, requires emphasis because during the formative years of data analysis and demography, procedures were so untrustworthy that in order to obtain unequivocal findings, large samples were required. With enormous mounds of raw data, the curves that trace the tops of these mounds differed inappreciably from mathematical curves. Technically speaking, given sufficient data, any reasonable curve estimator should yield an estimated curve that is indistinguishable from the curve that it estimates.

Yet how much data are enough depends not only on the variety of data gathered, but on the type of curve estimated. As has proven true of the personal computer (PC), demands keep pace with enhanced capacity to fulfill these demands The curve estimators of the past were much like older PC's. They sufficed only as long as the tasks required of them remained simple. For statistics, as tasks increase in scope, no concept is more central to the accomplishment of these tasks than is the distinction between estimated and measured curves. This is because complex data analysis applications draw heavily on both: modes of thinking that are unique to statistics; and modes of thinking germane to mathematical astronomy as well as other non-statistical, but quantitative, sciences.

## 6.5 Curve properties and symbols

The distinction between variates, variate values, and subjects is crucial to a clear understanding of statistical processes. When phrases like "the expectation $E(X)$ of a variate" are applied, the words "of a variate" are best thought of in reference to some particular curve that associates the values assumed by the variate with probabilities of these values' occurrence. The curves or functions that provide information about the distribution of one or more variates have as their arguments (for example the argument $x$ of $f(x)$) the values assumed by variates. There are many philosophical concepts that concern these variates, such as causality, smoothness, and knowledge. Yet the universe of discourse that concerns these concepts has increasingly called upon terminology that expresses the relationships between different types of curves and curve properties, as well as different ways of representing curves and properties.

When the symbol $f$ is included as a subscript of $E_f(X)$, this emphasizes that the combination of symbols $E_f(X)$ is mathematical shorthand notation for a characterization of some $X$ variate that associates $X$ with one particular property of the curve $f$. For example, $E_f$ can be thought of as the point at which a uniformly thick cardboard cutout of the region beneath $f$ and above the abscissa would balance, if the cutout were allowed to rotate within a plane. When applied in this sense, the word "property" denotes a number obtained from a curve to determine this balance point by means of executing a series of well-defined computational or mathematical operations, just as in a physical sense the balancing point of a cardboard cutout might be determined.

If the symbol $\mathbf{M}$ designates some particular matrix, then the symbol $|\mathbf{M}|$ represents a single number, the determinant of $\mathbf{M}$. Unlike $\mathbf{M}$, which is an organized collection of many numbers, $|\mathbf{M}|$ is always a scalar. Similarly, the curve $f$ is much more than a mere single number; while the curve property $E_f(X)$, like the matrix property $|\mathbf{M}|$, is a form of condensed information. As is true for the data content of some matrix in its entirety, as compared to this matrix's determinant, information is lost in the condensation process that yields $E_f(X)$. Consequently, besides being a representational form of shorthand, $E_f(X)$ is an abbreviation or summary. Hence it is suitable in some applications and inappropriate in others.

While curves and values assumed by variates are concrete entities, statistical *variates* are defined statistically or mathematically in terms of particular curves that describe a univariate or multivariate distribution. Largely for the reason that curves provide unambiguous information about variates and distributions, there has been an increasing tendency to reduce data analysis problems to quantitative problems that concern particular curve types or properties of curve types.

Besides the functions $f$ and $F$; regression curves, survival curves, $Q-Q$ plots, fractile diagrams, confidence bands, box plots, and dose response functions, are applied by demographers, quantitative epidemiologists and other statistically oriented researchers. What these curves and their properties are, and how they are related to each other, forms the warp and woof of statistical structure.

## 6.6 Variates, variables, and $E_f(Y|x)$ regression

In the interest of clarity, the terms "random variate," "stochastic variate," and the word "variate" itself, will be reserved exclusively for descriptions of those aspects of an observed subject that are in some way directly associated with probability, as indicated by the phrase, "the probability that a variate assumes a particular value." On the other hand, when a researcher chooses a value without reference to any randomization process or table of random numbers then in this and later sections the entity that has this value will be called a *variable*, not a variate.

Specifically, no variate will have its value determined in its entirety by the investigator, without what is defined in Appendix II as *randomization*. This is not to say that the probabilities with which statisticians work are unaffected by variables. When variables, rather than variates, influence probabilities, the probabilities involved are "conditioned" *on* the value or values of the variable, while proper usage refers to the probability *of* a variate value's occurrence.

The above distinction between variates and variables renders the phrase "the probability *of* a variable" self-contradictory. Instead, the probability of a variate $Y$'s value $y$, given the value $x$ of a variable, denotes a *conditional* probability. Thus a value of a variable is not itself associated with an $f$ or $F$-type curve directly and, consequently, a variable's value is seldom distinguished from the variable itself through the choice of uppercase or

## 6.6. Variates, variables, and $E_f(Y|x)$ regression

lowercase letters. However, in the statistical literature lowercase letters designate variables and their values much more frequently than uppercase letters.

The term "random variable" is sometimes taken to be a synonym of the single word "variate" and some authors feel that it is helpful to precede the term "variate" with the words random or stochastic. However to apply the term "random variable" in place of the term "random variate" makes it difficult to explain why it is the designation "multivariate analysis," not "multirandom variable analysis," that appears in the titles of many statistical texts.

Few real world studies concern merely a single number per experimental or observed subject. Consequently, were it not the case that a single variate is often accompanied by multiple variables—as distinguished from varia*tes*—as a field of study, univariate analysis would be overshadowed completely by its multivariate counterpart. This is definitely not the case today, when univariate analyses are commonly undertaken in applications where one or more variables provides information supplemental to the one variate whose conditional distribution, given variable values, is studied.

The word "multiple," as in "multiple regression," implies that a problem concerns more than one variable; for example the pair of variables: age and pack-years of cigarettes smoked. With regard to variates, the terms "bivariate" and "multivariate" analysis refer to those investigations that involve two, in the case of bivariate, and two or more, in the case of multivariate, variates. Hence, it is best to avoid the phraseology "multiple variates," since it is helpful to reserve the term "multiple" for service as a modifier of the term "variable."

When variables assume qualitative values, such as smoker/nonsmoker, these different values are called *levels*. Unfortunately, the subscript notation $x_1$ and $x_2$ is often called upon alternatively to distinguish two different variables, such as height and weight, or to determine two different numerical levels assumed by a single variable. However, when notation such as $|x_1, x_2$ occurs within an expression such as the argument of a density curve, the symbols $|x_1, x_2$ always designate that part of the argument that pertains to two distinct variables, such as height and weight, and definitely not to two values, such as 6 and 6.5 feet or 60 and 65 lbs.

In general, the shorthand method for distinguishing variates from variables is based on the symbolic placement of the symbol for the value $x$ of a variable after a "slash" symbol, as in $f(y|x)$. Here $|x$ corresponds to the phrase, "given the value $x$." For example, the condensed shorthand notation $E_f(Y|x)$ designates a property of the curve $f(y|x)$ that potentially can change as $x$ changes. Based on this notation, the expression $z = E_f(Y|x)$ represents a useful variety of what is called a *population regression curve*.

## 6. Variates, Variables and Regression

Just as there are silent letters in vernacular language, some symbols are silent—in the sense that they play an indirect role in statistical notation. Of the four letters that form the notation $E_f(Y|x)$, the symbols $f$ and $Y$ act as silent characters that indicate which density and variate are referred to by other symbols. For example, $z = E_f(Y|x)$ describes: (1) the relationship that the variable $x$'s value has; (2) with a $z$ ordinate value that is the value attained by the particular property $E$ of the curve $f$, where, in turn; (3) $f$ describes the conditional distribution of the variate $Y$.

The thick diagonal line shown is Figure 6.1 marks part of a common $z = E_f(Y|x)$ regression curve. Also shown are two conditional $f$-type curves. Over two values of the single variable $x$, specifically, $x_1$ and $x_2$, the expected values, $E_f(Y|x_1)$ and $E_f(Y|x_2)$, determine the $z$ coordinates of the regression line. The density curve whose property $E$ is evaluated given $x_1$ is shown by dashes, and its counterpart, whose $E$ property determines $z$ at $x_2$, is shown as a solid curve.

The curve $z = f(y|x)$ involves three entities; $z, y$, and $x$; while, strictly speaking, $z = E_f(Y|x)$ describes the relationship between the quantities $z$ and $x$. This underscores the role that a regression curve like $z = E_f(Y|x)$ plays as a bridge between, on the one hand, curves like $f$ and $F$ whose values are related in one way or another to probability and, on the other, curves like a single time series, or coordinates of a planet's path through the sky. The latter two curves have ordinates that, like the $z$-axis of the regression curve, are expressed in the same units as the variate $Y$'s values.

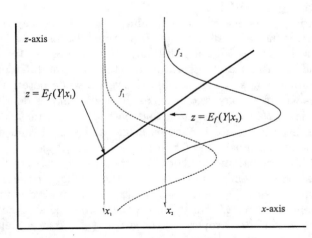

**Figure 6.1.** $E_f(Y|x)$ regression curve illustrated in detail at $x = x_1$ and $x = x_2$.

## 6.7. $\mu(x)$, $E_f(Y|x)$ and regression alternatives

Statistical problems often are posed in terms of comparisons between two or more regression curves like $z = E_f(Y|x)$ and $z = E_g(Y|x)$, where $f$ and $g$ designate two, possibly distinct, density curves. For the same purpose, the notation $z = E_1(Y|x)$ and $z = E_2(Y|x)$ is sometimes applied; for example, to help compare how the expected value, $Y$ variate packs-of-cigarettes-smoked-per-year curve property is related to the $x$ variable age: (1) for men; and (2) for women. The choice between the alternatives, $z = E_f(Y|x)$ and $z = E_1(Y|x)$, as representations of the same curve, illustrates the latitude of statistical shorthand. When it is not important to keep in mind that regression curves being compared correspond to two different populations, one with density $f$ and another with density $g$, then the notation $z = E_1(Y|x)$ and $z = E_2(Y|x)$ can serve as well as $z = E_f(Y|x)$ and $z = E_g(Y|x)$.

Notation like $z = E_f(Y|x)$ has an important conceptual advantage over notation like $y = E_f(Y|x)$. The reason why the ordinate-axis is labeled by some letter other than $y$ in Figure 6.1 is illustrated by the $Y$-variate family size. For this $Y$-variate, the lower case italicized letter $y$ will ordinarily denote a value assumed by some family-size measurement. However, the $E$-curve property of a density that described the Y-variate's distribution could assume a decimal or fractional value like 3.5. As a consequence, the notation $y = E_f(Y|x)$ might seem to imply that some individual family actually has 3.5 members. Calling upon the notation $z = E_f(Y|x)$ in place of $y = E_f(Y|x)$ stresses that here the symbol $z$ designates the value assumed by a curve property, not by a variate.

## 6.7  $\mu(x)$, $E_f(Y|x)$ and regression alternatives

Due to the many possible different permutations and combinations of statistical symbols, it is important to keep sight of representational basics. For example, the letter $E$, like the six letters MEDIAN, is taken to represent a particular curve property. Consequently, $z = E_{(f+g)/2}(Y)$, despite having a subscript that describes the average of two density curves $f$ and $g$, is merely the symbolic way of representing some single curve property. On the other hand, if $f$ were a density that is nonzero only over variate values that are less than -1, and if $g$ were a density that is nonzero only over positive variate values, then MEDIAN$_{(f+g)/2}(Y)$ could be defined to equal any negative value greater than -1. In this example, that MEDIAN$_{(f+g)/2}(Y)$ is not unique, but $E_{(f+g)/2}(Y)$ is unique, illustrates just one of many quirks of particular curve properties.

The notation $z = E(Y_1|x, u, v)$ represents a relationship that involves one variate, $Y_1$, and three variables; $x, u$, and $v$. In other words it denotes a multiple regression curve. For example, expected packs of cigarettes smoked per year might be related to age $x$, yearly income $u$, and education level $v$. Here again, as is the case of the expression $z = E_{(f+g)/2}(Y|x)$, the equation $z = E(Y_1|x, u, v)$ represents a single regression curve.

The number of distinct values assumed by a variable is rarely specified explicitly when a symbolic representation like $z = E(Y_1|x, u, v)$ appears in print. Many current packaged statistical computer programs take a similar approach when they allow a user to deal with variates and variables as wholes, without requiring that attention be paid to the individual elements, in other words, values of a data vector.

Instead of its real name—location parameter—it is handy to think of $\mu$ as a *relocation parameter*. Suppose standard density curve $g$ describes the distribution of a standard variate $Z$. For any nonzero value of the parameter $\mu$ the curve $g(y - \mu)$ can be thought of as the relocated cousin of $g(y)$ that is moved over $\mu$ units from its original location. How much a curve is relocated can be a function, such as $\mu(x)$, of one or more variables. For example a curve like $\mu(x) = \alpha + \beta x$ often serves as a component of $Y = \alpha + \beta x + \sigma Z$.

At first glance the formula $Y = \alpha + \beta x + \sigma Z$ does not seem to have an obvious connection with the concept of location parameter. However, suppose that for $\sigma = 1$ the quantity $\alpha + \beta x$ is subtracted from both sides of the equation $Y = \alpha + \beta x + Z$ and equation sides are reversed. This yields the relationship $Z = Y - (\alpha + \beta x)$. The substitution of $y - (\alpha + \beta x)$ for variable value $z$ within $g(z)$ designates that the curve has been relocated or, to apply the more common term, "translated." The translation, expressed as a function of $x$, is disclosed by the regression model $\mu(x) = \alpha + \beta x$.

Curves usually link single numbers to other single numbers. Consequently a parametric model like $\mu(x)$, and some nonparametric counterpart of this model, like $E_f(Y|x)$, both help bridge the gap between complex concepts that involve variates such as $Y$, and curves that merely associate number pairs. On the other hand, there can be subtle, yet important, differences between parametric and model-free curves.

For example, since curves can attain distinct relative maxima, model-free $\text{YMODE}_f(Y|x)$ curves need not be single valued. $\text{YMODE}_f$ denotes the single or multiple valued curve property, mode, of the conditional curve $f$. Along these lines an additional possibility is now often contemplated. The important class of statistical procedures subsumed by the term "switching regression" is based on the replacement of a single regression model like $\mu(x) = \alpha + \beta x$ by multiple distinct model components, such as for the two-component case, $\mu_1(x) = \alpha_1 + \beta_1 x$ and $\mu_2(x) = \alpha_2 + \beta_2 x$. For example, these twin models are applied in economic and other areas of study interested in dualities, such as bullish, as parametrized by $(\alpha_1, \beta_1)$,

## 6.7. $\mu(x)$, $E_f(Y|x)$ and regression alternatives

and bearish, as parametrized by $(\alpha_2, \beta_2)$, markets (Quandt, 1958, 1972; Quandt and Ramsey, 1978).

Switching regression is based on the mixture *models* that are discussed in Appendix I. Because formal modeling procedures are applied, switching regression is a parametric procedure. On the other hand, YMODE$_f(Y|x)$ regression curves are property-based. Hence the distinction between YMODE$_f(Y|x)$ and switching regression illustrates the distinction between parameter-based and property-based approaches.

Switching regression and YMODE$_f(Y|x)$-based approaches usually yield distinct curves. To see why these approaches differ, assume so much data are available that all differences between estimated curves and the curves they estimate are imperceptible. Suppose also that the unique mode of a standard curve g equals zero and that this standard curve describes both the residuals for which $Y - (\alpha_1 + \beta_1 x)$ is $g$-distributed, and those for which $Y - (\alpha_2 + \beta_2 x)$ is $g$-distributed where, to make things even simpler, $\beta_1 = \beta_2 = 0$.

For example, for $\zeta = 1.25$, suppose data from two Normal distributions are mixed as described in Appendix I.4 so that density representation $[f(x) + h(x)]/2$ discloses that the population under consideration contains two types of variates, those with density $f(x) = \phi(y - \zeta)$, and those with $h(x) = \phi(y + \zeta)$. Then the rightmost mode of this mixture density is 1.10, not the value of parameter $\zeta$, here 1.25. The modes of individual curves can equal corresponding mixture-component location parameters. Nevertheless, curve overlap will almost always cause these modes to move together so that they lie between, but not on top of, their location parameter counterparts. With sufficient overlap, mode distinctiveness will disappear completely.

The expression $Y = \alpha + \beta x + \sigma Z$ is often called a regression model. This terminology tends to obscure the fact that two curves, $\mu(x) = \alpha + \beta x$, as well as $Z$'s inverse cdf, density, or cdf, are often modeled. For this reason, here and in later chapters, $Y = \alpha + \beta x + \sigma Z$ will be called a *data generation model*, while $\mu(x) = \alpha + \beta x$ will be called the regression model that represents the parametric curve $\mu(x)$.

There are other unclear regression curve classifications. For example, consider the useful curve $G^{-1}F(x)$ that is often estimated by what is called a Q-Q plot. In a very simple case, $G^{-1}$ is the standard Normal inverse cdf $\Phi^{-1}$, while $F$ is estimated by an $\hat{F}$ curve called a sample cdf or, alternatively, by a cumulative polygon curve estimator. Leaving the issues of defining, selecting and computing some curve estimator $\hat{F}$ of $F$ aside, still leaves the question: Is the Q-Q plot curve $G^{-1}F(x)$ like the curve $\mu(x) = \alpha + \beta x$, in the sense that $G^{-1}F(x)$ is a regression curve?

While $\mu(x) = \alpha + \beta x$ describes only a proper subset of a distribution's features, for most choices of $G^{-1}$, knowledge of $G^{-1}F(x)$'s shape can be applied to determine both $F$ and $f$. Hence, since unlike the curve $\mu(x) = \alpha + \beta x$, the curve $G^{-1}F(x)$ discloses all distribution features, $G^{-1}F(x)$ differs from commonly encountered regression curves that abstract distributional information. A second, and perhaps more pronounced difference between $G^{-1}F(x)$ on the one hand, and a curve like $\mu(x) = \alpha + \beta x$ on the other, stems from the question of model validity. For any continuous variate $X$ there is no question that $G^{-1}F(x)$ exists and is well-defined. On the other hand, the validity of a model like $\mu(x) = \alpha + \beta x$ is an open question. Consequently, it makes a great deal of sense to apply a curve like $G^{-1}F(x)$ to check model validity; for example by substituting different alternative choices for $G$.

When it is viewed as a curve uncertainty examination tool, an important characteristic of the function $G^{-1}F(x)$ is that, unlike kurtosis and goodness of fit criteria, $G^{-1}F(x)$ is a curve, not a scalar quantity. Kurtosis, for example, is a single curve property, and hence when a density $f$'s kurtosis equals zero, this is not a sufficient condition for $f$'s Normality. On the other hand, in the sense that it provides detailed and comprehensive evidence, an estimate of $G^{-1}F(x)$ is like a fingerprint. Of course, also like the blurring of a fingerprint, any blurring produced by the curve estimation process will lower $G^{-1}F(x)$'s value substantially.

The differences between the full view provided by $G^{-1}F(x)$, and $\mu(x) = \alpha + \beta x$'s partial view, are clear-cut. A nonparametric estimate of YMODE$_f(Y|x)$, or its $E_f(Y|x)$ or MEDIAN$_f(Y|x)$ counterparts, can validate some model like $\mu(x) = \alpha + \beta x$. In the sense that they are associated with one and only one parameter, estimates of the three curves $E_f(Y|x)$, MEDIAN$_f(Y|x)$, and YMODE$_f(Y|x)$ are location specific. Regarding scale, consider the curve formed as the reciprocal of the derivative of $G^{-1}F(x)$, specifically, $[\phi\Phi^{-1}F(x)]/f(x)$. If $F$ actually is represented accurately by some general Normal cdf, and $f$ is described by the general Normal model $\phi([x-\mu]/\sigma)/\sigma$ then at every value of $x$ the function $[\phi\Phi^{-1}F(x)]/f(x)$ equals scale parameter $\sigma$. The curve $[\phi\Phi^{-1}F(x)]/f(x)$ is related to a proper subset of the parameters within the density model $\phi([x-\mu]/\sigma)/\sigma$; here only to $\sigma$ and not to $\mu$.

The curve $[\phi\Phi^{-1}F(x)]/f(x)$ can be generalized by replacing estimates of marginal $F$ and $f$ curves with their conditional counterparts to estimate the curve $[\phi\Phi^{-1}F(y|x)]/f(y|x)$. As discussed by Tarter and Lock (1993, Chapter 9), the curve $[\phi\Phi^{-1}F(y|x)]/f(y|x)$ can be for scale what a nonparametric counterpart of $\mu(x) = \alpha + \beta x$ can be for a location parameter that changes in value as $x$ changes. However, unlike most location parameter-based estimators, procedures based on $[\phi\Phi^{-1}F(y|x)]/f(y|x)$ by-

## 6.7. $\mu(x), E_f(Y|x)$ and regression alternatives

pass the selection of a particular scale-associated curve property like Var or ADM. Yet, as is the case for the right hand side of $\mu(x) = \alpha + \beta x$, the curve $[\phi\Phi^{-1}F(y|x)]/f(y|x)$ discloses how a parameter changes as the variable $x$ changes.

Because the function $[\phi\Phi^{-1}F(y|x)]/f(y|x)$ is based on nonparametric estimators of $F(y|x)$ and $f(y|x)$, it should be classed as a nonparametric procedure, and yet what estimates of $[\phi\Phi^{-1}F(y|x)]/f(y|x)$ actually estimate is the parameter $\sigma$. The catch here is that $\phi\Phi^{-1}F(z)]/f(z)$ will be identically equal to $\sigma$ if, and only if, $f$ and $F$ are Normal densities and cdf's, respectively. However, this often is a blessing in disguise. It is a blessing because the curve $[\phi\Phi^{-1}F(y|x_0)]/f(y|x_0)$, when estimated at any particular value $x_0$ of the variable $x$, provides a graphical check of Normality, while, given Normality, for choices $y_0$ of the variate value $y$, the function $\sigma(x) = \phi\Phi^{-1}F(y_0|x)]/f(y_0|x)$ discloses, when it is constant valued, the property known as *homoscedasticity*.

# CHAPTER 7
# Mixing Parameters and Data-generation models

## 7.1 An introduction to data-generation models

No single curve type provides the ideal framework about which all statistical problems can be formulated. For a continuous variate, at a specific point $x_1$, it is curve evaluation $F(x_1)$, not $f(x_1)$, that equals a probability. On the other hand, even though the cdf of a continuous variate is connected more directly with the important concept, probability, than its pdf counterpart, since no parametric model for any Normal cdf $\Phi$ exists, the "Gaussian" $F$ cannot be computed with perfect accuracy. This is true even though models for all Normal $f$-type curves exist and can be expressed concisely enough for $f(x_1)$ to be computed easily at any $x_1$.

When no single curve is universally preeminent among its peers, it is hardly surprising that distinct curve properties and model parameters all have their individual niches. The number of these niches increases as the number of variates whose values are obtained from a given sample subject increase. In the univariate case, curve properties like AD, $E$, and Var can be exhibited clearly. For example, the curve property $E$ can be thought of as the balancing point of a curve $f(x)$, as well as the difference between two areas defined in terms of either the curve $F(x)$ or the curve $xf(x)$. Unfortunately, the multivariate case cannot proceed much further along these lines. Hence additional approaches are needed.

The next four chapters emphasize multivariate data and univariate data accompanied by one or more variables. In order to interpret multivariate and multiple regression techniques, emphasis will be shifted away from curve properties and towards model parameters. In particular, consideration will be given to models that describe how variates are thought to have been generated in environmental, occupational, and many other areas of study.

There are many applications where variates are simulated. For example, environmental science researchers often model risk as a product of variates like the volume of air taken into the lungs during one typical breath, or the amount of a heavy metal in a gram of cooked fish. Viewed from a statistical perspective, air, lead, and mercury variates have distributions and, in turn, these distributions are described by many individual curves. Among these is the function $F^{-1}$, a curve that can help shed light on a wide variety of data-generation approaches.

In the sense they can be visualized clearly, the AD, $E$, and Var curve properties discussed in the previous chapters resemble Descartes' parabola latus rectum length. Unfortunately, with one notable exception, other curve properties and model parameters are linked only with great difficulty to geometric concepts. As was the case for the parameter first discussed by Descartes, this exception concerns an interpretation as a specific line segment's length. However this geometric interpretation applies only to one particular statistical model, that of the *bivariate Normal density*. In the sense of applying to only one model, *correlation geometry* has less scope than the geometric interpretations presented in earlier chapters.

The algebraic model side of Descartes' linkage of geometry to algebra can be called upon to compensate for geometric limitations. In later chapters, this side will help examine the important parameter $\rho$ in non-Normal cases where the algebraic interpretation of correlation and other curve properties is based on expressions within which $F^{-1}$ plays a central role.

It is handy to call the model for a pdf or a cdf curve that describes the distribution of one or more continuous variates, a *probability model*. This helps distinguish these models from *data-generation models* within which some standard inverse cdf curve often serves as a central component. For example, the popular technique called *Model I analysis of variance (ANOVA)*, is based on a data-generation formula expressed in terms of a single inverse cdf curve. On the other hand, two or more inverse cdf's are called upon to simulate *Model II" ANOVA data*, where each inverse cdf yields an artificial "component" variate.

Mathematical and statistical packaged programs provide random number generation capability that can be called upon to simulate ANOVA and many other types of variates. This allows a data-generation model's formula to serve as a recipe whose basic ingredients include one or more computer generated random numbers. Inverse cumulative distribution functions whose arguments are these numbers, synthesize a variate $Y$'s values. In particular, suppose $Z$ is a variate with a uniform (in other words, "rectangular") density $I_{[0,1)}(z)$ whose values are obtained by calling a random number generation function or subroutine provided within a packaged computer program. Then, assuming $F$ is differentiable, $F^{-1}(Z)$ designates a random variate whose density is $f$, the derivative of $F$.

For example, suppose a researcher can approximate the standard Normal inverse cumulative, $\Phi^{-1}$, to a high degree of accuracy, and a formula that describes how the variate $Y$ is thought to have been generated contains a term denoted by the error symbol $\varepsilon$ When this latter term is taken to be a Normal variate with density $\phi([y-\mu]/\sigma)/\sigma$ then, for some particular choice of location parameter $\mu$ and scale parameter $\sigma$ as far as computer simulation or *Monte-Carlo* experiments are concerned, $\varepsilon$ can be taken to designate the component $\mu + \sigma\Phi^{-1}(Z)$. Such a data generating model serves as a recipe that combines two ingredients: (1) a random number $Z$ generated by some computer program and; (2) an algorithm that determines inverse cdf values of the curve $\mu+\sigma\Phi^{-1}(z)$.

## 7.2 Error, regression, and probit, models

There are many ways to generate artificial samples besides calling upon some inverse cdf. For example, generation procedures can be based on the $f$-type of curve. However, when a density curve $f$ serves this purpose, at least two random $Z$ numbers are required to produce a single variate like $Y$. For this reason, when Tocher (1963) describes the process by which simulated values written on slips of paper are selected from a hat or gold fish bowl, he calls the $F^{-1}$ type of curve the top hat function. (Actually, since $Z$ has a uniform density $I_{[0,1)}(z)$, the curve $F^{-1}(1-z)$ would serve Tocher's purposes just as well as $F^{-1}(z)$ itself.)

To see what $F^{-1}(z)$ or $F^{-1}(1-z)$ have to do with data generation, consider the era that immediately predated PC-based computation. At that time statistical models often were formulated as models with an *error term* denoted as $\varepsilon$. Unfortunately, the Greek letter $\varepsilon$ of the formula $\varepsilon = \mu + \sigma G^{-1}(Z)$ has many notational and conceptual disadvantages. For one thing, as statistical terminology has evolved, it has become commonplace for Greek letters to be reserved for the representation of either model parameters or for a specialized density, like the standard Normal curve $\phi$. Even were an acceptable symbolic substitute for $\varepsilon$ found, the problem, defining exactly what an error is, would remain a formidable task.

There are many current interpretations of the concept "probability," and it is still not easy to define what a random uniformly-distributed variate like $Z$ precisely is, or what is meant when two such variates are taken to be independent and identically distributed. Consequently, it seems unwise to add "error" as an additional concept. Mention of an error implies that there is some true or ideal entity about which errors can be measured.

While this makes sense in some situations, there are others where a single path to truth is strewn with stumbling blocks.

The formula $\varepsilon = \mu + \sigma G^{-1}(Z)$ does contain two symbols that designate parameters, $\mu$ and $\sigma$ It also contains $G^{-1}(Z)$, where this variate assumes values equal to the curve $G^{-1}$ evaluated at the variate $Z$'s value. Thus $G^{-1}(Z)$ is a variate—not a parameter. (The evaluation of any inverse cdf curve within the unit interval, (0, 1), is a curve property known as a *quantile* or, when expressed in whole numbers such as 50, a "percentile" like the $50^{th}$ percentile or population median.)

The formula $\mu(x) = \alpha + \beta x$ can be thought of as a way of expressing one type of parameter, here a location parameter $\mu$, as a model based on two other parameters, $\alpha$ and $\beta$. On the other hand, when $\mu + \sigma G^{-1}(Z)$ designates a model component, the upper case letter $Z$ indicates that $\mu + \sigma G^{-1}(Z)$ is a variate, not a parameter. A model component's subcomponents often are expressed in terms of parameters. The curve $G^{-1}$ can be involved, and a variate identified by the upper case letter $Z$ is one commonly encountered subcomponent.

A sort of symbolic musical chairs is played by the symbols for expected value, exponential, and error. In contrast to the three letters, Var, called upon to designate the *variance* curve property, the three letters, "Exp", never represent the *expected value* curve property. Instead, $E$ represents this property in general, and a lower case $e$, as in the curve $e(x)$, designates the length-of-life-remaining regression curve. This and an additional variation on this theme, the representation $e_i$, for expected number, will be defined and discussed in several later chapters. At this point however, it is worth noting that all of these different usage's of the letter $e$ leave no $e$-designated symbolic chair left for the term *error*.

As illustrated by Figure 6.1, the expression $E_f(Y|x)$ denotes the path, locus, or trace of some sequence of curve properties that change as the conditioning variable x changes, while the symbolic representation $\mu(x)$ designates a different variety of curve. A subscripted symbol like $\mu_x$ or, equivalently, the curve notation $\mu(x)$, represent some path, locus, or trace, such as $\mu(x) = \alpha + \beta x$. The important idea is that the model $\mu(x) = \alpha + \beta x$ designates that some straight line describes the $x$-variable's effects on a location parameter component of a density or cdf model. In general, since regression models can be formulated without calling upon the concept, error, it seems preferable to think of an equation whose left hand side is some variate $Y$ as a data-generation model, not a regression model.

There are several reasons why it is not the case that all regression functions can be represented as some composite of elementary functions $x^k$, $\sqrt[k]{x}$, $e^{kx}$, $k^{-1}\log(x)$, $sin(kx)$ and $k^{-1}sin^{-1}(x)$. For example, since no finite number of elementary functions can be assembled to determine $\Phi$, strictly speaking, no model for a regression curve identical to $\Phi$ exists. In his paper entitled, "Why I prefer logits to probits," Berkson (1951) argues that a tractable substitute for a $\Phi$-shaped regression curve's inverse often should be considered in place of the curve $\Phi$ itself. Statisticians who deal with functions that assume an inverse s-shape, as does the function $\Phi$'s inverse; often approximate the *probit function*, the standard Normal inverse cdf, $\Phi^{-1}(z) + 5$. A logit can be thought of as one such approximation, although a rough one, based on the function $ln(z/[1-z])$ which—when scaled properly—looks, in the dark with the light behind it, somewhat like $\Phi^{-1}(z)$. (As discussed in Section 4.3, while over its body portions, the logistic is bell-shaped, it has more in common with the Laplace tail regions than with the Normal model over its tails.)

(A value assumed by the $\Phi^{-1}$ function has sometimes been called a "Normal deviate." By adding 5 to evaluations of $\Phi^{-1}$, early data analysts avoided negative numbers. Any modern computer in the hands of a scientist who is comfortable with negative number arithmetic renders the probit function's definition with the extra 5, at best, quaint.)

## 7.3 Regression and data-generation models

In applications where the expectation curve property $E$ is modeled, the notation $E_f(Y|x)$ indicates through the subscript $f$ that the expectation property is defined in terms of some density $f$. Yet, an explicit model is often assumed for $E_f(Y|x)$, for example $E_f(Y|x) = \alpha + \beta x$, without the complete explicit specification of any model for $f$ like the Normal.

A model for $\mu_x$ describes how it is presumed that the parameter $\mu$ within the symbolic representation of the variate $Y$'s density changes as $x$ changes. On the other hand, $E_f(Y|x)$ describes how a particular curve property is translated by a changed value of $x$. Neither $\mu_x$ nor $E_f(Y|x)$ discloses how probable values of $Y$ are, or exactly how $Y$ has been generated.

The difference between a variate $Y$'s value and a counterpart $\mu_x$ or $E_f(Y|x)$, is called a *residual*. Residuals, in the model fitting sense, are generalizations of univariate quantities called *deviations*. However, while by convention deviations are defined as distances between variate values and some curve property like $E(X)$, residuals, on the other hand, quantify departures of variate values from some true or fitted model.

The $\mu(x)$ component within data-generation model $Y = \mu(x) + \sigma F^{-1}(Z)$ is often called a *predictive* or *prediction* model. When in such cases the term $\mu(x)$ or its estimator is called the portion of the variate $Y$ that is "explained" by the $x$ variable, in a similar fashion the model component $\sigma F^{-1}(Z)$ is called the portion of the variate $Y$ *unexplained* by the $x$-variable.

The popularity of models based on the curve $\mu_x$, such as the $Y = \mu_x + \sigma F^{-1}(Z)$, stems primarily from a dual form of condensed notation. Because traces or loci of parameters like $\mu_x$, or of curve properties like $E_f(Y|x)$, are both mere parts of a larger whole, the formulation of a model for a curve like $\mu_x$ is a less ambitious project, typically, than the determination of model for curves like $f$ or $F$ in their entirety. This is because both curves $f$ and $F$, unlike the curve parameter locus or trace $\mu_x$, describe a variate's distribution completely.

For example, suppose **x** is vector valued, and the generation of $Y$ is related to these multiple x-variables. Then in most cases, there are many other characteristics of the curve $f$ besides $E_f$, such as the curve $f$'s single or multiple modes (in other words, extrema locations). Alternatively, its median, can be functionally related to the variables that constitute $x$.

Regarding parameters, as opposed to curve properties, when a curve like $f$ or $F$ is modeled, besides location parameter $\mu$ other parameters will often be functionally related to **x**. However, despite this wide latitude of choice, it is obviously easier to assess how the single curve property $E_f$ or the single parameter $\mu$ is related to vector $x$, than it is to determine how $f$ or $F$—taken as wholes—change as $x$ changes. Here an $f$ or $F$ curve can be taken to be analogous to the whole spectrum of white light, while individual parameters and curve properties constitute light waves of particular frequencies.

Many properties of a pdf curve, or many parameters of this curve's model, change as functions of $x$. Nevertheless, there has been only one curve property, other than the location-related properties mean, median and mode, whose estimation has generated a substantial literature. For some scale-related property such as the variance, the term "homoscedasticity" designates that the scale-related curve is horizontal, while the term "heteroscedasticity" indicates that some non horizontal curve describes how a scale parameter or, depending on the application, a curve property like AD or the standard deviation, changes as $x$ changes.

## 7.4 Probability, proportion, and data-generation models

The distinction between the two terms, "probability" and "proportion," will now be explained in terms of random number generation. Again, suppose $Z$ represents a variate whose values are obtained by calling a random number generation function or subroutine provided within a packaged computer program, and also suppose that $F^{-1}(Z)$ designates a random number whose pdf is $f$, the assumed continuous derivative of $F$.

## 7.4. Probability, proportion, and data-generation models

When the term $F^{-1}(Z)$ has the capital letter $Z$ as its argument, the $F^{-1}(Z)$ notation designates that it is not just the value of a single random number, $z$, that is transformed. Instead, usage based on the capital letter $Z$ provides a clue that all $Z$ variate values are transformed in some specific way. Today's packaged computer programs call upon this convention when they define a variate label, such as $Y$, and then permit one program statement to be specified, such as $Y' = \log(Y)$, which denotes that a new variate $Y'$ is produced by transforming all $Y$-variate values of sampled data.

The **S⁺** function, runif; the *Mathematica* function, Random; the **MATLAB** and **C⁺** functions, rand; as well as the **MATHCAD** function, rnd(1), provide random number generation capability for obtaining random $Z$-variates. Here for practical purposes all values from 0 to 0.999999.. can be taken to be *equally likely*, and no other values are ever assumed by $Z$. By calling a function like rnd(1), today many other convenient computational tools are available that can begin the process by means of which simulated variates are generated. For example, the program *Mathematica* offers a "numerical function" called "Sign[$x$]" that returns the value 1 whenever the argument $x > 0$, and returns the value -1 whenever arguments are negative. (Notice that *Mathematica's* Sign function is undefined for $x = 0$.)

*Mathematica* and other programs provide "conditional" functions such as "If[$x, y, z$]," if $x$ is true then $y$, and if $x$ is not true then $z$. For nonzero arguments, the Sign[$x$] function can be thought of as the special case of the conditional, If[$x > 0, 1, -1$]. However, the function, If[$x > 0, 1, -1$] is not quite the same as the Sign[$x$] function because, although when $x$ is positive, If[$x > 0, 1, -1$] equals 1, and when $x$ is negative or zero the function If[$x > 0, 1, -1$] equals -1, unlike If[$x > 0, 1, -1$], the function Sign is undefined for a 0 argument.

Sign[$x$] and If[$x > 0, 1, -1$] functions often appear within products designed to change the sign of an expression. For example, forming the product of the function If[$x > 0, 1, -1$] times the expression $x$ will yield the absolute value of the quantity $x$ which, of course, is the quantity designated by the shorthand representation $|x|$, while the function $\{|x|+x\}/\{2x\}$ equals 1 for any positive $x$, 0 for any negative $x$, and is undefined for $x = 0$.

Computer programs offer a built-in function known as the *Heaviside step function*, that for nonzero x, is identical to $\{|x|+x\}/\{2x\}$, and returns the value 1 when $x = 0$. Since this is a useful function, it is unfortunate that the same Greek letter, $\Phi$, is called upon in the computational literature to denote the Heaviside function that is called upon elsewhere to denote the standard Normal cdf. To avoid confusing these curves, here, and below the italicized Greek letter $\mathit{\Phi}$ represents the Heaviside function shown in Figure 7.1, and non-bold, non-italicized, $\Phi$ will continue to designate the standard

96    7. Mixing Parameters and Data-generation models

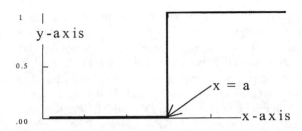

Figure 7.1. The Heaviside function with argument $(x - a)$.

Normal cdf. (In actuality the Heaviside function is the cdf $F_2$ discussed in Chapters 2 and 3.)

The indicator special curve, $I_{(a,b]}(x)$ that is shown in Figure 7.2 is not often a built-in function. However, for $b > a$, the function $I_{[a,b]}(x)$ can be defined in terms of the Heaviside function by the relationship $I_{[a,b]}(x) = \Phi(x - a)\Phi(b - x)$. Here, as before, by substituting a parenthesis for a bracket, $I_{(a,b]}(x)$ is defined to equal $I_{[a,b]}(x)$ everywhere except for the argument $x = a$, at which point $I_{(a,b]}(x) = 0$. Similarly $I_{[a,b)}(x) = I_{[a,b]}(x)$ when $x \neq b$, while $I_{[a,b)}(b) = 0$ and, finally, $I_{(a,b)}(x) = I_{[a,b)}(x)$, except when $x = b$ or $x = a$, while $I_{(a,b)}(b) = I_{(a,b)}(a) = 0$.

Suppose the letter $p$ denotes a real number whose value is between zero and one. In addition, suppose the command rnd(1) returns a random number $Y$ between 0 and 1, including 0 but never the value 1 exactly. Then $\Phi(\text{rnd }(1)-p)$ can be called upon to serve as a "switching function," in the sense that $\Phi(\text{rnd }(1)-p)$ returns the value 0 whenever rnd (1) ¡ $p$ and returns the value 1 otherwise. Looked at from the reverse point of view, the "probability" that the value 0 will be returned equals $p$, and the probability that 1 will be returned equals 1-$p$.

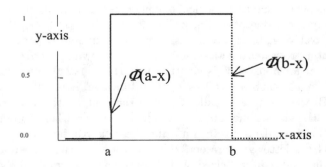

Figure 7.2. The indicator function $y = \Phi(x - a)\Phi(b - x)$.

## 7.4. Probability, proportion, and data-generation models

The definition of probability $p$ in terms of the function rnd(1) is "circular." It is circular because what rnd(1) actually does depends on what is meant by the term "equally likely," where the two words—equally likely— imply that were the variate $Y$ generated as rnd(1), then the pdf that describes the distribution of $Y$ equals $I_{[0,1)}(y)$. The area of study, *random number generation*, concerns procedures by which random variates with density $I_{[0,1)}(y)$ can be obtained, and has been the topic of many papers, including those discussed by Jansson (1966). Later chapters discuss this topic further and elaborate upon the important concept, "identical and independently distributed," that is often abbreviated as iid.

The simulation of independent variates with pdf $f(y) = I_{[0,1)}(y)$, and the interpretation of the $p$ term within $\Phi(\text{rnd }(1)\text{-}p)$, helps to clarify the concepts, *probability* and *proportion*. Suppose, for example, that the function $\Phi(\text{rnd }(1)\text{-}p)$ is *called* only once when a computer program is executed. Then to simulate a single variate that assumes the value 0 with probability equal to $p$, the formula that contains $\Phi(\text{rnd }(1)\text{-}p)$ would be computed a single time whenever $\Phi(\text{rnd}(1)\text{-}p)$ is called once.

The crucial idea here is that the number of times $\Phi(\text{rnd}(1)\text{-}p)$ is called upon in order to yield a variate value has no bearing on the probability that the $\Phi(\text{rnd}(1)\text{-}p)$ formula will return the value 0 or the value 1 which, throughout all these calls, equals $p$, where $0 < p < 1$. On the other hand, the proportion of times 0 or 1 is actually returned can—and usually does— change from sample to sample. When a computer program calls upon $\Phi(\text{rnd}(1)\text{-}p)$ only once, either the value 0, or the value 1—and no other value—will be returned. Thus, even when $p$'s value is between 0 and 1, the proportion of times that 0 is returned equals either 0 or 1. (A proportion is distinguished from a probability like $p$, by the symbol $\hat{p}$ in exactly the same way that the curve $\hat{f}(x)$, estimated from a sample, is set apart from the $f(x)$ curve that describes the population from which members of the sample were selected.)

When a sequence of random values is selected from a table of random numbers, each member of this sequence can be applied, in essence, to flick a single, or multi-pole, switch. As far as most statistical simulation experiments are concerned, how this switch operates is described by some generalization of the expression $\Phi(\text{rnd}(1)\text{-}p)$ that: indexes; selects; or provides the subscript of, a specific population member or subject. On the other hand, among variates whose common distribution is described by the curve $f(y)$, it is the measurement obtained from this subject that yields the variate $Y$'s, value.

A generalization of the simple expression $\Phi(\text{rnd}(1)\text{-}p)$ can help explain the relationship between the generation of mixture data, on the one hand, and the generation of non-mixture data, on the other. Suppose rnd(1) or some other command yields a random variate with density $I_{[0,1)}(y)$. In other words, suppose all values between 0 and 1, including 0 but *not including* 1, are equally likely. Then, for $N > 1$, the formula

$$\sum_{j=1}^{N} j I_{[(\{j-1\}/N),(j/N))}(\operatorname{rnd}(1))$$

is a generating function that yields a random positive integer between 1 and $N$. (The cdf that describes the distribution of this variate is a step function with risers over the sequence that begins with 1 and ends with $N$, where the height of each riser's equals $1/N$.)

A function called *ceil* with argument $x$ assumes a value equal to the least integer $\geq x$, while the *floor* function evaluated at $x$ equals the largest integer $\leq x$. Consequently, determining

$$\sum_{j=1}^{N} j I_{[(\{j-1\}/N),(j/N))}(\operatorname{rnd}(1))$$

is equivalent to rounding the product of $N$ times $\operatorname{rnd}(1)$ to the largest integer $\leq 1 + N \times \operatorname{rnd}(1)$, by calling upon the floor function evaluation $\operatorname{floor}(1 + N \times \operatorname{rnd}(1))$. Equivalently, this computation can be performed by rounding the product $N([1 - \operatorname{rnd}(1)]$ to the least integer $\geq N \times [1 - \operatorname{rnd}(1)]$; in other words, to the integer $\operatorname{ceil}(N \times [1 - \operatorname{rnd}(1)])$.

When some particular coefficient $j$ of the $I$ function within

$$\sum_{j=1}^{N} j I_{[(\{j-1\}/N),(j/N))}(\operatorname{rnd}(1))$$

is replaced by $j^2$, or some other function of the particular $j$, this substitution resembles the computation of $F^{-1}(Z)$ at a random variate value $Z = z$. In general, suppose the symbol $j$ is changed to $h(j)$. Also assume that within this sum, no other integer originally yields, or is changed to yield, the value $h(j)$. Then the probability of obtaining $h(j)$ is still equal to $1/N$. But were both of the two coefficients, $j$ and $(j + 1)$, changed by the function $g$ to equal the same number, $K$, and no other coefficients changed by $g$ in this way, then the probability of obtaining the changed value K would be $2/N$. This process illustrates why the $F^{-1}$ symbol within $F^{-1}(Z)$ represents a spreading function that compresses the random values assumed by $Z$ when the slope of $F$ is high, and spreads out these numbers when $F$ approaches a horizontal line. The latter situation, of course, occurs typically for values placed along one or more pdf curve tails.

## 7.5 The generation of contagious model and mixture model data

Suppose the DDT blood levels of a sample of migrant field workers are measured, and that these workers belong to the two categories: (1) parents, in other words, first generation workers and (2) their children, many of whom are second generation workers (Tarter and Silvers, 1975). However, since generation membership has not been ascertained at the time measurements are made, the variable, generation membership, is the sort of *lurking variate* that often motivates mixture model choice.

This section will illustrate how a data-generation model can be applied to simulate a $c = 2$, dual mixture component sample. Suppose $\phi(y; \mu, \sigma) = \phi([y - \mu/\sigma]/\sigma)$, and that $\mu_0$ and $\mu_1$ denote two distinct location parameters of two Normal pdfs, $\phi(y; \mu_0, \sigma)$ and $\phi(y; \mu_1, \sigma)$, respectively. Also suppose the 0 or 1 value of switching function $\Phi(\text{rnd}(1)-p)$ determines the subscript $_0$ or $_1$ that distinguishes $\mu_0$ from $\mu_1$; while, for a given sample, the probability $p$ of component membership is constant even though, from sample to sample, the proportion of sample members from one or another mixture component varies. One way to interpret what switching function $\Phi(\text{rnd}(1)-p)$ does, is to regard the location parameter $\mu$ as a variate whose distribution is described by a specialized cdf-type curve. This cdf is a two-step step function with risers at $\mu_0$ of height $p$ and at $\mu_1$ of height $1 - p$.

Here mixture data simulation is a two stage process where; as a first step, the choice is made between two alternative location parameter values; say $\mu_0$ as distinguished from $\mu_1$, within $\phi(y; \mu_0, \sigma)$ and $\phi(y; \mu_1, \sigma)$. In the general case, where $c$ may be greater than 2, for example where there may be three or more generations of workers, suppose $p_j, j = 1, ..., c$; where

$$\sum_{j=1}^{c} p_j = 1,$$

are $c$ "mixing parameters" that individually specify the "allocation probabilities" of data points to mixture components, while $a_0 = 0$ and $a_j = a_{j-1} + p_j, j = 1, ..., c$. Then the

$$\sum_{j=1}^{c} j I_{[a_{j-1}, a_j)}(\text{rnd}(1))$$

function can be executed to select a component in the same way that the $\Phi(\text{rnd}(1)-p)$ function can perform the same role in the $c = 2$ case. Then, as a second step, a variate that has the appropriate density $\phi(y;\mu_j,\sigma)$; where $j = 1, ...,$ or $c$; can be generated.

The term "identically distributed with density $f$" implies that the same data-generation procedure applies to all data points. For example, mixture variates are often taken to be identically distributed in the sense that a mixture density $f(y)$ equals

$$\sum_{j=1}^{c} p_j \phi(y;\mu_j,\sigma),$$

or equals some alternative *weighted average* of members of some alternative sequence of $f$-type curves. In the multiple-generation worker example, $\phi(y;\mu_1,\sigma)$ might describe the distribution of parent exposures and $\phi(y;\mu_2,\sigma)$ might describe the distribution of offspring exposures.

Even when $\mu_1 \neq \mu_2$, in the absence of information that discloses which generation characterizes which worker, variates are taken to be identically distributed. However, were worker generation membership information obtained, and were $\mu_1 \neq \mu_2$, then variates would no longer be iid. Before this information was added, the same curve;

$$\sum_{j=1}^{c} p_j \phi(y;\mu_j,\sigma)$$

where $c = 2$; described the distribution of all variates. The added information disclosed which variate's curve is represented in terms of $\mu_1$ and which is represented in terms of $\mu_2$.

We now have increased the palette of model parameter types from the simple pair, $\mu$ and $\sigma$, to the trio: $\mu$; $\sigma$; and $p$. Occasionally, rather than the Latin letter $p$, the Greek letter $\pi$ denotes the mixing parameter. However, to avoid confusing a mixing parameter with the number $\pi = 3.1415926536...$, the Latin lower case letter $p$, not the Greek letter $\pi$, is called upon in this book to denote a mixing parameter.

Many statistical theorems, such as the Central Limit Theorem (CLT) help study identically distributed data. However, suppose that two or more mixing parameters $p_j, j = 1, ..., c$; say $p_u$ and $p_v$; are nonzero, and that the two corresponding location parameters, $\mu_u$ and $\mu_v$, are unequal. Then it is unlikely that the curve property $E_f(Y)$ equals any individual location parameter. Instead, for any mixture-weighted sum of densities $f_m$ expressed in terms of individual location parameters; $\mu_j, j = 1, ..., c$; the curve property $E_f(Y)$ typically equals

$$\sum_{j=1}^{c} p_j \mu_j.$$

## 7.5. The generation of contagious model and mixture model data

Therefore, even though some constant times the average of $f_m$-distributed variates has a density that approaches some Normal density, the location parameter of this Normal density typically will be

$$\sum_{j=1}^{c} p_j \mu_j,$$

a quantity that is unlikely to be equal to any individual $\mu_j, j = 1, ..., c$.

Although mathematically similar, mixtures and *contagious* model-based expressions are rarely studied by Bayesian analyses. This is true even of contagious models that generalize the discrete valued index of a mixing parameter, so that some parameter like $\mu$ is taken to itself have a distribution. In specialized applications, this variate's density is modeled in terms of its own parameter sequence. Finite mixtures, where $c < \infty$, are formulated in terms of parameters like $\mu_j, \sigma_j; j = 1, ..., c$; as well as the mixing parameters $p_j; j = 1, ..., c - 1$; that are all taken to be fixed, but unknown, values. For both contagious models and for finite mixtures, this single set of parameter values is assumed to apply as a whole to the process that generates sampled data. In effect, some indicator or Heaviside-like function simply blanks out all but the appropriate parameter values. Like many human genes, they are there, but unused.

In the sense, for example, that it provides the index value 1 of the parameter $\mu_1$, a variate will only have a distribution described by a simple density like $\phi(y; \mu_1, \sigma) = \phi([y - \mu_1]/\sigma)/\sigma$ in the hypothetical case where a lurking variate shows itself. However, many realistic data sets are based on pairs that consist of two random numbers, where the first pair member helps determine the value of a parameter like $\mu_1$'s subscript, and the second provides the $Z$ for a simulated variate like $Y = \mu_1 + \sigma \Phi^{-1}(Z)$ that is based on this choice of subscript.

The Bayesian point of view is based on past studies that, taken together, have disclosed a prior distribution for the parameter $\mu$ or another single parameter's, or set of parameters', values. For example, suppose, for $b > a$, it is known solely that the parameter $\mu$ must be between two values, $a$ and $b$, and no particular value on, or within, this interval is deemed a better choice than any other. Then the density that describes this prior distribution will be uniform pdf $I_{[a,b]}(\mu)/(b - a)$, while, given $\mu$, the random variate $Y$'s distribution might be represented by the expression $f(y - \mu)$.

Often, unlike finite mixture or contagious model representation, a model for $f(y - \mu)$ like the single component Normal is taken to be valid. In rare situations, Bayesian methods could be based on some mixture like

$$f(y) = \sum_{j=1}^{c} p_j \phi(y; \mu_j, \sigma)$$

or even some contagious model. However, were this done, it would be necessary to provide a prior distribution that describes the distribution of all parameters such as, $p_j; j = 1, .., c-1$; as well as $\mu_j, \sigma_j; j = 1, ..., c$. In the contagious model case, some parameter is conceptualized as a continuous variate. In its turn, the density of this variate might be modeled in terms of one or more parameters. Specifying prior distributions for these parameters would be a daunting problem.

A $c$-component mixture model generalizes a simple expression like $\phi([y-\mu_1]/\sigma)/\sigma$, Such generalizations can help describe first-and-second-generation migrant field-worker mixtures, and other complex distributions, where what is unknown is a lurking variate's values. Knowledge of these values would simplify what had been a mixture-decomposition problem so that statistical procedures based on the assumption of a simple model like $\phi([y-\mu_1]/\sigma)/\sigma$ could be applied. Such procedures might, or might not, be Bayesian.

Contagious modeling procedures differ from finite mixture decomposition solely in terms of the sort of extension of a curve like $\phi([y-\mu_1]/\sigma)/\sigma$ that is selected. For a variety of reasons, a parameter like $\mu$ or $\sigma$ might be assigned its own density curve which, in the case of $\mu$ might even itself be Normal. When the Normal location parameter is itself taken to be Normally distributed, this will not affect the statistical process substantially, but this process will concern a completely new form of density when, for example, a scale parameter to taken to have its own non-degenerate distribution. Nevertheless, despite the seeming complexity of contagious models, in reality they, like finite mixture models, are merely ways of gaining representational scope. Based on this enhanced model, either Bayesian or frequentist techniques can be applied.

While the Bayesian perspective, like the formulation of a contagious model, asserts that one or more parameters has its/their own distribution, the purpose served by a sample is to change the *prior* opinion of this distribution to an "a posteriori" opinion. According to the frequentist approach, even when a variate whose distribution is described by a contagious model is studied by calling upon repeated samples, there is one sole density curve, for instance $f$, involved. On the other hand, as repeated samples are scrutinized, the Bayesian approach considers a sequence of density curves, where each curve describes an opinion of some parameter, or set of parameters, within $f$'s representation. The distributions described by these curves are almost always distinct. Indeed, these distributions will remain unchanged when a sample provides no additional information. To many, this makes considerable sense, since it is appealing to regard knowledge-accretion as a process that begins with the placement of a horizontal line on a tabula rasa, such as $I_{[a,b]}(\mu)$ or $I_{[a,b]}(\mu_1 - \mu_2)$, and ends with a single vertical line over some specific value, such as $\mu = 98.6°$ or $\mu_1 - \mu_2 = 0$ units.

# CHAPTER 8
# The Association Parameter $\rho$

## 8.1 Response, key, and nuisance, variates

In previous chapters, curve types, such as $f$, $F$, and $F^{-1}$; parameters, such as location $\mu$, scale $\sigma$ and mixing parameter $p$; and curve properties, such as AD, $E$, and Var have all been discussed from a theoretical point of view. This chapter deals with distinctions that have an applied flavor. For example, the difference between random variates on the one hand, and conditioning-explanatory or exogenous variables on the other, is based on subject matter interpretation. While definitions of parameters and curve properties can be expressed mathematically, in terms of concepts such as that of a curve and its derivative, the distinction between response, key, and nuisance variates can be explained best in terms of a given investigation's purpose.

Unfortunately, like the dual definitions of a parameter as a model component and as a variate, the word "independent" appears in the statistical literature with two distinct personas: Usage (1) is the $f(y_1, y_2) \equiv f(y_1)f(y_2)$ relationship described in Appendix II; usage (2) distinguishes specific regression components. For example, the arguments of regression curves are sometimes called independent variables where, for example, if a regression curve estimates a curve like $E_f(Y|x)$, then $x$ is called an independent variable.

There is little, or at best an indirect connection between usage (1) and usage (2). Consequently, usage (1) will be called upon exclusively, while the terms: "response, key and nuisance" will help accomplish usage (2) goals. In particular, a "response variate" is what a researcher would like to affect (for example survival time or amount of profit.) On the other hand, a *key* or *explanatory* variate or variable is the means by which the response variate's value can be altered or predicted. It is typical for a scientist to focus his or her expertise on the study of one or more key variates or variables.

A *nuisance* variate or variable is one that must be taken into account when models or other more general curve representations are formulated. Yet it is not of primary interest in its own right—either as a means or end—in the sense that it is neither a key nor response type of variate or variable. For example, since health, environmental, and demographic statistical investigations never have as their goal making people healthier by making them younger, age is commonly a nuisance variate but seldom, if ever, is age a key variate.

Since variables are associated with probability only indirectly, in the sense that a variable's value affects a response variate's density given this variable's value, it follows that when prospective studies are conducted, at least one variate must be the target of investigation whose chances of occurrence are at issue. The words "key" and "nuisance" can modify either the term "variate," or the term "variable," in either prospective, or non-prospective—in other words—*retrospective*, studies. *Observational data* often yields the values of variates, while a setting listed as part of an experiment's protocol provides a variable's value.

Two branches of data analysis methodology deal with nuisance variates and variables, and yet do so in different ways. In branch (1), experiments are designed so that analysis of variance (ANOVA), and other specialized methodologies such as analysis of covariance (ANCOVA), are applied to factor out nuisance factors and covariates. In branch (2) investigations, observational, and experimental data sets that cannot be gathered in a form suitable for branch (1) adjustment, are studied with partialing or adjustment procedures.

Branch (1) methodology is subdivided according to a variety of ANOVA models and designs. For example, when it follows an acronym like ANOVA, the word "model" refers to a data-generation model that describes how key and nuisance, variable and variate, values affect one or more response variate values. The word "design" refers to the allocation of key and nuisance variable values to experimental subjects. For example, univariate Model I ANOVA associates a response variate with variables, not with other variates. The design's goal is to assure that variables can be dealt with individually. Since this model assumes that the values of variables are selected non-randomly by the researcher, Model I ANOVA-based inferences apply solely to these researcher selected values.

Design quality depends on the careful assignment of subjects. Even for so simple a model as Model I, no amount of computing power can undo the damage caused by an incautious design. For this reason, despite and in some cases because of advances in computational capability, the field of study *design of experiments* remains an important statistical subspecialty.

Branch (2), like branch (1), contains many subdivisions. However, there has yet to be as much attention paid to the similarities and differences between various branch (2) techniques as there has been to the similarities among, and differences between, ANOVA and related procedures. In the health and the social sciences, information concerning human beings cannot be gathered commonly from a designed experiment but, instead, exists only in an observational form. Consequently, scientists must often rely on observational data—hence the attention paid in later sections to the similarities and differences between the curves and parameters that underlie branch (2) techniques.

## 8.2 The association parameter $\rho$

In this and later sections the capital letter $Z$ represents a random variate generated by the procedure based on the rnd(1) discussed in the previous chapter, or some function of rnd(1). To clarify how $Z$ is applied, it is handy to call upon several shorthand conventions. For instance, in statistical books and papers the symbol $\sim$ is often placed between a symbol that specifies a variate, and some sequence of symbols that specify a particular distribution. Unfortunately, in many mathematical publications the $\sim$ symbol denotes the term "approximately equal." Consequently, in order to avoid this ambiguity, the inclusion symbol $\in$, as in $Z \in f$, will designate that some variate, $Z$, has a distribution described by a specific density curve, here, for $Z \in f$, represented by the letter $f$.

For example, were $Z$ a standard Normal variate, then $X = \mu + \sigma Z \in \phi((x-\mu)/\sigma)\sigma$ specifies that $\mu + \sigma Z$ is a Normal variate with scale parameter $\sigma$ and location parameter $\mu$. The middle bar within the symbol $\in$ helps underscore that $\mu + \sigma Z$ and $\phi((x-\mu)/\sigma)/\sigma$ are two distinctly different statistical entities, in the sense that $\mu + \sigma Z$ is not a component part of the expression $\phi((x-\mu)/\sigma)/\sigma$, quite the contrary, since $\mu + \sigma Z$ is related more closely to the Normal distribution's inverse cdf than it is to the density curve $\phi((x-\mu)/\sigma)/\sigma$.

Whenever $\mu + \sigma Z \in \phi((x-\mu)/\sigma)\sigma$ for a data-generation model $X = \mu + \sigma Z$ and a density model $\phi((x-\mu)/\sigma)/\sigma$ it is clear that the symbols $\mu$ and $\sigma$ within $X = \mu + \sigma Z$ correspond to the symbols $\mu$ and $\sigma$ within $\phi((x-\mu)/\sigma)/\sigma$. However, in the case of the parameter that is represented by the Greek letter $\rho$, the connection between this symbol as a data-generation model component, and as a density model component, is less obvious.

Four distinct interpretations of the symbol $\rho$ will be discussed in this chapter. Besides parametric interpretations as density, and data generation, model components; interpretations as geometric and *global* curve properties will be considered. (The term "global" refers to an expression that discloses how sums or integrals determine a property's value.)

To generate independent standard Normal variates, $Z_1$ and $Z_2$, as a first step the rnd(1) function can be called twice to obtain two pseudo-independent uniform variates $U_1$ and $U_2$. (When called upon to generate a random number, the number actually provided by computer programs is *pseudo-random*. Since it is based on a finite sequence of arithmetic operations, a pseudo-random simulated variate value has only as many properties that match those of true random numbers as is computationally practical.)

Each of a pair of pseudo-independent rnd(1) uniform variates, $U_1$ and $U_2$, can be transformed individually to create a second variate pair, for example $Z_1 = \Phi^{-1}(U_1)$ and $Z_2 = \Phi^{-1}(U_2)$. There are procedures, such those suggested by Tarter (1968), that can approximate $\Phi^{-1}$ accurately. (An approximation is needed because no composite-of-elementary-functions model for $\Phi^{-1}$ can exist (Rosenlicht,1975)).

While there is no formula that allows $Z_1 = \Phi^{-1}(U_1)$ and $Z_2 = \Phi^{-1}(U_2)$ to be calculated exactly for some choice of $U_1$ and $U_2$, there is an alternative exact approach based on the two functions,

$$Z_1 = \sin(2\pi U_2)\sqrt{-2 \log_e U_1}$$

and

$$Z_2 = \cos(2\pi U_2)\sqrt{-2 \log_e U_1}.$$

These formulas, called collectively, the *Box Muller transformations* (1958), take two independent uniform variates as input, and produce two independent Normal variates as output.

A simple procedure resembles Box and Muller's but, instead of yielding independent variates, it will take two independent Normal variates $Z_1$ and $Z_2$ as input, and generate two dependent variates. These variates will be dependent for any choice of *association parameter* $\rho$ that assumes a nonzero value between -1 and 1. Specifically, for such a $\rho$, the pair of data-generation models, $W_1 = \mu_1 + \sigma_1 Z_1$ and $W_2 = \mu_2 + \sigma_2(\rho Z_1 + \sqrt{1-\rho^2} Z_2)$, produces dependent Normal variates. (Here $\rho$ is referred to as an *association parameter* so that in later sections it can be called upon in situations where the curve property correlation does not exist.)

Beside variates that are Normally distributed, when they are based on independent variates $Z_1$ and $Z_2$, the equations $W_1 = \mu_1 + \sigma_1 Z_1$ and

$$W_2 = \mu_2 + \sigma_2(\rho Z_1 + \sqrt{1-\rho^2} Z_2)$$

## 8.2. The association parameter ρ

can be adapted to yield many alternative types of dependent variates. Nevertheless, later sections will stress the particular case where $Z_1$ and $Z_2$ are chosen to be independent Normal variates. This choice assures that the variate pair $W_1$, $W_2$ jointly are bivariate Normal with population correlation coefficient $\tilde{\rho}$ where, in this Normal special case, the curve property $\tilde{\rho}$ is numerically equal to the association parameter $\rho$.

By means of the data-generation formulas like $W_1 = \mu_1 + \sigma_1 Z_1$ and

$$W_2 = \mu_2 + \sigma_2(\rho Z_1 + \sqrt{1-\rho^2} Z_2),$$

bivariate and multivariate Normal distributions can be defined without specifying the joint Normal density explicitly. This is fortunate because multivariate

$$f(x,y) = [\exp(-h(x,y))]/[2\pi\sigma^{(x)}\sigma^{(y)}\sqrt{1-\rho^2}],$$

where

$$\begin{aligned} h(x,y) &= ([x-\mu^{(x)}]/\sigma^{(x)})^2 - 2\rho([x-\mu^{(x)}]/\sigma^{(x)})([x-\mu^{(y)}]/\sigma^{(y)}) \\ &+ ([x-\mu^{(y)}]/\sigma^{(y)})^2\}/\{2(1-\rho^2), \end{aligned}$$

can be expressed compactly only by calling upon vector and matrix notation. Like the $\sigma$ divisor of $\phi$ that is required as part of the definition of univariate model $\phi((x-\mu)/\sigma)/\sigma$, bivariate density models often are assembled in a way that, at least at first glance, seems to include redundant terms. On the other hand, unlike other bivariate generalizations of Laplace, extreme value, logistic, Cauchy, and lognormal univariate models, many specialized bivariate distributions can be defined in a concise fashion by substituting the appropriate univariate variates for $Z_1$ and $Z_2$ in the pair of relationships, $W_1 = \mu_1 + \sigma_1 Z_1$ and $W_2 = \mu_2 + \sigma_2(\rho Z_1 + \sqrt{1-\rho^2} Z_2)$.

Despite their simplicity, caution is necessary when data-generation models like $W_1 = \mu_1 + \sigma_1 Z_1$ and $W_2 = \mu_2 + \sigma_2(\rho Z_1 + \sqrt{1-\rho^2} Z_2)$ are applied. Difficulties can be traced to two sources. The first concerns the interpretation of $\rho$, both as an association parameter and as the type of correlation coefficient curve property discussed in the next section. The second limitation is due to the effect that the term

$$\rho Z_1 + \sqrt{1-\rho^2} Z_2$$

has on discrete, Poisson-like variates.

When a bivariate Normal distribution is said to have a *population correlation coefficient* $\rho$, the symbol $\rho$ denotes a curve property that will be discussed in detail below. Unfortunately, unlike the symbolic distinction between the Latin letter $E$ that denotes a curve property and the Greek letter $\mu$ that designates a location parameter, there is no simple symbolic distinction between $\rho$ as curve property and as parameter. Thus, for many models, parameter-with-property mismatch can be confusing.

For example, the logistic density and the corresponding "logit" inverse cdf, $F^{-1}(y) = \log(y/[1-y])$, is not defined so that there is an exact match between the association parameter $\rho$ within

$$W_2 = \mu_2 + \sigma_2(\rho Z_1 + \sqrt{1-\rho^2} Z_2),$$

and the curve property, correlation. Yet an advantage of

$$W_2 = \mu_2 + \sigma_2(\rho Z_1 + \sqrt{1-\rho^2} Z_2),$$

is that this data-generating relationship can be applied to simulate bivariate data based on Cauchy distributed $Z_1$ and $Z_2$, even though the curve property, correlation, does not exist for any of the resulting bivariate densities obtained in this way.

A mismatch between the model parameter denoted by the Greek letter $\rho$ and a curve property like the correlation coefficient, is a danger sign that is similar to the division of the term $y^2$ by what, in the Physics and Optics literature, would seem to be an extra constant 2 within the exponent of the Normal density curve's model

$$\phi(y) = e^{-y^2/2}/\sqrt{2\pi}.$$

Within the statistical and computational literature, the adaptation of standard models so that parameters match curve properties is limited to the Normal $y^2/2$ argument. Consequently, there often is a mismatch between parameters and properties that disclose association, correlation, and scale. For example, were the scale parameter $\sigma$ to appear within a model for the logistic density $f(x) = (4\sigma)^{-1} \text{sech}^2\{(x-\mu)/(2\sigma)\}$, then the symbol $\sigma$ would not be numerically equal to the standard deviation property of this density curve.

Before considering the distinctions and similarities between regression models and bivariate data generation-models in further detail, an additional limitation of the data-generation model pair, $W_1 = \mu_1 + \sigma_1 Z_1$ and

$$W_2 = \mu_2 + \sigma_2(\rho Z_1 + \sqrt{1-\rho^2} Z_2),$$

should be mentioned. While, from a theoretical point of view, independent Poisson $Z_2$ and $Z_1$ variates can be substituted into the formulas that define dependent variates $W_1$ and $W_2$, the generation of dependent Poisson variates does raise a difficult practical problem.

## 8.2. The association parameter $\rho$

Poisson variates can be simulated in several different ways. For example, suppose that, starting with the half open interval $[0, e^{-\lambda})$, the index-generating formula

$$\sum_{j=1}^{N} j I_{[(\{j-1\}/N),(j/N))}(\text{rnd}(1)) - 1,$$

is modified so that, for $N \geq j > 1$, the $j^{th}$ interval, $[\{j-1\}/N, j/N)$, within this formula's subscript is replaced by an interval of length equal to $e^{-\lambda}\lambda^j/j!$ that starts where the previous interval leaves off. Then, as $N$ increases, this modified generator will determine a variate whose distribution approaches the distribution of a Poisson variate.

Alternatively, consider that if $x$ represents a non integer real number, then the shortest interval—bordered by integers—that contains $x$ is the interval $(\text{floor}(x), \text{ceil}(x))$. Consider the sequence, $U_1, U_2, .., U_y$, with the last, here the $y^{th}$, member $U_y$, that is comprised of variates obtained by calling rnd(1) repeatedly until the inequalities,

$$\text{floor}(\sum_{j=1}^{y} [-\log U_j]) \leq \lambda$$

and

$$\text{ceil}(\sum_{j=1}^{y} [-\log U_j]) \geq \lambda,$$

are both satisfied. Then the integer $y$ that is determined in this way is the value of a Poisson variate $Y$ such that $Y \in e^{-\lambda}\lambda^y/y!$.

*Dependent* Poisson variates cannot be simulated as easily as their independent Poisson variate counterparts. For example, suppose that in order to yield a pair of variates $W_1$, $W_2$ based on association parameter $\rho$, independent Poisson $Z_2$ and $Z_1$ are generated and then substituted into the expressions $W_1 = \mu_1 + \sigma_1 Z_1$ and $W_2 = \mu_2 + \sigma_2(\rho Z_1 + \sqrt{1-\rho^2} Z_2)$. The $W_1$ and $W_2$ variates will then differ from the $Z_1$ and $Z_2$ variates substantially. They are different because a given set of parameter $\mu_1, \mu_2, \sigma_1, \sigma_2$, and $\rho$ values will rarely yield variates $W_1$ and $W_2$ that, like the original Poisson $Z_2$ and $Z_1$, are integer-valued.

Generally speaking, one discrete variate will often depend on a second discrete variate. Yet these variates cannot be simulated easily by calling upon an expression like $W_2 = \mu_2 + \sigma_2(\rho Z_1 + \sqrt{1-\rho^2} Z_2)$. Although two integer-valued indices or "discrete" variates, such as the number of deaths in a census tract and the number of cases of a given disease in the same tract, can be closely related, because of the limitations of the formula

$$W_2 = \mu_2 + \sigma_2(\rho Z_1 + \sqrt{1-\rho^2}Z_2),$$

it is more difficult to study these discrete variates' relationship than it is to study the relationship between pairs of Normal, or even between pairs of Cauchy, variates. Similarly, almost nothing is known about parametric approaches that concern the relationship between a Normal variate and a Poisson variate or, heaven forbid, a Cauchy variate and a Poisson variate.

## 8.3 Conditional, joint and marginal, notation

Many notational conventions that apply to variates, variables, and the relationships between them, are based on distinctions between conditional, joint, and marginal, density curves. For example, consider the simple *discrete uniform* density that assigns equal probability to the values 1, 2, 3, 4, 5, and 6, and otherwise equals 0, where here, when this curve describes the number of dots on the top side of a rolled cube, it is a pair of dice, not a single die, that is rolled.

Suppose a red die and a green die are thrown and the following three different classifications of probability are compared: (1) the chance of rolling a 2 with the red die and a 5 with the green die; (2) the chance of rolling a 2 with the red die given that a 5 is known to have been rolled with the green die or; (3) the chance of rolling a 2 with the red die without any prior knowledge of the roll of the green die.

Why marginal probabilities are germane to type (3) rolls is illustrated by a gambling table laid out to describe the possible outcomes of type (1) dice rolls. Marginal probabilities can be visualized as the result of sweeping (integration or summation) processes, in which joint probabilities are combined across a row or down a column, and where the integral or sum tallied in this way is recorded along either a side, or a bottom, *margin*.

The subjective or, as it is called by Savage (1968), "personalistic" interpretation of probability, concerns an individual's confidence that a statement is true—for example, that it will rain at some future date. It is obviously easier to be confident about simple occurrences than about complex occurrences, and thus marginal probabilities often assess simpler statements than are assessed by their joint probability counterparts. The probability of rain today is a simpler concept than is the probability that it will both rain today and rain tomorrow as well. This is in accord with the simple fact that sweeping processes *accumulate*, rather than reduce, whatever is being swept.

## 8.3. Conditional, joint and marginal, notation

Once it is known that a given day is rainy, it is often important to consider the *conditional probability* that it will also rain the next day. Similarly, it may also be of interest to know that the conditional probability of rolling a 6 with a red die, given that a 6 has been rolled by a green die, is equal to the marginal probability of rolling a 6 with a red die, without any concern with, or reference to, the variate that corresponds to its green counterpart.

The outcome of a red die throw (or tomorrow's weather) given prior information about a green die's throw (or that it has rained today), is represented in terms of the set of symbols that follow a slash sign in the *conditional density* notation. For example, the $|x_1, x_2, \ldots$ symbols within $f(y|x_1, x_2, \ldots)$ help to pin down which particular $f$-type curve is of current interest. Therefore as discussed in Section 6.6, $x_1, x_2, \ldots$ are equivalent to the symbol for a density's subscript, for example the subscript of the symbol $f_{x_1, x_2 \ldots}$.

A subscript of a subscript, the first for $f$ itself and the second of some variate value like $x_1$, are difficult both to print and to read. Consequently, conditional densities are represented by placement after a slash, within the parentheses that follow an unsubscripted $f$ symbol, of what might otherwise appear as a subscript.

The placement of the symbols $x_1, x_2, \ldots$ within $(y|x_1, x_2, \ldots)$ leaves the space beneath a symbol $f$ free to serve a purpose other than the listing of variables. For example, suppose two (possibly distinct) curves both represent conditional pdf's that refer to the same variate and variables. Then the subscripts $_1$ of $f_1$ and $_2$ of $f_2$, as in $f_1(y|x_1, x_2, \ldots)$ and $f_2(y|x_1, x_2, \ldots)$, often distinguish these curves.

Suppose, for a specific $x_1, x_2, \ldots$ sequence, a joint density curve is multiplied by a constant so that the resulting curve integrates to 1 over the $y$-axis. Then the curves $f_1(y|x_1, x_2, \ldots)$ and $f_2(y|x_1, x_2, \ldots)$ that are formed in this way represent two different multidimensional functions that can be said to have been sliced at the same given values, $x_1, x_2, \ldots$ In particular, when a *single* letter, say $y$, is placed before the symbol for the word "given" (the slash, |), then this placement designates that the slices at $x_1, x_2, \ldots$ here represented by the symbols $f_1$ and $f_2$, are one dimensional. Hence slices $f_1$ and $f_2$ describe *univariate conditional* distributions.

The expressions, $f_1(y_1, y_2|x_1, x_2, x_3)$ and $f_2(y_1, y_2|x_1, x_2, x_3)$ denote the two curves, $f_1$ and $f_2$, where, for example, $f_1$ might be a bivariate Normal density and $f_2$ a bivariate Cauchy density. Notation that expresses $f_1(y_1, y_2|x_1, x_2, x_3)$ and $f_2(y_1, y_2|x_1, x_2, x_3)$ also discloses that both of these curves describe conditional *bivariate* distributions of the variates $Y_1$ and $Y_2$, given the three variables $x_1, x_2, x_3$. In other words, two-dimensional slices are taken in the sense that any model for $f_1(y_1, y_2|x_1, x_2, x_3)$ would be called a *bivariate conditional model*.

A lowercase subscripted form of a letter like $y$ often represents the value assumed by a variate. By convention, the uppercase form of this same letter, with the same subscript as the lowercase letter, represents the particular variate or sequence of variates, say the sequence whose members are $Y_1$ and $Y_2$, whose distribution is described symbolically, for instance, by the density $f_1(y_1, y_2|x_1, x_2, x_3)$.

The usage of lowercase and uppercase letters also applies to data-generation models. For instance, since the expression

$$\mu + \sigma Z \in \phi((x - \mu)/\sigma)/\sigma$$

asserts that a variate $\mu + \sigma Z$ has a nonstandard Normal density, and the lowercase letter $x$ serves as the argument of $\phi((x - \mu)/\sigma)/\sigma$ it follows implicitly that the uppercase letter $X$ represents $\mu + \sigma Z$.

When $Z \in \phi(z)$, the expression $X = \mu + \sigma Z \in \phi((x - \mu)/\sigma)/\sigma$ is a shorthand characterization of variate $X$. It classifies $X$ in two ways: (1), as a variate generated by taking a standard Normal variate, multiplying it by $\sigma$ and then adding $\mu$; and (2), as a variate whose density can be modeled by $\phi((x - \mu)/\sigma)/\sigma$. The implicit notation $N(x; \mu, \sigma)$, of the expression $X = \mu + \sigma Z \in N(x; \mu, \sigma)$ serves the same notational purpose. The expression $N(x; \mu, \sigma)$ sometimes represents a distribution in its entirety and sometimes designates a model for this distribution's density. Here and below, $N(x; \mu, \sigma)$ represents the general Normal model $\phi((x - \mu)/\sigma)/\sigma$.

In some references the distribution of a variate, such as $X$, is represented as "dist $X$." For example, dist $X \equiv$ dist $Y$ indicates that the distribution of $X$ is identical to the distribution of $Y$. If two variates are identically distributed, any curve (for example an $f$-type of curve) that happens to describe the distribution of the first variate will serve equally well as a description of the second variate's distribution. As was true of the letter $f$, in order to describe conditional densities, expressions such as $X = \mu + \sigma Z \in N(x; \mu, \sigma)$ as well as the

$$X = \mu + \sigma Z \in \phi((x - \mu)/\sigma)\sigma$$

notation, are often generalized to indicate which parameter is related to which particular sequence of variates and/or variables.

For example, $Y|x \in N(y; \sigma|x, \beta)$ denotes (1) that for a given value of the variable $x$, a variate $Y$, conditioned on this $x$, is Normally distributed and (2) that this Normal distribution is described by a Normal model with a scale parameter $\sigma$. Here the placement of $\sigma$ indicates that this parameter is unrelated to $x$'s value. On the other hand, in this example the parameter $\beta$ will often help describe how the variable value $x$ affects the location of the density curve.

## 8.3. Conditional, joint and marginal, notation

The rule, one variate, one density, implies that the conditional density is taken to be a *single* entity that is translated from one position to another. For a given variable value, where this curve actually is, often depends on a location parameter's value as determined, wholly or in part, by $\beta$. To indicate this form of dependence symbolically, an alternative to $N(y; \sigma | x, \beta)$ written as $N(y; \mu_x, \sigma)$ is called upon in situations where the symbol $\mu_x$ designates a function $\mu(x) = h(x; \beta)$. (It is often necessary to distinguish an $X$-variate's location parameter from a $Y$-variate's location parameter. When a lowercase letter is placed within superscript parentheses in later sections, the symbols $\mu^{(x)}$ and $\mu^{(y)}$ will serve this purpose.)

The two curves, $f$ and $F$, can be thought of as descriptors of a distribution in its entirety. When a parameter is considered to characterize $f$ or $F$ as wholes, it will be called a *broad-brush* parameter. Like $f$ and $F$, broad-brush parametric expressions $\mu(x)$ or $\mu_x$ are taken to be single entities. By convention, whenever the representation $\mu_x$ merges the concept *parameter* with the concept *curve*, curve interpretation dominates. For example, if $x_1 \neq x_2$, then $\mu(x_1)$ together with $\mu(x_2)$ designate two evaluation-type properties of the single curve $\mu(x)$, not two distinct parameters. For this reason, the notation $\mu(x_1)$, $\mu(x_2)$ differs substantially from each of the paired symbols $\mu_1, \mu_2; \mu^{(1)}, \mu^{(2)}$; and $\mu^{(x)}, \mu^{(y)}$, because, for example, the symbols $\mu^{(x_1)}$ and $\mu^{(x_2)}$ designate two distinct parameters, not the one parameter $\mu$ of $\mu(x_1)$ and $\mu(x_2)$.

Based on the pair of parameters $\alpha$ and $\beta$, the model $\mu_x = \alpha + \beta x$ is often called upon to represent the curve $\mu_x$. There is no theoretical reason why an expression like $\mu_{x,z} = \alpha_z + \beta_z x$ should not continue this process one step further. Perhaps, for example, $z$ designates the value of a nuisance variable that alters the slope and $y$-intercept of a regression line. (Regression methodology that concerns $\alpha_z + \beta_z x$ is presently statistical terra-incognita, although, thanks to modern computational and representational techniques, it may not be long before $\alpha_z + \beta_z x$-based explorations take place.)

Expressions $h(x; \beta)$ and $h(x; \beta_1, \beta_2)$ denote functions based on one or two parameters, respectively. In much the same way that the letter $i$ represents an arbitrary subscript of a variate, such as $Y_i$, a generic symbol like $\theta$ often denotes an arbitrary parameter that could, for instance, be any one of the parameters more definitively denoted by symbols such as $\mu$ and $\sigma$ that designate specific parameter types.

Location, scale and association parameters each have their own role to play in terms of a particular form of uncertainty about a curve. The role is determined by the adjoining symbols which surround the symbols that represent these parameters within a density, cdf, or data-generation model. On the other hand it is sometimes useful to designate a parameter, pure and simple. For this purpose, within a sequence like $\theta_1, \theta_2, ..., \theta_{10}$ of $f(x; \theta_1, \theta_2, ..., \theta_{10})$, the symbol $\theta$ represents a parameter when no indication whether $\theta$ is a scale, location, or other specific parameter type is needed.

## 8.4 The sample and the population correlation coefficient

Suppose $\{(X_j, Y_j)\}$ ; $j = 1, 2, ..., n$; represents a sample of iid-bivariate-Normal $f$-distributed variate pairs with values $\{(x_j, y_j)\}$; $j = 1, 2, ..., n$; while $\bar{x}$ represents the arithmetic average

$$n^{-1} \sum_{j=1}^{n} x_j$$

computed from $X$ variate values, and $\bar{y}$ represents the comparable average of $Y$ variate values. Then

$$n^{-1} \sum_{j=1}^{n} (x_j - \bar{x})(y_j - \bar{y})$$

is the most commonly applied estimator of the covariance-composite $\rho\sigma^{(x)}\sigma^{(y)}$ assembled from the scalar-valued $\rho, \sigma^{(x)}$ and $\sigma^{(y)}$ parameters within the formula for the bivariate Normal density curve $f$. The combination of parameters $\rho\sigma^{(x)}\sigma^{(y)}$ equals the curve property of the bivariate Normal density $f$ that is defined as

$$\text{Cov}(X_j, Y_j) = \int_{-\infty}^{\infty} \int_{-\infty}^{\infty} (x - E(X))(y - E(Y))f(x, y)dxdy,$$

where

$$E(X) = \int_{-\infty}^{\infty} \int_{-\infty}^{\infty} xf(x, y)dxdy \quad \text{and} \quad E(Y) = \int_{-\infty}^{\infty} \int_{-\infty}^{\infty} yf(x, y)dxdy.$$

(Even though the curve property Cov is well-defined for any bivariate Normal $f$, it is not necessarily the case that $\text{Cov}(X_j, Y_j)$ is well-defined for some other distribution.)

Each $X_j$ variate among the bivariate Normal variates

$$\{(X_j, Y_j)\}; j = 1, 2, ..., n;$$

has a marginal distribution described by a density $\phi([x_j - \mu^{(x)}]/\sigma^{(x)})/\sigma^{(x)}$. Similarly, each $Y_j$ variate of $\{(X_j, Y_j)\}$; $j = 1, 2, ..., n$; has a distribution

## 8.4. The sample and the population correlation coefficient

described by a density $\phi([y_j - \mu^{(y)}]/\sigma^{(y)})/\sigma^{(y)}$. The sample mean statistics $\bar{x}$ and $\bar{y}$ are common estimators of location parameters $\mu^{(x)}$ and $\mu^{(y)}$, while modifications of $s$ that serve as estimators of scale parameters $\sigma^{(x)}$ and $\sigma^{(y)}$ commonly are

$$s^{(x)} = \sqrt{n^{-1}\sum_{j=1}^{n}(x_j - \bar{x})^2} \quad \text{and} \quad s^{(y)} = \sqrt{n^{-1}\sum_{j=1}^{n}(y_j - \bar{y})^2},$$

respectively. For many bivariate data sets a fraction is formed with a numerator equal to the estimator of $\rho\sigma^{(x)}\sigma^{(y)}$, the *cross product average*,

$$\text{CPA}^{(x,y)} = n^{-1}\sum_{j=1}^{n}(x_j - \bar{x})(y_j - \bar{y}).$$

This fraction's denominator is set equal to $s^{(x)}s^{(y)}$ and then the Pearson's r statistic $\text{CPA}^{(x,y)}/(s^{(x)}s^{(y)})$ is computed. The statistic $r = \text{CPA}^{(x,y)}/(s^{(x)}s^{(y)})$ is often referred to as the *Pearson coefficient of correlation* or the *product moment correlation estimator*.

Among many alternative statistics that estimate association parameters, Pearson's r stands out in terms of the high place it has retained in the esteem of theoretical and applied statisticians. This is due to both its own advantages, and to the disadvantages of alternative procedures. Nevertheless, because of the important roles that the covariance curve property and the association parameter $\rho$ both play in many statistical investigations, "rank correlation" and "tetrachoric correlation" alternatives to r have also been studied extensively.

To construct a rank correlation coefficient, a sample of $n$; $(X, Y)$ variate pairs, is separated into $X$ variate and $Y$ variate subsets. Suppose within each subset no two variates assume the same value. Then within each subset the variate with the largest value can be given the rank 1, the one with the next largest value the rank 2, and so forth. For each $j = 1, ..., n$; the $j^{th}$ variate pair's difference, $D_j$, can then be computed between the $X$ variate's, and the corresponding $Y$ variate's ranks. These $n$ differences can then be squared and the results all added together to compute SDS, the *sum of squared rank differences*. The rank correlation coefficient counterpart of Pearson's $r$ statistic is then defined to equal $r_s = 1 - 6(\text{SDS})/(n[n^2 - 1])$.

Pearson's r statistic assumes values between -1 and +1. Yet today, the estimators of the population correlation coefficient that are calculated by many packaged computer programs can, in rare instances, be either greater than 1 or less than -1. This difference arises because the statistics these programs compute take advantage of the fact that when $\sigma^{(x)}$ and $\sigma^{(y)}$ are estimated, $(X, Y)$ variate pairs are separated into $X$-variate and $Y$-variate subsets. While the computation of

$$n^{-1} \sum_{j=1}^{n} (x_j - \bar{x})(y_j - \bar{y})$$

requires knowledge of *both* the $j^{th}$, $x$, and the corresponding $j^{th}$, $y$, values, if the $x$-value is known, but the $y$-value is unknown, then this $x$-value will still contribute to an estimator of $\sigma^{(x)}$, even though it doesn't contribute to either the numerator of r, or r's denominator component that estimates $\sigma^{(y)}$. When there are many missing values in a data set, efficiency can be increased by individual estimation of the $\rho, \sigma^{(x)}$, and $\sigma^{(y)}$ components that constitute $\rho, \sigma^{(x)}$ $\sigma^{(y)}$.

The above *missing value* procedure can be thought of as an attempt to retain as much information as possible. On the other hand, like the rank correlation coefficient, tetrachoric correlation replaces exact values with reduced or abstracted data. For example, suppose instead of its rank, a researcher determines whether or not each $X$-variate value is above or below the value assumed by the sample median, the point that separates the ordered sample values into two halves. $Y$ variate values are treated similarly, and in this way data points are subdivided among four quadrants depending upon their coordinate's value being larger or smaller than the $X$ and $Y$ sample medians, respectively.

The number of points both of whose $x$-values and $y$-values are greater than the corresponding medians, in other words—those in the first quadrant—is added to the number in the third quadrant. The sum of these two numbers is divided by $n$, the number of points in all four quadrants. This step yields a third number, the first candidate value. This is repeated to yield a value equal to the total points in the second and fourth quadrants, divided by $n$. The larger of the two candidate values is called the *tetrachoric r*.

## 8.5 Correlation geometry

Two variates that are legitimately characterized as independent have a joint density that must be identically equal to the product of the marginal densities of these two variates. As a consequence, statistical dependence and independence is determined by the structure of three curves, considered in their entirety. Yet, both the association parameter $\rho$ and covariance, $\text{Cov}(X, Y)$, assume individual numerical values. For a given distribution

## 8.5. Correlation geometry

the values of $\rho$ and $\mathrm{Cov}(X,Y)$ are measures of one—out of a great many—facets of variate dependence. However, for bivariate Normally-distributed variates it does follow that $\rho = 0$ and $\mathrm{Cov}(X,Y) = 0$ are both necessary and sufficient conditions for the independence of the two variates, $X$ and $Y$. In addition, for bivariate Normal $X$ and $Y$, the parameter $\rho$ equals curve property $\mathrm{Cov}(X,Y)$ whenever the product of $\mathrm{Var}(X)$ times $\mathrm{Var}(Y)$ equals 1.

For non-Normal distributions, even when only two variates are involved, dependence does not necessarily imply that a parameter $\rho$ within a representation of a joint density curve, or even that the $\mathrm{Cov}_f(X,Y)$ property of this curve, is nonzero. Additional complexities arise because Pearson's important statistic r is usually taken to be an estimate of the curve property

$$\mathrm{Cov}(X,Y)/\sqrt{\mathrm{Var}(X)\mathrm{Var}(Y)},$$

and not considered to be an estimator of the Normal curve's parameter $\rho$, or an estimator of the parameter $\rho$ within the data-generation model

$$W_2 = \mu_2 + \sigma_2(\rho Z_1 + \sqrt{1-\rho^2}Z_2).$$

To see what these complexities are, it is useful to keep in mind that data-analysis computations often are opposite to the step-sequence that generates artificial data. For instance, to generate a variate $Z$, $Z$'s value is multiplied by the scale parameter $\sigma$ and—finally—$\mu$ is added to this product. This pattern continues when dependently-distributed data are generated. The two standard independent variates, $Z_1$ and $Z_2$, are assumed to be mixed together to form $X = Z_1$ and

$$Y = \rho Z_1 + \sqrt{1-\rho^2}Z_2$$

*first*, and then only after this has been accomplished are the relationships $X' = \mu^{(x)} + \sigma^{(x)}X$ and $Y' = \mu^{(y)} + \sigma^{(y)}Y$ called upon to attach the location and scale parameters, $\mu^{(x)}, \sigma^{(x)}, \mu^{(y)}$ and $\sigma^{(y)}$, to the variates, $X$ and $Y$.

Suppose variates are generated by the above process, and an elementary function is found to serve as a model for the density of the resulting variate $(X',Y')$, where $\rho$ is included among this density model's parameters. Then the correspondence between a parameter, such as $\rho$, and a curve property of the density model of the variate pair $(X,Y)$; such as the covariance $\mathrm{Cov}(X,Y)$, or correlation

$$\mathrm{Cov}(X,Y)/\sqrt{\mathrm{Var}(X)\,\mathrm{Var}(Y)},$$

is not guaranteed.

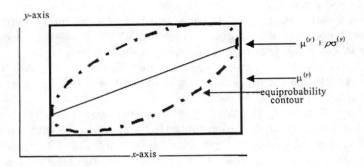

Figure 8.1. Bivariate normal parameter.

For example, if either or both $Z_1$ and $Z_2$ are Cauchy-distributed, then $\text{Cov}(Z_1, Z_2)$ is ill-defined and

$$\sqrt{\text{Var}(Z_1)\,\text{Var}(Z_2)}$$

is infinite. Yet, despite the Cauchy density's correlation property's poorly defined value, to obtain bivariate data the simulation formulas, $X = Z_1$ and

$$Y = \rho Z_1 + \sqrt{1-\rho^2}\, Z_2,$$

can be applied despite a parameter-with-property mismatch. The symbol $\rho$ can designate the association parameter within

$$\rho Z_1 + \sqrt{1-\rho^2}\, Z_2$$

even though, in the Cauchy $Z_1$ and $Z_2$ case no correlation curve property that matches $\rho$ exists.

Unlike the Cauchy case, in the bivariate Normal case there is a simple geometric interpretation of the curve property associated with the bivariate Normal parameter $\rho$. Consider, for example, the elliptical Normal density contour shown in Figure 8.1. The theoretical regression line that is shown in this picture has a slope equal to $\rho$ times the ratio of the two scale parameters of the $Y$-variate to that of the $X$-variate, respectively.

It is not unusual to treat a key variate $X$ as if it were a variable whose value forms part of a regression model. This approach can sometimes be justified because of an important relationship. This relationship links $\tilde{\mu}^{(y)}(x)$, the curve that describes the $Y$ response variate's conditional distribution's Normal density model's location parameter, to the four Normal

## 8.5. Correlation geometry

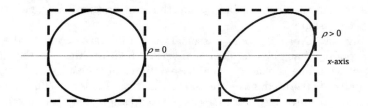

Figure 8.2. Contours of transformed variates.

marginal density model parameters $\mu^{(x)}, \sigma^{(x)}, \mu^{(y)}$ and $\sigma^{(y)}$. Specifically suppose $(X, Y) \in f(x, y)$, where

$$f(x,y) = [exp(-h(x,y))]/[2\pi\sigma^{(x)}\sigma^{(y)}\sqrt{1-\rho^2}]$$

and

$$\begin{aligned}h(x,y) &= ([x-\mu^{(x)}]/\sigma^{(x)})^2 - 2\rho([x-\mu^{(x)}]/\sigma^{(x)})([x-\mu^{(y)}]/\sigma^{(y)}) \\ &+ ([x-\mu^{(y)}]/\sigma^{(y)})^2\}/\{2(1-\rho^2),\end{aligned}$$

Then in this bivariate Normal case the three relationships

$$\tilde{\mu}^{(y)}(x) = \mu^{(y)} + (\rho\sigma^{(y)}/\sigma^{(x)})(x - \mu^{(x)}), E(Y|x) + \mu^{(y)} = (\rho\sigma^{(y)}/\sigma^{(x)})(x - \mu^{(x)}),$$

and

$$E(Y|x) = E(Y) + [Cov(X,Y)/Var(X)][x - E(X)],$$

are all valid.

Within an ellipse, the line segment called the *major-axis* is defined as the longest chord that can be drawn between two points that both lie on the ellipse. However, if a segment of some $\tilde{\mu}^{(y)}(x)$ regression line were located within the ellipse shown in of Figure 8.1, then this segment would not lie on the major-axis of the ellipse. Instead, it would pass through the point that has a $y$ coordinate equal to ordinate $\mu^{(x)} + \rho\sigma^{(y)}$ and an $x$ coordinate equal to the value $\mu^{(y)} + \sigma^{(x)}$ in the abscissa of Figures 8.1 and 8.2. In Figure 8.1 the circle is the largest figure of this type that is enclosed completely by the square with side of length 2.

The point where the ellipse touches the rightmost vertical side of the square is exactly $\rho$ units above the $x$-axis. The $\rho = 0$ and $\rho > 0$ cases are illustrated in Figure 8.2. As $\rho \to 0$, the ellipse broadens to a circle. As $\rho \to 1$, a cigar-shaped ellipse narrows until, as if to fit into the square's corner, it degenerates to a straight line segment. When $\rho \to -1$, not $+1$, the same phenomenon occurs, except that the limiting chord has a negative, not a positive, slope.

The bivariate Normal model's association parameter is connected with a curve property that discloses degree of variate relationship. Yet the slope of the regression line is exactly equal to $\rho$ only when the two variates that are Normally distributed have the same scale parameters. This means that even in the bivariate Normal case, a regression line's slope only is proportionate to the strength of an association—as opposed to being merely an indicator of the presence or absence of a bivariate relationship—when variates are scaled identically.

In general, even in the bivariate Normal case, strength of variate relationship is due primarily to the fitting of an ellipse into a corner tightly, and not due to the slant of a line. In simulation applications, it is obviously far easier to deal with independent than with dependent variates. Hence caution is called for regarding a false sense of security that can be afforded by a regression line whose slope is slight.

For Normal distributions, the two parametric and two property-based interpretations of the symbol $\rho$ are compatible. For other distributions, each interpretation has its own strengths and weaknesses. In particular, geometric equiprobability contour interpretation provides useful information even when contours are not perfectly elliptical.

Many univariate density curves are multi-modal. Consider, for example, a halved bell-shaped curve where each half subtends an area equal to 0.5. If halves are drawn apart to form a new curve, the curve will have a median curve property that can be defined equal to any point between the halves. Similarly, a bivariate density could conceivably have as many different contour-based correlations as there are distinct $\rho$ parameters in some mixture model. There is even more reason to conceptualize multiple variate associations than multiple medians. We ask: "What is the association between variates $X$ and $Y$? rather than, What *are* the associations between these variates?" due to habit, not necessity. Curves, like faces that morph on a computer screen as a function of time, can morph as a function of a key or nuisance variate or variable's value so that variate associations change.

A global curve property like $\text{Cov}(X, Y)$ is usually assumed to be single-valued. On the other hand, many geometric curve properties linked to variate association, can, like those illustrated in Figures 8.1 and 8.2, be multiple-valued or even change continously. For example, consecutive pieces of a bivariate Normal can be grafted together so that association increases as a function of either the $X$-variate's or the $Y$-variate's values and, in this sense, a sequence of elliptical curve segments morphs from circular to linear, segments.

## 8.5. Correlation geometry

A computational procedure for studying association constancy is described in Tarter, Lock and Ray (1995). Local estimates of the scale parameter of the $Y$-variate given the $X$-variate modify the traditional bands that enclose a regression line. Thus, when there is little or no variation about the line for large values of $X$, but much more variation for small values, bands first are broad then narrow as $x$ increases.

Generally speaking, such bands provide wiggle room for uncertainty about some putative true curve. Conventional bands fan out at both edges in accord with sparseness of data over tail regions of the multivariate Normal density, where, in the Normal case this fanning out process is taken to be symmetric. On the other hand, it can easily be the case that changes in variate values are closely related for large values of a variable, but loosely related for small values.

# CHAPTER 9
# Regression and Association Parameters

## 9.1 The curse of dimensionality

Difficulties compound as the number of variables and variates increase. For example, in our daily lives we deal routinely with a few people who are close to us, in the sense that their personal complexities are known. We get to know other individuals only in a one-dimensional sense of personal interactions involving one, or at most a handful of variates. Obviously, prejudices result from simplistic univariate judgments. Both the vernacular use of the word "bias" and the application of this term in statistics are connected with what has been called the *curse of dimensionality*. (According to Scott (1992) this term was applied first in Bellman's classic 1961 text.) In statistics, when we can deal with the complexities of curves in their entirety, rather than with abstracts of these curves, this can reduce statistical biases to acceptable levels.

In this chapter, the curse of dimensionality is viewed in terms of multiple-variable and multivariate models that contain the two parameter sequences, $\theta_1, \theta_2, ..., \theta_p$ and $\omega_1, \omega_2, ..., \omega_q$. The first parameter classification determines multivariate structure. The second helps assemble individual variables to form one or more regression curves that, in turn, condition one or more curve properties or model parameters.

For example, with the exception of the geometric interpretation of $\rho$ that was considered at the end of the last chapter, $\rho$ is found among a list of $\theta_1, \theta_2, ..., \theta_p$ parameters. In the univariate case, when homoscedasticity is assumed, this implies that $\sigma$ is a member of the sequence $\theta_1, \theta_2, ..., \theta_p$, none of whose values are related to any variable's value. On the other hand, a location parameter's value often is modeled in terms of some sequence or subsequence of $\omega_1, \omega_2, ..., \omega_q$ parameters. Hence, when it is model parameters, not curve properties that are studied; $\omega_1, \omega_2, ..., \omega_q$ are components within the representation of some broad-brush (global) parameter-with-curve composite like $\mu(x)$.

In previous sections, the term "variate" was distinguished from the term "variable," and key variates and variables were distinguished from nuisance variates and variables. Much can be learned about statistical usage from the connections and differences between the variate/variable distinction and the key/nuisance distinction. For example, a model that describes a response variate's relationship to a key variable—given a nuisance variate—is quite different from a model that concerns a response variate's relationship to a key variate given a nuisance variable (or even a nuisance variate).

To help interpret the geometric illustrations shown at the end of this chapter, it is convenient to discuss a model's structure in terms of an implicit model of the density of $u$ distinct variates given $v$ distinct variables. Again, for the purposes discussed below, the two Greek letters $\theta$ and $\omega$ distinguish two varieties of model parameters.

The first two letters of the abbreviation "iid" correspond individually to the Greek letters $\theta$ and $\omega$. For example, it is rare in simulation studies to generate an entity like the parameter $\rho$ as a function of key or nuisance variable values, while a location parameter like $\mu$, or even a scale parameter like $\sigma$, is often related to variable values. Expressed verbally, the parameter $\rho$ usually is assumed to characterize bivariate structure and to play no role in the determination of curve identity. The parameter $\rho$ helps disclose variate dependence or independence. Because, in this way, it helps define the structure of a single joint density function, $\rho$ is taken, almost without exception, to represent a $\theta$-type parameter.

The characteristic that distinguishes broad-brush $\theta$-type from $\omega$-type parameters is that, unlike any $\theta$, each $\omega$ serves as a part of some regression equation that combines $x_1, x_2, ..., x_v$ variables in order to describe a model's structure. Suppose a curve, like some density $f$, is expressed as a composite of two other curves, $g$ and $h$. For example, $g$ might be $\phi$, the standard Normal density model while, on the other hand, as part of this model's argument a second function, $h$, is expressed in terms of a particular parameter sequence, say $\omega_1, \omega_2, ..., \omega_q$. In other words, the density's model is expressed as a function, $g$, of a second function, $h$, where this second function is expressed in terms of the parameters $\omega_1, \omega_2, ..., \omega_q$.

In this chapter, $h$-type functions like $\mu_1(x)$ and $\mu_2(x)$ are generalized and taken to be functions of one or more $x_1, x_2, ..., x_v$ arguments assembled with the aid of $\omega$-type parameters. Special attention will be given to functions that are said to be *linear* in these parameters. Besides location-determining parameters that are designated as such by the Greek letter $\mu$, another closely related type of parameter is prefixed by the term "threshold" and denoted by the Greek letter $\tau$. For instance, for some value of $\tau$ it may be the case that $Y' = \log(Y - \tau)$, not $Y$ itself, is described by a

## 9.1. The curse of dimensionality

model which is based on a location parameter expressed as a function of variable $x_1, x_2, ..., x_v$ values strung together by $\omega$-type parameters.

The letter $\tau$ designates the variate $Y$'s value's lower limit. In many practical applications, as $x_1, x_2, ..., x_v$ values change, there is as little reason to include $\tau$ among the list of $\omega$-type parameters that are unrelated to these values as there is to assume some scale parameter $\sigma$ is unrelated to all $x_1, x_2, ..., x_v$. There are also instances when there is some truncation point below which a piece of a density is, in effect, lopped off. Both this truncation point, and the threshold parameter $\tau$ can be modeled in terms of a linear or other function of $\omega$-type parameters. However, for reasons of simplicity, $h$ is often thought of solely as a curve that determines some location parameter's value. For example, $\theta_x = \mu_x$ is a function of the variable $x$ such that $\theta_x = h(x; \omega_1, \omega_2, ..., \omega_q)$ and

$$h(x; \omega_1, \omega_2, ..., \omega_q) = \sum_{j=1}^{q} \omega_j x^{j-1}.$$

The expression $f(y_1, y_2, ..., y_u; \theta_1, \theta_2, ..., \theta_p | x_1, x_2, ..., x_v; \omega_1, \omega_2, ..., \omega_q)$, where the letters $p$ and $q$ represent positive integers, describes a density model implicitly. (Note that here the subscript letter $_p$ designates an integer index, not a probability or a proportion.) There are so many density model alternatives that it is handy to call upon

$$f(y_1, y_2, ..., y_u; \theta_1, \theta_2, ..., \theta_p | x_1, x_2, ..., x_v; \omega_1, \omega_2, ..., \omega_q)$$

notation in order to focus attention on the components of the individual argument $(y_1, y_2, ..., y_u; \theta_1, \theta_2, ..., \theta_p | x_1, x_2, ..., x_v; \omega_1, \omega_2, ..., \omega_q)$.

Whenever it is assumed that variables $x_1, x_2, ..., x_v$ do not affect the values of the parameters $\theta_1, \theta_2, ..., \theta_p$, these symbols are not followed by any arguments. On the other hand, inclusion of the parenthesized $x_1, x_2, ..., x_v$ of $\theta_j(x_1, x_2, ..., x_v); j = 1, 2, ..., p$; denotes the possibility that these parameters change as one or more $x_1, x_2, ..., x_v$ change. The $\theta_1, \theta_2, ..., \theta_p$ symbols do not themselves disclose where a slice is taken. This role is played by the sequence $x_1, x_2, ..., x_v$, whose members are assembled to form a function like $\theta_j(x_1, x_2, ..., x_v)$. If $\theta_j(x_1, x_2, ..., x_v)$ is modeled by an expression based on the parameter values $\omega_1, \omega_2, ..., \omega_q$, then these values, together with the coordinates of the slice, $x_1, x_2, ..., x_v$, determine $\theta_j$'s value.

Before further discussing the symbols that appear after the slash sign, one more notational convention is a useful preliminary. Concerning the subscripts of Greek letters, when a symbol such as $p$, that is chosen from the middle of the alphabet, appears as the subscript of a parameter such as $\theta$ or $\mu$ then this subscripted symbol represents a scalar, not a curve.

On the other hand, one or more letters, such as $x$, chosen from the end of the alphabet, designate variables when they appear as a subscript of $\theta$ or $\mu$. For example, $\mu_p$ denotes the $p^{th}$ location parameter while $\mu_x$ denotes a curve $\mu(x)$ that is often modeled in terms of $\omega$ parameters.

The expression $f(y_1, y_2, ..., y_u; \theta_1, \theta_2, ..., \theta_p | x_1, x_2, ..., x_v; \omega_1, \omega_2, ..., \omega_q)$ specifies two details of the structure of the density $f$. It denotes symbolically that the $p$ parameters $\theta_1, \theta_2, ..., \theta_p$ are taken to be unrelated to the $v$ variables $x_1, x_2, ..., x_v$. It also discloses that $q$ parameters, here $\omega_1, \omega_2, ..., \omega_q$, serve as $\alpha$—intercept-type or $\beta$—slope-type regression, or more complex parameter components, within explicit models that also contain variable value $x_1, x_2, ..., x_v$ components.

The meaning of the phrase *identically and independently distributed* (iid), which is also discussed in Appendix II, can now be illustrated by an example of the representation

$$f(y_1, y_2, ..., y_u; \theta_1, \theta_2, ..., \theta_p | x_1, x_2, ..., x_v; \omega_1, \omega_2, ..., \omega_q).$$

For example, for $u = 2$ suppose that the variate values $y_1$ and $y_2$ denote the number of mature ants and the number of ant larvae, respectively, within a glass-enclosed ant colony. The variable $x_1$ represents the temperature of the colony which, until later in this chapter, is assumed to be a constant value throughout the experiment. In addition, suppose $x_2$, soil moisture level, is also constant, as are $x_3$, light level, and all other control settings, including the last setting $x_v$, which represents the $v^{th}$ variable value, the surrounding air humidity level. When the values $x_1, x_2, ..., x_v$ are constant throughout the experiment, it follows that the $u = 2$, $Y$ variates, $Y_1$ and $Y_2$, whose joint conditional density is $f(y_1, y_2; \theta_1, \theta_2, ..., \theta_p | x_1, x_2, ..., x_v; \omega_1, \omega_2, ..., \omega_q)$, are identically distributed.

When a single unsliced curve $f$ describes all relevant aspects of some variates' distribution, and these variates are said to be iid, then $f$ is taken to be a unique curve, in the sense that the conditional density, say

$$f_c(y_1, y_2, ..., y_u; \theta_1, \theta_2, ..., \theta_p | x_1, x_2, ..., x_v; \omega_1, \omega_2, ..., \omega_q),$$

of the $Y_1, Y_2, ..., Y_u$ variates is taken to be identical to the joint density $f(y_1, y_2, ..., y_u; \theta_1, \theta_2, ..., \theta_p)$. Here the distribution of the $Y_1, Y_2, ..., Y_u$ variates is taken to be the same no matter what values are assumed by the $x_1, x_2, ..., x_v$ variable values. In this situation, knowledge of the single curve $f(y_1, y_2, ..., y_u; \theta_1, \theta_2, ..., \theta_p)$ is tantamount to complete statistical information.

A choice must often be made concerning joint versus conditional models. Treating observed variate values as $x_1, x_2, ..., x_v$ values is tantamount to taking some $\theta_1, \theta_2, ..., \theta_p$ parameter that serves as an association parameter,

## 9.1. The curse of dimensionality

for example $\rho$ and exchanging it for one or more $\omega_1, \omega_2, ..., \omega_q$ regression model parameters, for example a line's slope, $\beta$. Whether a variate's pdf should be modeled as multivariate Normal density, and *correlation-based* methods called upon or, alternatively, whether a model based on regression parameters $\omega_1, \omega_2, ..., \omega_q$ should be assumed, is decided by processes that concern model validity and project need considerations.

One of the most basic of rules that helps to provide guidance in this matter, is that regression focuses upon the issue, whether or not one or more response variates depends upon a sequence of key variables. Here the term "depends" simply refers to linkage and does not necessarily imply cause and effect. That the variables involved are themselves mutually related is of less importance in regression applications than it usually is in correlation-based analyses and in specialized regression modifications such as ANOVA and ANCOVA. Naturally these interrelationships will affect parameter estimators. But the job of a regression-type analysis is to predict a response in its totality, not to compare the contributions made by individual variables.

ANOVA and ANCOVA procedures are designed to help tease effects apart. Most of these procedures can be formulated as specialized regression techniques that are designed to deal with the interrelationships among specialized variables. They accomplish for these variables what the partialling techniques discussed extensively below are designed to accomplish for key and nuisance variates, as opposed to key and nuisance variables. Even the procedures collectively referred to by the term "stepwise regression" focus upon dependence, not interdependence. This basic idea will be elaborated upon in later chapters that consider topics such as partial correlation, nonparametric age-adjustment, ANOVA, ANCOVA, and nonparametric regression.

When $x_1, x_2, ..., x_v$ are non-Normal researcher-selected, and thus are not observed values, it is obviously impossible to justify the validity of any multivariate joint density model where $x_1, x_2, ..., x_v$ are taken to be the values assumed by variates. Yet, as will be discussed in the next chapter, it is foolish to call upon regression rather than correlation-based approaches if, in doing so, adjustment for nuisance variate contamination is performed incautiously. Uncertainty about curve representation is not reduced by simply assigning variates to the role of regression variables. As we shall see, the compatibility of certain regression and correlation methods, when data are actually multivariate Normal, does not extend safely very far beyond this useful distribution. Hence, when data are observed values, one of the worst things a researcher can be uncertain about is the multivariate Normality of his or her data.

In controlled experiments in which settings are assigned to variables by a researcher, the iid assumption is either atypical or, alternatively, this assumption is applicable only in the "null hypothesis" case illustrated in Appendix II. Either (1) it will be impossible to keep all variables $x_1, x_2, ..., x_v$ constant due to the nature of the many nuisance variables whose values cannot be set exactly; (2) the effects that various combinations of $x_1, x_2, ..., x_v$ have on $y_1$ and $y_2$ is the point of the experiment; or (3) both (1) and (2) are true. Data are seldom iid universally throughout a designed experiment and, consequently, the main thrust of regression and related ANOVA and ANCOVA methodology is the study of non-iid data.

## 9.2  Multiple variable interdependence

To illustrate the role played by the $\theta_1, \theta_2, ..., \theta_p$ parameters, suppose temporarily that the ant colony population-size variates actually are iid. This assumption implies that a bivariate model symbolically represented as $f(y_1, y_2; \theta_1, \theta_2, ..., \theta_p)$ describes the distribution of the ant experiment data. Parameters like $\theta_1, \theta_2, ..., \theta_p$ of $f(y_1, y_2; \theta_1, \theta_2, ..., \theta_p)$ can play many roles within parametric models. For example, in the ant experiment, $\theta_1$ might represent some location parameter represented as $\mu_1$. There will usually be two location parameters within $f$'s parametric model, $\mu_1$, the location parameter of the marginal density of the mature-ant colony-size variate, and $\mu_2$, that of the larval colony-size variate's distribution's marginal density.

There obviously will be variation in ant population size from colony to colony about, or from, the location parameter values $\mu_1$ and $\mu_2$. In addition to the location parameters $\theta_1 = \mu_1$ and $\theta_2 = \mu_2$, a model for $f$ would contain typically two "scale" parameters, for example $\theta_3 = \sigma_1$ and $\theta_4 = \sigma_2$. Besides scale parameters, there would be one or more parameters that designate the degree of association between any two $Y$-variates whose distribution is described by any bivariate or multivariate model-based representation of a curve $f$. For example, in the ant experiment, perhaps it might be one or more association parameters, for example, $\theta_5 = \rho$ that quantifies the relationship between larval and adult ant colony sizes.

As discussed in the previous chapter, often $\theta_5 = \rho$ is the sole parameter that quantifies association. For example, in the bivariate Normal case the equality $\rho = 0$ implies the absence of variate relationship. Conversely, for variates like the two components of ant colony size, the closer the value of $\rho$ is to the number 1, the more perfect the correspondence between these two components' sizes. Were inverse relationships encountered, the closer the value of $\rho$ is to the number -1, the closer an increase in one variate will correspond to a decrease in its counterpart.

It would be convenient if only one concept was needed to deal with all forms of association, much in the way that the parameter $\rho$ performs

## 9.2. Multiple variable interdependence

a crucial role in the bivariate Normal special case. Yet, for a variety of reasons, parametric models deal with dependence and, as will be explained in the next three chapters, interdependence, in several different ways. One reason for model complexity is that any single association parameter, such as $\rho$ merely describes how two variates are related. For instance $\rho$ often quantifies the relationship of response, $Z_1$, and key, $Z_2$, variate components, while another type of association parameter that is encountered in studies based on canonical correlation, describes the relationship between two linear combinations of variates.

When $p = q = 2$, an example of $\omega_1, \omega_2, ..., \omega_q$ symbols placed after the slash sign, together with $\theta_1, \theta_2, ..., \theta_p$ symbols placed before the slash sign that appears within $f(y; \theta_1, \theta_2, ..., \theta_p | x_1, x_2, ..., x_v; \omega_1, \omega_2, ..., \omega_q)$, is $\theta_1 = \alpha, \theta_2 = \sigma, \omega_1 = \beta_1$, and $\omega_2 = \beta_2$; where a single response variate, such as total ant colony population-size, $Y$, is described by the data-generation model

$$Y = \alpha + \beta_1 x_1 + \beta_2 x_2 + \sigma Z,$$

and $Z$ is a standard Normal or some other continuous variate whose density is assumed known. Because the *additive linear model*

$$Y = \alpha + \beta_1 x_1 + \beta_2 x_2 + \sigma Z$$

is a popular way for a variate like $Y$ to be associated with two variables like $x_1$ and $x_2$, it is worth considering exactly what this model asserts about variate $Y$. The pair of graphs shown in Figure 9.1 can help interpret an important characteristic of any additive linear model.

To simplify matters, assume that $\alpha = 0$ and $\beta_1 = \beta_2 = 1$. Under these assumptions, for any value of the constant $c$, the curve $\beta_1 x_1 + \beta_2 x_2 = c$ is a line with slope equal to -1. Now consider the conditional density curve $f(y|x_1, x_2)$. As far as the values taken on by this curve are concerned, as long as $\beta_1 x_1 + \beta_2 x_2 = c$, where $c$ is a constant, it makes no difference what the values of $x_1$ and $x_2$ are individually. This is indicated by the two callouts $f(y|c_1)$ and $f(y|c_2)$ on the right-hand side of Figure 9.1.

The thick-dashed diagonal line on the left hand side of Figure 9.1 coincides with the dashed bell-shaped curve on the right-hand side of this figure. Any combination of $x_1$ and $x_2$ that totals $c_1$ corresponds to the same $f$-type curve. Thus, exactly how the variate $Y$ is distributed is determined solely by the sum, $c_1$, in the sense that, from a statistical point of view, other than this total, what $x_1$ and $x_2$ are individually is irrelevant.

The $\beta_1 x_1 + \beta_2 x_2 = c$ model makes sense in circumstances where variables $x_1$ and $x_2$ are commensurable. For example, suppose $x_1$ and $x_2$ are, respectively, a father's and a mother's yearly income, and the $Y$ variate is the amount spent on their child's college education. Since the equation

$\beta_1 x_1 + \beta_2 x_2 = c$ represents a weighted family income, it is quite reasonable to combine $x_1$ and $x_2$ in some simple fashion in order to obtain the single value $c$.

Unfortunately, there are many situations where it stretches credulity to assert, for example, that there are parameters $\beta_1$ and $\beta_2$ for which an increment in the corresponding value of $x_1$ is counterbalanced by some exact decrement in the value of $x_2$. Validity is particularly problematic when the two values $x_1$ and/or $x_2$ are measured by an instrument that cannot be depended upon to yield a reliable value. Is the decrease in some true $x_1$-value or, alternatively, some measured $x_1$-value, compensated for by some increased true or measured $x_2$-value? What if the true values actually do exert their influence solely as a weighted sum, but the measured $x_1$-values and $x_2$-values differ in their proximity to true values?

When a model $Y = \alpha + \beta_1 g(x_1) + \beta_2 h(x_2) + \sigma Z$, where $g$ and $h$ are user-specified single-valued functions, is substituted for the model $Y = \alpha + \beta_1 x_1 + \beta_2 x_2 + \sigma Z$, the nature of the trade-off curve is changed, but the existence of a valid trade-off-function still remains to be corroborated. Consider, for example, the

$$Y = \alpha + \beta_1 g(x_1) + \beta_2 h(x_2) + \sigma Z$$

generalization of $\beta_1 x_1 + \beta_2 x_2 = c$ to $\beta_1 g(x_1) + \beta_2 h(x_2) = c$, described by Figure 9.2. When functions $g$ and $h$ change the linear shape of the curves shown in Figure 9.1 to yield the thick dashed and solid curves shown in

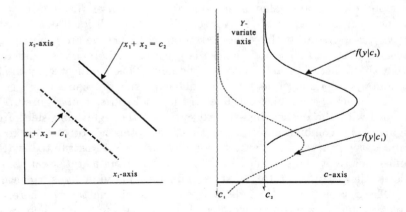

Figure 9.1. When a reqression function $x_1 + x_2 = c$ yields different values of $c$, different density curves $f(y|c)$ are selected.

## 9.2. Multiple variable interdependence

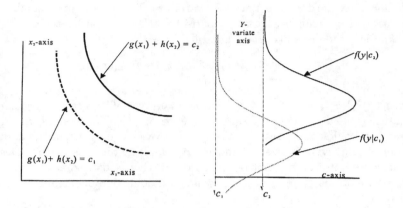

**Figure 9.2.** As a regression function $g(x_1) + h(x_2) = c$ yields different values of $c$, diferent density curves $f(y|c)$ are selected.

Figure 9.2, it has been assumed that not only is the distribution of $Y$ still constant over one such curve, but that it is constant for all such curves.

Even for a single choice of $c$, and for such a commonplace pair of variables as height and weight, it is hard to see why a variate $Y$ that depends on these two variables would have the same distribution for a constant $c$ determined by the height–weight index $\beta_1 g(x_1) + \beta_2 h(x_2) = c$. When alternative functions $g$ and $h$ are tried within $\beta_1 g(x_1) + \beta_2 h(x_2)$, exactly where $f$ is constant is typically changed. But for $g$ and $h$ there still supposedly exists a family of curves in the $x_1, x_2$ space such that each individual curve, in its entirety, determines a single $f$-type curve of the $Y$ variate. For example, as far as the height $x_1$–and weight $x_2$–variables are concerned, some short and heavy individual would have to be the exact statistical counterpart of some tall and thin individual.

The constancy assumption over a set of curve values becomes more, not less, difficult to justify as the number of key and nuisance variables increases. Because they determine a single value of $c$, the three variables $x_1, x_2$ and $x_3$, together with the three curves $g_1, g_2$ and $g_3$ of the model component $\alpha + \beta_1 g_1(x_1) + \beta_2 g_2(x_2) + \beta_3 g_3(x_3)$, imply that the values assumed by all points on some specific three-dimensional surface have the same effect on the variate's distribution. For example, were $g_1(x) = x, g_2(x) = x$, and $g_3(x) = x$, then $f$-type curve constancy along a line, as illustrated by Figure 9.1, would be generalized to constancy at any point on the plane

$$\alpha + \beta_1 x_1 + \beta_2 x_2 + \beta_3 x_3 = c.$$

While it is hard to see how two variables, like height and weight, can be traded off against each other perfectly, it's even harder to justify "tri-variable" trade-offs like that quantified by

$$\alpha + \beta_1 g_1(x_1) + \beta_2 g_2(x_2) + \beta_3 g_3(x_3).$$

Thus, no matter how much curve shapes are altered by applying $g_1(x) \neq x, g_2(x) \neq x$, and $g_3(x) \neq x$, or how many variables are used within an additive linear model, the same sequence of $f(y|c)$ curves shown on the left-hand side of Figures 9.1 or 9.2 apply. In fact, even when the assumption of additivity is relaxed and more complex functions of variables are considered, it still may not be the case that there exists some choice of the three functions $g_1, g_2$, and $g_3$ for which a model component like

$$\alpha + \beta_1 g_1(x_1, x_2, x_3) + \beta_2 g_2(x_1, x_2, x_3) + \beta_3 g_3(x_1, x_2, x_3) + \sigma Z$$

is a valid representation. There simply may not be any choice of functions; together with the four parameters $\alpha, \beta_1, \beta_2, \beta_3$, that can describe how the variables $x_1, x_2$, and $x_3$ affect the response variate $Y$'s distribution and, after so doing, leave a quantity $\sigma Z$ that is unrelated to these three variables.

Uncertainties of the above type can be somewhat reduced by the simple expedient of increasing the number of regression model terms so that $q$ of the sequence $g_1, g_2, g_3, ..., g_q$ is far greater than the number of variables. Unfortunately, this leads to the important curse of dimensionality consideration that is known as *overfitting*. As general curve estimation techniques that can deal with densities as wholes improve, it is becoming less and less attractive to treat variates as if they were variables and increase $q$ rather than to deal with complex multivariate densities directly.

## 9.3 Logit and linear models

Model restrictiveness illustrated in the previous section for continuous variates, also applies to any dichotomously-valued variate $Y$ that assumes either of two possible binary values, e.g., 0 and 1. For example, a particularly dangerous form of curve uncertainty concerns of the well-known logit model

$$\log[P/(1-P)] = \alpha + \beta_1 g_1(x_1) + \beta_2 g_2(x_2) + \beta_3 g_3(x_3).$$

Here, in the sense that the cdf that describes this variate's distribution is a two-step step function, the response variate is a two-valued special case of a discrete variate.

Suppose that within the $\log[P/(1-P)]$ side of the logit relationship, the uppercase letter $P$ denotes the chance that $Y$ will assume the value 1,

## 9.3. Logit and linear models

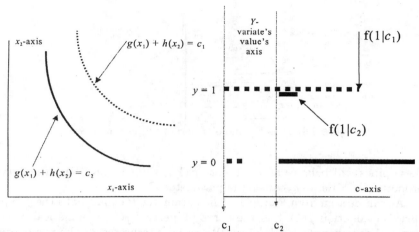

Figure 9.3. As a logit reqression function $g(x_1) + h(x_2) = c$ yields different values of $c$ different density curves $f(y|c)$ are selected.

rather than the value 0. Because the symbol $P$ appears within the model term, $\log[P/(1-P)]$, and because

$$\log[P/(1-P)] = \alpha + \beta_1 g_1(x_1) + \beta_2 g_2(x_2) + \beta_3 g_3(x_3)$$

lacks any $\sigma Z$ term following the model component

$$\alpha + \beta_1 g_1(x_1) + \beta_2 g_2(x_2) + \beta_3 g_3(x_3),$$

at first glance the logit model seems different from the density/data-generation and regression models that were discussed in the previous section. However, in the logit/logistic case there is a close connection between a density model and a regression model.

Figures 9.3 and 9.4 illustrate how a single elementary function representation can simultaneously serve as two types of model, a logistic density model and a logit data-generation model. Figure 9.3 is the logit model analog of Figures 9.1 and 9.2. Figure 9.4 demonstrates why the quantity $P$ can be interpreted either as a value attained by an $f$-type curve or, alternatively, as the curve property $E(Y)$ over which this $f$-type curve, a curve that takes on the value $1-P$ at 0 and the value $P$ at 1, will balance.

As far as the dual interpretation of $P$ is concerned, suppose the two solid vertical lines shown in Figure 9.4 that rise above the base each represent, respectively, different lengths of the same uniformly heavy pipe or metallic strip. Were the configuration to rotate, it follows that for any $0 < P < 1$

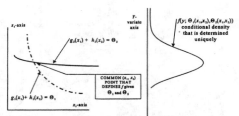

Figure 9.4. $P$ as a value of $f$ and of regression curves.

these pipes will balance at the point $(E(Y) = P, 0)$, since here the curve property $E_f(Y)$ always equals the parameter $P$.

As can be seen from Figure 9.4, the parameter $P$ equals both the $f$-type curve's evaluation $f(1)$, and the curve property $E_f(Y)$'s numerical value. Hence the expression $\log[P/(1-P)]$ can be interpreted as the result produced by substituting density-curve-evaluation, $f(1) = P$, for the argument $y$ of the logit function inverse cdf $\log[y/(1-y)]$. (Because $P/(1-P)$ denotes the chance of an event occurring over the chance of its non-occurrence, when it is evaluated at the value $y = f(1) = P$, the function $\log[y/(1-y)]$ is often called the "log odds ratio.")

Figure 9.4 illustrates why the regression function,

$$\log[P/(1-P)] = \alpha + \beta_1 g_1(x_1) + \beta_2 g_2(x_2) + \beta_3 g_3(x_3),$$

is often taken to be a curve that equates a specific curve property, here $\log[E_f(Y|x_1,x_2,x_3)/\{1-E_f(Y|x_1,x_2,x_3)\}]$, to the linear function

$$\alpha + \beta_1 g_1(x_1) + \beta_2 g_2(x_2) + \beta_3 g_3(x_3).$$

Similarly, the assumption of a *logistic regression* relationship asserts that

$$E_f(Y|x_1,x_2,x_3) = 1/\{1 + \exp[-(\alpha + \beta_1 g_1(x_1) + \beta_2 g_2(x_2) + \beta_3 g_3(x_3))]\},$$

a strange-looking, but nevertheless popular, assumption of a curve property's relationship to, in this particular example, the three variables $x_1, x_2,$ and $x_3$.

Figure 9.3 illustrates the geometry that underlies the logit model

$$\log[P/(1-P)] = \alpha + \beta_1 g_1(x_1) + \beta_2 g_2(x_2) + \beta_3 g_3(x_3),$$

and the "logistic model"

$$E_f(Y|x_1,x_2,x_3) = 1/\{1 + \exp[-(\alpha + \beta_1 g_1(x_1) + \beta_2 g_2(x_2) + \beta_3 g_3(x_3))]\}.$$

## 9.3. Logit and linear models

Unlike Figures 9.1 and 9.2, the response variate's distribution is described by an $f$-type curve that can be nonzero solely over the two values, $y = 0$ and $y = 1$. Nevertheless, as was illustrated in the Normal case by Figures 9.1 and 9.2, whenever logit or logistic regression is applied, constancy over each of a family of curves whose members resemble the two curves shown on the left side of Figure 9.3, is still assumed.

The equivalence of logistic regression based on the curve property $E_f(Y|x_1, x_2, x_3)$, and logit regression based on probability $P$, is convenient. Yet it should be kept in mind that both these regression models are merely two, out of a great number of potential alternative, left hand sides of a model regression model whose right hand side is either

$$1/\{1 + \exp[-(\alpha + \beta_1 g_1(x_1) + \beta_2 g_2(x_2) + \beta_3 g_3(x_3))]\}$$

or $\alpha + \beta_1 g_1(x_1) + \beta_2 g_2(x_2) + \beta_3 g_3(x_3)$.

Chambers and Hastie (1992) illustrate three commonly used (inverse) link functions for the binary family. One of these is the logit discussed above. The second is called a *probit curve*. However, this probit differs from the probit curve discussed in the early statistics literature because, sensibly, the Chambers and Hastie probit is identical to the inverse Normal cdf, not the older usage of the word "probit," $\Phi^{-1}(z) + 5$. The third link function is called the *complementary log(-log) model*.

Unlike the logit and probit inverse cdf curves, the third link function does not correspond to a symmetric $f$-type curve like the logistic or the Normal density. To see where symmetry or asymmetry might come into play, consider that

$$\log[P/(1-P)] = \log(P) - \log(1-P).$$

This equality suggests that $\log[P/(1-P)]$ should be taken to be a special case of the more general function $\log(P) - K[\log(1-P)]$. For any positive constant $K$, the function $\log(P) - K[\log(1-P)] \to +\infty$ as $P$ approaches 1 from below, and it approaches $-\infty$ as $P$ approaches 0 from above.

Consider the function $f_K$ that is defined to equal the derivative of $F$, where $F$ is the inverse of the function $F^{-1}(z) = \log(z) - K[\log(1-z)]$. Simple calculus procedures can be applied to show that for any positive $K$, $f_K$ is a density. Thus, no matter what positive value is selected for the parameter $K$, any

$$F^{-1}(z) = \log(z) - K[\log(1-z)]$$

is a legitimate inverse cdf. Another way of putting this latitude of $K$'s choice, is that the conventional right side of the logit model allows for the possibility that $P$ is functionally related to

$$\alpha + \beta_1 g_1(x_1) + \beta_2 g_2(x_2) + \beta_3 g_3(x_3),$$

but this logit model is based on the assumption that the parameter $K$ equals 1, no matter what $\alpha + \beta_1 g_1(x_1) + \beta_2 g_2(x_2) + \beta_3 g_3(x_3)$ happens to equal.

The term "logistic" appears in many different branches of the statistical literature. For example, in his classic book that deals with engineering applications, Hald (1952) states that "The growth curve with the equation

$$y = \lambda/(1 + \gamma e^{-\chi x})$$

is called the *logistic* curve." In the biological sciences, this curve has long been applied to describe growth processes that, at a small value of $x$, start off slowly, increase rapidly (while there are plenty of resources remaining), and then taper off to approach what, in Hald's notation, is denoted by the Greek letter $\lambda$. The capitol Greek letter $K$ of $\log(z) - K[\log(1-z)]$, can consequently be thought of as a *skewness* parameter that takes into account that the rate of early growth $R_1$, and $|R_2|$, the absolute value of the final, decline-in-growth, rate, often differ.

## 9.4 Dual regression functions

In previous sections the representational limitations of linear regression models were discussed in terms of common-sense ideas such as the equivalency of, for example, a short and fat person to a tall and thin person. The curve representation $f(y_1, y_2, ..., y_u; \theta_1, \theta_2, ..., \theta_p | x_1, x_2, ..., x_v; \omega_1, \omega_2, ..., \omega_q)$ and the curves shown in Figure 9.5, illustrate some of the strengths and some of the weaknesses of linear and other regression model parametric representations.

Suppose $u = 1, v = 2, p = 2$, and the $q > 5$ parameters $\omega_1, \omega_2, ..., \omega_q$ are divided into two subsets, $\omega_1, \omega_2, ..., \omega_\zeta$, and $\omega_{\zeta+1}, \omega_{\zeta+2}, ..., \omega_q$, where $1 < \zeta < q - 2$, and that $\theta_1$ is a location parameter $\mu$ which is assumed to be related to the variables $x_1, x_2$ through the intermediary $\omega_1, \omega_2, ..., \omega_\zeta$, parameters, say assembled to form the function $g_1(x_1) + h_1(x_2)$. However, in addition, suppose that the scale parameter $\theta_2 = \sigma$, like its location parameter $\mu$ counterpart, rather than appearing bare without the ( ) barrel of function notation, is *also* a non trivial function, say $g_2(x_1) + h_2(x_2)$, based on the $q - \zeta$ "post slash sign parameters" $\omega_{(\zeta+1)}, \omega_{(\zeta+2)}, ..., \omega_q$, within $f$'s model.

## 9.4. Dual regression functions

In the sense that $f(y_1, y_2, ..., y_u; \theta_1, \theta_2, ..., \theta_p | x_1, x_2, ..., x_v; \omega_1, \omega_2, ..., \omega_q)$ can be expressed validly as the conditional model special case

$$f(y_1; \Theta_1(x_1, x_2), \Theta_2(x_1, x_2)),$$

for some experimenter-determined values of the variables $x_1$ and $x_2$, dual relationships, $g_1(x_1) + h_1(x_2) = \Theta_1$ and $g_2(x_1) + h_2(x_2) = \Theta_2$ are often valid. Since location and scale parameters determine distinct features of a statistical model, it follows that for $f(y_1; \Theta_1(x_1, x_2), \Theta_2(x_1, x_2))$ it is not a single parameter but instead the *pair* of parameters ($\Theta_2(x_1, x_2)$ and $\Theta_2(x_1, x_2)$) that determines density curve uniqueness given the values of the variables $x_1$ and $x_2$.

Were merely a single parameter $\Theta_1$, as in a model of $f(y_1; \Theta_1(x_1, x_2), \theta_2)$, related to the slice determining coordinates $x_1, x_2$, then all points along the dashed curve shown on the left side of Figure 9.5 would correspond to the same density curve, such as the single curve shown on the right side of this figure. Correspondingly, were the second, rather than the first $\theta$ parameter, taken to be a function of $x_1, x_2$, then a common density $f(y_1; \theta_1, \Theta_2(x_1, x_2))$ would result from any choice of $x_1, x_2$ along, say, the solid curve. However, unlike the regression models discussed in previous sections, for $g_1, g_2, h_1$, and $h_2$, each different $x_1, x_2$ pair can correspond to a distinct density of the $Y_1$ variate. Consequently, whenever multiple parameters are linked to the $x_1, x_2$ variables, it is unlikely that different combinations of $x_1$ and $x_2$ determine the same conditional density curve

$$f(y_1, y_2, ..., y_u; \theta_1, \theta_2, ..., \theta_p | x_1, x_2, ..., x_v; \omega_1, \omega_2, ..., \omega_q).$$

Each of the curves shown on the left side of Figure 9.5 can be thought

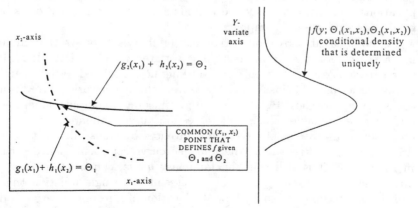

Figure 9.5. Dual regression functions, each determining a distinct conditional pdf parameter.

of as a single thread within the fabric of variate-with-variable relationships. Unless applied with the great care often taken when ANOVA and ANCOVA are applied, regression is a poor way to deal with nuisance variates and variables whose effects are to be removed, yet it can be an excellent way to tease apart the complex threads that join different combinations of the same sequence of variables to one or more response variates. Note that the thick solid and dashed curves of Figure 9.5 each join a pair of key variables together to determine $\Theta_1(x_1, x_2)$ and $\Theta_2(x_1, x_2)$. Variables are not removed. Instead they are woven together by dual regression functions.

Regression is called for when, instead of variates being adjusted or partialled, as described in the next chapters, the structure of a conditional is teased apart or, alternatively, a sequence of variables is to be joined together for the purpose of predicting the values of one or more response variates. Distribution dissection and the welding of variables can be based on roles parameters play. For instance, on the one hand, since $\theta_1, \theta_2, ..., \theta_p$ individually disclose location, scale, association, mixing, or threshold curve characteristics, regression techniques can help disclose whether, how, or to what degree, some combination of variables affects these different characteristics of a distribution.

Alongside parameters, different curve properties like the mean, median, or mode can be called into service and linked to some $h$-function based on $\omega_1, \omega_2, ..., \omega_q$ parameters. This provides the representational latitude to examine distribution structure. For example, distributional asymmetry can be discerned when a curve that estimates a response variate's median value is found to differ from a curve based on the population mean. Since the mode, unlike the mean and median, can assume multiple distinct values, regression approaches that target the mode curve property have many applications. For instance, they can provide a clue that a population is best treated as a mixture of components.

A mode-based regression curve that bifurcates suggests that mounting a search for one or more lurking variates is called for. If such a search is successful, then this once-lurking-now-disclosed variate can often serve as an additional variable within the framework of traditional statistical approaches. Before modern graphical and computational methods became available, almost all regression curve estimators were based on minor modifications of a technique invented by Gauss almost two hundred years ago. Yet, unlike the multiple curve representation described in Figure 9.5, it is still common for linear model construction to apply all of the parameters $\omega_1, \omega_2, ..., \omega_q$ to assemble a single function that is called upon to describe solely how the location of a response variate's distribution is linked to the variables $x_1, x_2, ..., x_v$. For example, $\mu_{Y|x} = \omega_0 + \omega_1 x_1 + \omega_2 x_2 + ... + \omega_v x_v$ often serves as a single superparameter or,

## 9.4. Dual regression functions

alternatively, the $E_f(Y|x_1, x_2, ..., x_v)$ curve-based property is taken to be the sole pertinent indicator of dependence.

Without checking by graphical and other techniques, data analyses based on superparameter $\mu_{Y|x} = \omega_0 + \omega_1 x_1 + \omega_2 x_2 + ... + \omega_v x_v$ turns a blind eye to the possibility that there does not exist a single function of all $v$ variables $x_1, x_2, ..., x_v$ that, unaided, can reduce the influence that these variables have upon $Y$'s conditional density $f$ to a scalar valued quantity. To see why this is the case, suppose dependence is constrained to only one scalar valued parameter that is taken to be related to all variable values. To define this scalar, the $\omega_1, \omega_2, ..., \omega_q$ are pressed into service to represent a curve within the space spanned by the possible values of $y_1, x_1, x_2, ..., x_v$. This is tantamount to the assumption that a single scalar value of this curve can be substituted for these v distinct variables without any loss in generality. There is supposed to be a knothole sized passage, one supervariable wide, within the slash symbol | of $f(y_1|x_1, x_2, ..., x_v; \omega_1, \omega_2, ..., \omega_q)$ through which all $v$-variables influence, or predict, variate distribution.

Powers of individual variables are often assembled to form linear combinations of polynomials. In other applications, the logarithms of some or all variable values, rather than these values themselves, are combined linearly to form a regression function. Unfortunately, however, the extent of literature devoted to alternatives to the right-side of $\mu_{Y|x} = \omega_0 + \omega_1 x_1 + \omega_2 x_2 + ... + \omega_v x_v$ can distract attention away from characteristics of regression methodology that are only indirectly connected to the generalization of $\omega_0 + \omega_1 x_1 + \omega_2 x_2 + ... + \omega_v x_v$. Instead, these concern the following two issues: (1) Why should there be a merely a single knothole (in the sense that there instead are likely to be two or more distinct functions that, together, describe variable–variate relationship)? (2) Why is the single parameter $\mu_{Y|x}$ or the single curve property $E_f(Y|x)$ the sole parameter or property predicted or determined by the variable values? In the ant experiment, as well as in many other experiments and observational studies, it is just as likely that a variate's scale parameter $\sigma$ is nontrivially related to $x_1, x_2, ..., x_v$, as it is for the corresponding location parameter to have this sort of relationship. Variables such as air or soil temperature can affect both the most typical number of ant larvae produced in an ant colony, as well as modify the variability of larvae production.

Consider, for example, a Poisson model that describes the distribution of some discrete variate such as ant colony size. Unlike the Normal, the Poisson model depends on a single parameter, and thus this one parameter plays the dual role of roughly determining both curve location and curve scale. Consequently, in the Poisson response variate case, both location and scale must either be related, or unrelated, to any key or nuisance variate or variable's value. Either the assumption that the variate is Poisson distributed must be generalized, or regression methods that take this duality into account must be applied.

# CHAPTER 10
# Parameters, Confounding, and Least Squares

## 10.1 Ideal objects

What actually does uncertainty about statistical curves refer to? Possible answers to this question can be divided into the two categories, models and parameters. The sharp division between some model-choice opening gambit and the estimation of parameters endgame has, over time, become blurred to the point of seeming artificial. Yet, even today, in comparison to model-based procedures, nonparametric approaches such as those discussed in the next chapter are often viewed as use-them-if-you-have-to or fall-back-position substitutes.

From a historical perspective, there seems to be a natural sequence that begins with the parameter $\mu$, as a natural runner-up, proceeds to $\sigma$, then on to $\rho$ and ends with either the mixing, threshold or, as discussed below, the truncation-type of parameter. No matter what its actual members are, or how they are ordered, this sequence without doubt has far fewer members than would be contained in the set consisting of all distinct univariate density curve models that have individually generated many statistical publications. Consequently, it seems far wiser to first decide which member, along the sequence that consists of a relatively small number of parameter-type alternatives, corresponds to what one wishes to know and, only then, proceed with the issues of model selection.

The crucial decision that it is the teasing apart of interrelationships, not the mutual effect of variables in conjunction with each other, raises the question, how should teasing apart be accomplished? The partial correlation approaches outlined in Chapter 12, and the age-adjustment approaches of Chapter 11, provide two possible answers to this question. Traditionally, the former class of approaches has been more popular than the latter. However, thanks to improved computational methodology, age-adjustment techniques now have many new potential uses. Both a life table that estimates an expected length of life curve and any one of the many varieties of age-adjustment procedures are based on curve properties, not on model parameters. These techniques differ substantially from regression procedures that estimate $\mu_{Y|x}$. This chapter discusses the general points of view that underlie the difference.

Curve property and model parameter distinctiveness can be traced back at least to 360 BC when, in reference to all beds and tables, the *Republic* Socratic dialog included the assertion: "But there are only two ideas of them—one the idea of a bed, the other of a table" (translation by B. Jowett). This is the likely published origin of the concept that there exists an ideal bed and an ideal table about which, borrowing statistical jargon, all observed beds and tables vary. More than two millennia later this viewpoint still could influence eighteenth and nineteenth century statisticians to the degree that the location parameter $\mu$ and regression models of $\mu_{Y|x}$ were taken to be ideals about which data points vary.

There is an all but invisible serpentine argument here. Location, as in location parameter, of course has more "there-ness" than scale, association, mixing, threshold, or truncation. Points, for example, do not vary about a scale or an association parameter. Hence, in the eighteenth and nineteenth century it seemed natural to connect the concept of location with the concept of perfection; and to leave location's parametric sibs relegated to a category, called collectively, measures of imperfection and/or error.

The thought processes of Laplace, Gauss, and other Normal and regression curve innovators may have been in tune with their times, but when these times were viewed from the vantage point of our discordant era, according to E. T. Bell (*The Search for Truth*, 1934, p. 193) "On the whole the Eighteenth Century seems to have been one of the most cocksure scientifically of modern times, running close second to the Nineteenth." Throughout most of this 200 yearlong period, especially when applied in reference to a resting or moving physical object, the meaning of the term "observed" seemed perfectly clear. For example a substantial proportion of seventeenth and eighteenth century mathematical literature concerned the particular curve, called a *catenary*, that can be formed by holding a chain by its two ends, where one such curve, whose lowest point has coordinates $(0, q)$, can be expressed as

$$y = (q/2)[\exp(x/q) + \exp(-x/q)].$$

Of course, no chain itself is really a segment of a perfectly flexible, uniformly dense, and inextensible cable. Nevertheless, it would be splitting hairs to call the concept of catenary into question, in the sense of

## 10.1. Ideal objects

doubting the existence of some standard catenary with a model component $[\exp(x) + \exp(-x)]/2$ that can be generalized to represent a model based on the expression $(q/2)[\exp(x/q) + \exp(-x/q)]$. Because they describe a parameter, and not the position of a tangible object like a link of a chain, what a true $\mu_{Y|x}$ curve actually is, is often less than perfectly clear. Hence, after illustrating one reason for this lack of clarity, by considering data points that surround one particular curve like $\mu_{Y|x}$, the remainder of this chapter will discuss an even more questionable assumption which concerns the individual variables that serve as regression curve arguments.

Early data analysis texts often discussed data-generation models that combined some formula for the curve $\mu_{Y|x}$ with what were then called *error terms*. Many of these books contained chapters that dealt with what was then called the *theory of errors*. As the word "error" suggests, some backbone curve or model is error-free and, in this sense, resembles Plato's ideal table or bed.

Since a location parameter is not always surrounded by observed values, what location actually means requires considerable elaboration. For example, within the lognormal cdf model $F(y) = \Phi(\log[y - \tau])$, the symbol $\tau$ represents a location parameter, since whenever the value of $\tau$ changes, the curve $F$ is translated by exactly the difference between the new and the previous $\tau$-parameter values. However, since no real number is the log of a negative number, no value of the variate $Y \in \Phi(\log[y - \tau])$ can be less than $\tau$. As a consequence, when values of the threshold/location parameter $\tau$ of this variate's model are taken to be functionally related to a variable $x$, then all differences between observed and explained values are positive.

In circumstances where it is given that data points vary about, and not above, some ideal $\mu_{Y|x}$ curve, and given that some linear or other model describes a distribution's $\mu_{Y|x}$ backbone, a major limitation still pertains to the $x_1, x_2, \ldots, x_v$ components which distinguish exactly where ribs are attached to this backbone. These components differ in a crucial respect from $Z_1, Z_2$, and $Z_3$ independent variates. Even when $x_1, x_2, \ldots, x_v$ actually designate variate, not variable, values, it is unlikely that the variates being studied can be treated as independent variates. Were $x_1, x_2, \ldots, x_v$ thought of as ribs then they would not be as neatly stacked as $Z_1, Z_2$, and $Z_3$. A more apt analogy would be the loosely attached blades of a poor quality penknife. Due to this loose attachment of variables, it is rarely the case that the mere inclusion of a third variable, say $X_{nui} = x_3$, with its own regression coefficient $\beta_{nui}$ within an error-based regression model, helps a researcher to cope thoroughly with the contaminating effects of this variable.

The subscript "*nui*" helps to distinguish nuisance variates from other types of variates. In particular, suppose that $X_{nui}, Z_1, Z_2$, and $Z_3$ are four independent variates applied to simulate the three variates:

$$X_1 = \rho_1 X_{nui} + \sqrt{1-\rho_1^2} Z_2;$$
$$X_2 = \rho_2 X_{nui} + \sqrt{1-\rho_2^2} Z_3;$$

and the response variate

$$Y = \beta_1 X_1 + \beta_2 X_2 + \beta_{nui} X_{nui} + \sigma Z_1.$$

Then whenever $\rho_1 \neq 0$ and $\rho_2 \neq 0$, the two key variates

$$X_1 = \rho_1 X_{nui} + \sqrt{1-\rho_1^2} Z_2 \quad \text{and} \quad X_2 = \rho_2 X_{nui} + \sqrt{1-\rho_2^2} Z_3$$

are both related to the value assumed by the nuisance variate $X_{nui}$. (The reason why the $\sqrt{1-\rho_1^2}$ and $\sqrt{1-\rho_2^2}$ terms are included in these equations will be considered in the next section in detail.) For the purposes of this section it suffices to note that $Z_2$ designates the part of the key variate, $X_1$, that is unrelated to the nuisance variate $X_{nui}$. Correspondingly, $Z_3$ plays this same role for the key variate $X_2$, while variates $Z_2$ and $Z_3$ are independent, and hence unrelated, both to each other and, again thanks to assumed independence, to $X_{nui}$.

Substitution of the first two equations into the third yields

$$\begin{aligned} Y = & \beta_1(\rho_1 X_{nui} + \sqrt{1-\rho_1^2} Z_2) + \beta_2(\rho_2 X_{nui} + \sqrt{1-\rho_2^2} Z_3) \\ & + \beta_{nui} X_{nui} + \sigma Z_1, \end{aligned}$$

an equation that discloses how $Y$ could be assembled from the independent variates $Z_1, Z_2$, and $Z_3$ based on the coefficients $\beta_1\sqrt{1-\rho_1^2}$ and $\beta_2\sqrt{1-\rho_2^2}$ of $Z_2$ and $Z_3$, respectively. This formula shows clearly that even when the values of $\beta_1$ and $\beta_2$ are specified, the values assumed by the regression coefficients $\beta_1\sqrt{1-\rho_1^2}$ and $\beta_2\sqrt{1-\rho_2^2}$ of the independently distributed, non-$X_{nui}$ contaminated variates $Z_2$ and $Z_3$, can only be determined by calling upon additional information concerning two additional parameters, $\sqrt{1-\rho_1^2}$ and $\sqrt{1-\rho_2^2}$.

Suppose we add the ideal knife to Plato's ideal bed and ideal table. Is a Swiss Army knife less perfect than an ordinary kitchen knife because the Swiss Army knife has a screwdriver blade? It is certainly less "knife-like" than the kitchen knife. Indeed, when a cheap imitation has a screwdriver blade that catches on the knife blade when either is opened, the kitchen knife does indeed seem preferable.

Without $\sqrt{1-\rho_1^2}$ and $\sqrt{1-\rho_2^2}$, even were it the case that some process discloses all four parameter values, $\beta_1, \beta_2, \beta_{nui}$ and $\sigma$, of the model $Y =$

## 10.1. Ideal objects

$\beta_1 X_1 + \beta_2 X_2 + \beta_{nui} X_{nui} + \sigma Z_1$, the disclosed values are far from ideal. They are deficient because it is the products $\beta_1 \sqrt{1-\rho_1^2}$ and $\beta_2 \sqrt{1-\rho_2^2}$, not $\beta_1$ and $\beta_2$ alone, that permit the independent, and in this sense ideal, $Z_2$ and $Z_3$ variates to be assessed in terms of their effect on the interdependence of the response variate $Y$ and the two key variates $X_1$ and $X_2$. In many applications, solely because of nuisance variate $X_{nui}$'s contaminating effects on $Y, X_1$, and $X_2$, the key variates might misleadingly appear to affect variate $Y$. In particular, it might appear that changes in key variate $X_2$'s values cause changes in $Y$'s values, even though changes are induced entirely by the key variate's nuisance component.

The symbol $X_{nui}$ represents an observed, not an ideal, quantity. It is simplistic to assume that somehow the $\beta_{nui} X_{nui}$ term of model $Y = \beta_1 X_1 + \beta_2 X_2 + \beta_{nui} X_{nui} + \sigma Z_1$ draws all nuisance contamination out of the quantities $X_1$ and $X_2$. Omission of the $\beta_{nui} X_{nui}$ term within $Y = \beta_1 X_1 + \beta_2 X_2 + \beta_{nui} X_{nui} + \sigma Z_1$ can, of course, also cause difficulties. (Without this term, the only way that $X_{nui}$ is left to affect $Y$ is through the variates $X_1$ and $X_2$.)

The mere inclusion of $\beta_{nui} X_{nui}$ does not guarantee that parameter estimation procedures can winnow the effects of $X_{nui}$ from those parts of $X_1$ and $X_2$ that are attributable to $X_{nui}$. Unless, $X_{nui}, X_1$, and $X_2$ are adjusted to form quantities like $Z_1, Z_2$, and $Z_3$, which for multivariate Normal variates are ideal in the sense of being independent (and in this sense pure), the model $Y = \beta_1 X_1 + \beta_2 X_2 + \beta_{nui} X_{nui} + \sigma Z_1$ provides three interconnected avenues of $X_{nui}$ influence: (1) via $\beta_1 X_1$; (2) via $\beta_2 X_2$ or; (3) through the more direct but not necessarily exclusive route via the term $\beta_{nui} X_{nui}$.

Statistical investigations that take $x_1, x_2, \ldots, x_v$ components to be ideal in the sense of being non-interrelated, disregard the distinction between regression methodology, as a way of linking variables together, and regression's partial correlation, age-adjustment, and ANOVA and ANCOVA counterparts. Within the immense ANOVA and associated design of experiment literature, the bulk of space has been devoted to the concept of *confounding*. According to Kendall and Buckland (1960, p. 590), "Confounding is often a deliberate feature of the design but may arise from inadvertent imperfections." The purpose of much of the deliberation that precedes planned experiments is to confound only factors, in other words nuisance and key ANOVA variables whose combined influence on the response can be taken to be inconsequential.

Despite the apparent simplicity of certain ANOVA analyses, in order to cope with what Kendall and Buckland refer to as imperfections, design criteria, as well as assumption-checking procedures, must be implemented carefully. The error-based regression pathways are as much, if not more, subject to these imperfections, as are ANOVA representations. No amount of computing strategy or computer time can patch all flaws of an incautious ANOVA design. Similarly, confounding control via $\mu_{Y|x} = \omega_0 + \omega_1 x_1 + \omega_2 x_2 +, \ldots, +\omega_v x_v$ is a research pathway of last resort that should be regarded as a computational substitute that will hopefully yield the same results that could also be obtained by implementing the partial correlation procedures described in Chapter 12.

## 10.2 Linear data-generation models and mixture models

The pair of data-generation models, $X_1 = \rho_1 X_{nui} + \sqrt{1-\rho_1^2} Z_2$ and $X_2 = \rho_2 X_{nui} + \sqrt{1-\rho_2^2} Z_3$, does not contain any location or scale parameter components. Nevertheless, to help interpret these components, a feature of this model, the $\sqrt{1-\rho^2}$ multiplier of $Z_2$ that appears after the + sign within

$$(\rho Z_1 + \sqrt{1-\rho^2} Z_2),$$

provides a hidden message for standard Normal $Z_1$ and $Z_2$.

Suppose the association parameter $\rho$ within the linear data-generation model

$$W = \mu + \sigma(\rho Z_1 + \sqrt{1-\rho^2} Z_2)$$

equals 1/2 exactly. For comparison purposes, also consider a simple mixture model for a pdf, $Pf(x)+(1-P)g(x)$, where here $f$ and $g$ are themselves distinct density functions. (The capitol letter $P$ represents a mixing parameter to avoid the confusion of this symbol with both the lower case Greek letter $\rho$ and the lowercase letter $p$ representing the index of a broad brush parameter like $\theta_p$.) Because the area beneath any density must equal 1, when $P = 1/2$, it follows that the multiplier of $g(x), (1-P)$, also equals 1/2. Yet for the data-generation model $W = \mu + \sigma(\rho Z_1 + \sqrt{1-\rho^2} Z_2)$, when $\rho = 1/2$, $\sqrt{1-\rho^2} \neq 1/2$; instead, it equals $\sqrt{3}/2$.

A clue to the reasoning that leads to the data-generation model component $\sqrt{1-\rho^2}$, occurs among the operations from which a dependent Normal variate $W$, where $W = \mu + \sigma(\rho Z_1 + \sqrt{1-\rho^2} Z_2)$, can be simulated. The variates $Z_1$ and $Z_2$ contribute to a weighted sum based on the association parameter $\rho$ and, only when this sum is computed, are the scale and location parameters, $\mu$ and $\sigma$, put in play. Thus, two pure (in the sense of being *standard* Normal) variates are combined together and—only then— is the result scaled and, as a final step, situated at the desired location parameter value.

The sequence of steps, from association to scale to location, resembles the computational steps taken to simplify an expression like $4 + 3(5)^2$.

## 10.2. Linear data-generation models and mixture models

Any power such as $^2$, takes precedence over multiplication and addition. Similarly, within $\mu_2 + \sigma_2(\rho Z_1 + \sqrt{1-\rho^2}Z_2)$ the association parameter $\rho$ takes precedence over its scale, $\sigma$, and location, $\mu$ counterparts, in the sense that $\rho Z_1 + \sqrt{1-\rho^2}Z_2$ is computed as a first step.

To merit the designation, *association parameter*, a parameter's value must have nothing to do with location or scale. Correspondingly, no matter what value the parameter $\rho$ assumes, curve property $\text{Var}(\rho Z_1 + \sqrt{1-\rho^2}Z_2)$ equals 1 for any independent $Z_1$ and $Z_2$ that have unit variances. In particular, the analysis of multivariate Normal data depends on this distinctiveness. On the other hand, binary, or qualitatively valued key variables are studied by ANOVA and the age-adjustment techniques that are discussed in the next chapter. Even though they superficially resemble correlation procedures which target continuous key variates, univariate ANOVA methodology concentrate on model location and/or scale parameters to handle issues that multivariate procedures handle by assessing association parameters like $\rho$.

To see why $\text{Var}(\rho Z_1 + \sqrt{1-\rho^2}Z_2) = 1$, consider the formulas

$$\text{Var}(a_1 X + a_2 Y) = a_1^2 \text{Var}(X) + a_2^2 \text{Var}(Y),$$

and $\text{Var}(a_1 X + a_2) = a_1^2 \text{Var}(X)$, that are discussed at length in many elementary level statistics texts and reviewed in later chapters. These general formulas imply that the variance of the $W$ variate,

$$W = \mu + \sigma(\rho Z_1 + \sqrt{1-\rho^2}Z_2),$$

equals $\sigma^2$, the scale parameter $\sigma$ squared.

Unlike

$$\text{Var}(\rho Z_1 + \sqrt{1-\rho^2}Z_2),$$

the variance of the variate $W' = \mu + \sigma(\rho Z_1 + (1-\rho)Z_2)$ equals $\sigma^2[2\rho(\rho-1)+1]$. For example, for the parameter choice $\rho = 1/2$, it follows that $VarW' = \sigma^2/2$. Because the variance of the weighted sum $W'$ of $Z_1$ and $Z_2$ components is related to $\rho$—were it $W' = \mu + \sigma(\rho Z_1 + (1-\rho)Z_2)$, not $W$, that is applied to simulate a response variate $W$—then here the $\rho$ association parameter would help determine, at least in part, the curve property Var, a property that so often connotes scale.

Suppose $Z$ is a variate with density $I_{[0,1)}(z)$ and $g$ is the density of the random variate $Y' = G^{-1}(Z)$, such that the median and $E$ curve properties of $g$ are both zero, and the finite $\sqrt{\text{Var}}$ and AD properties of $g$ are $SD_g$ and $AD_g$, respectively. Then if $Y = \mu + \sigma G^{-1}(Z)$, the $\sqrt{\text{Var}}$ and AD curve properties of $Y$ about the value $\mu$ are $\sigma SD_g$ and $\sigma AD_g$, respectively. Therefore, when a Normal or any other variate is obtained such that $\mathrm{Var}(\rho Z_1 + \sqrt{1-\rho^2}Z_2)$ is unrelated to the parameter $\rho$, not only the variance and standard deviation, but also the $\sigma\mathrm{Ad}_g$ curve property of this variate's distribution, will be free from a relationship to the parameter $\rho$.

Statistical methods can be allocated to subcategories, some of which are concerned with location and scale parameters and others concerned primarily with generalizations of the association parameter $\rho$. For these $\rho$-based methods, location $\mu$ and scale $\sigma$ parameters are taken to be what are called *nuisance parameters*. Within simulation procedures designed to yield data that are dealt with appropriately by conventional statistical analyses, these nuisance parameters would not play their assigned roles if, instead of obtaining dependent Normal or other $Z_1$ and $Z_2$ variates from the model component $\mu+\sigma(\rho Z_1 + \sqrt{1-\rho^2}Z_2)$, it was the generation model component $\mu+\sigma(\rho Z_1 +(1-\rho)Z_2)$ that actually yielded data which matched the data under study.

## 10.3 Parameter distinctiveness

The next chapter will consider statistical approaches that do not rely on model-related assumptions. These procedures tend to be more general, and therefore less prone to validity issues and difficulties, than the regression procedures discussed here or the correlation approaches considered in Chapter 12. This raises an issue that is basic to curve uncertainty reduction. What does the price paid to alleviate validity-related troubles actually pay for?

One answer is that our human experience familiarizes us with concepts that correspond individually to particular parameters. This answer raises still another basic question. What precise characteristics distinguish parameters like $\rho, \mu$, or $\sigma$ from each other? Although there is considerable latitude in the symbolic representation of parameters, by convention distinct Greek letters denote distinct types of parameters. However, what the term *distinct* means requires considerable explanation.

The symbolic context of parameters, such as $\rho, \mu$, and $\sigma$ distinguishes these three Greek symbols from, let us say, the three symbols $\theta_1, \theta_2$, and $\theta_3$. For example, when some combination of symbols like $(x-\mu)/\sigma$ is the only place where the parameter $\mu$ appears in a density $f$ model's formula, this pattern defines $\mu$ to be a location parameter. The only time the location parameter $\mu$ appears in an $f$-type or cdf model's formula, the symbol $\mu$ must be separated from the symbol $x$ by a negative sign. Correspondingly, a parameter is referred to as a scale parameter when it multiplies a standard variate, say $Z$, within a data generation model. A parameter is referred to as an association parameter when it serves the role played by $\rho$ within

## 10.3. Parameter distinctiveness

the weighted sum that defines the variate $W$, as in $W = \mu + \sigma(\rho Z_1 + \sqrt{1-\rho^2} Z_2)$.

Because they play a dual role, threshold parameters, unlike their location, scale and association counterparts, are difficult to define. For example, suppose $Y' = \log(Y - \tau)$, is a Normal variate with location and scale parameters $\mu$ and $\sigma$ respectively. Then threshold parameter $\tau$ serves as part of the argument of the log function, while it also serves implicitly as a part of the indicator function $I_{[\tau,\infty)}(y)$ that helps define the lognormal variate in such a way that $Y - \tau$ cannot be negative.

When a parameter $\tau'$ helps determine the index of an indicator function like $I_{[\tau',\infty)}(y)$ that multiplies a density model, but does not serve within a function like $\log(Y-\tau)$, it acts as a *truncation*, not a threshold, parameter. However, in order to conform to the requirement that the area beneath any density equals one, it is necessary to multiply a pre-truncation pdf model by an appropriate quantity that, in essence, inflates the truncated curve piece. In general, this quantity will be a function of model parameters such as $\mu$ and $\sigma$ and, in the bivariate and multivariate contexts, association parameters such as $\rho$. Consequently, one of the major impediments that complicates parametric statistical analyses is that, unlike the three parameters $\mu, \sigma$, and $\rho$, curve truncation can be disentangled from other parameters only with great difficulty.

A lognormal or other curve with a threshold parameter $\tau$ of $\log(Y-\tau)$ can also be truncated in such a way that the lower truncation point is larger than $\tau$ (Tarter and Lock (1993), section (7.7)). Were this a density model that is truncated at both ends, then dual truncation parameters would appear within the curve's formula.

Many univariate model should include a location, a scale, and a threshold, as well as one or two truncation parameters. Proceeding still further, a univariate mixture formed of two distinct doubly truncated lognormal components has a formula that contains as many as 11 distinct parameters, 5 for each component, and a mixing parameter. The bivariate generalization of such a curve calls into question the wisdom of representing any bivariate distribution by a truncated density model. The curse of dimensionality is leveled with a vengeance against both the parametric and the model-free world views.

The number of distinct parameter types, such as: location; scale; mixing; threshold; truncation and; association, is only limited by manageability issues. However, within $f$'s argument

$$(y_1, y_2, \ldots, y_u; \theta_1, \theta_2, \ldots, \theta_p | x_1, x_2, \ldots, x_v; \omega_1, \omega_2, \ldots, \omega_q),$$

the number $p$ of theta parameters $\theta_1, \theta_2, \ldots, \theta_p$ is a quantity that increases rapidly as $u$, the number of modeled variates, increases. Each individual variate can be associated with its own location, scale, and threshold parameters, while each variate pair can have its own association parameter. The number of distinct combinations of dual variates compounds in rough proportion to $u^2$, where, $u$ designates the number of distinct variates $Y_1, Y_2, \ldots, Y_u$. This is true even for the most manageable of multivariate models, the multivariate Normal model.

## 10.4 Representational uniqueness and model fitting

Two distinct sets of parameter values within a univariate mixture model can represent the same curve. When, as is so often the case, it causes computational difficulties, parameter and curve property uniqueness is a bothersome problem. Yet, when duality provides two alternative ways to express the same thing, this latitude of choice can be a blessing in disguise. For instance, parameters like $\theta_1, \theta_2, \ldots, \theta_p$ and $\omega_1, \omega_2, \ldots, \omega_q$ are more than mere place-holding symbolic components within an expression, model, or representation. Instead, these two different Greek letters provide a guide to roles played by these different types of parameters. This leeway or looseness of parametric representation provides room for useful alternative parametrization choices.

For example, the symbols $\alpha$ and $\beta$ within the equation $y = \alpha + \beta x$ are parameters often connected to the properties of a line, $y$-intercept and slope, respectively. In the sense of specifying how these properties affect the value of $y$, these parameters specify the structure of a curve. Notice that $y = \alpha + \beta x$ is a symbolic representation, not a line. This distinction underlies the subtle but crucial difference between parameters and properties. A property of a specific curve, such as the point over which it would balance, can exist whether or not the curve is modeled. That it can be easier to establish the existence of a single curve property, than to corroborate model validity, motivates a good deal of today's statistical thinking.

In the context of simple analytic geometry, alternative forms of parametric representation of the same curve are well-known. For instance a two-point formula for a line, or, alternatively, the slope-intercept formula $y = \alpha + \beta x$, can suit the needs of specific applications. Generally speaking, that there can be two alternative models for a single curve is no more surprising than the existence of two different portraits of the same person.

A famous painting by a Belgian artist shows a smoker's pipe. Underneath the painted part of the picture is a legend which translates as "This is not a pipe." In a humorous way, the artist is making the important statement that a two- or even three-dimensional painted image of an object is

## 10.5. Model-fitting considerations

not actually the object that is depicted. This same distinction applies in the quantitative sciences as validly as it applies to the arts. In particular, a curve's symbolic representation is distinct from the curve itself.

While variable values must follow the slash sign, | , in

$$f(y_1, y_2, \ldots, y_u; \theta_1, \theta_2, \ldots, \theta_p | x_1, x_2, \ldots, x_v; \omega_1, \omega_2, \ldots, \omega_q),$$

there is some latitude permitted when variate value interrelationship is represented. Variate values can even be taken to be "pseudo-variables" and placed after, instead of before, the slash sign. The trade-offs between treating an observed variate as a variate or a pseudo-variable depends on the purposes to which models are put, and the harm that can be done by a curve-and-representation mismatch. For example, a curve representation can be incorrect from a theoretical perspective but be amenable to robust estimation procedures. In the long run, the representation that underlies a robust procedure may be superior to alternative representations that provide a more ideal fit, but are frail, in the sense that frailty denotes the antonym of the term "statistical robustness."

Models and other expressions are merely introductory chapters of any data-analysis story. In the physical sciences, accurate models are often deduced from basic laws. Yet in the demographic, biomedical, economic, and social sciences such pure deductive approaches have been of only moderate value. Instead, mathematically derived models are often combined with inductive techniques that blend data into the representational mix. Chief among the many reasons why such combinations are complex is that even when an accurate parametric model is available, parameters must be estimated.

When poor parameter estimators are substituted for the parameters of some putative ideal model, la Plato, such a substitution can do more harm than that caused by the insertion of good estimators into a curve representation that differs from the ideal model. Compromises of this sort are particularly important when conventional estimators of the ideal model's parameters are sensitive to even small quantities of pathological data.

## 10.5 Model-fitting considerations

A basic feature of the multivariate Normal model is that *each* variate is related to any other single variate by merely a single number where, for example, for both $i$ and $j \leq u$, if $i \neq j$ the $i^{th}$ variate's relationship with the $j^{th}$ variate is determined by the scalar $\rho_{ij}$. Gumbel (1961) as well as Mardia, Kent, and Bibby (1979), sought alternative multivariate models that

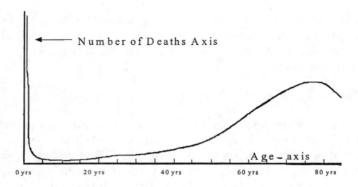

Figure 10.1. A bathtub function.

could match the multivariate Normal in terms of the simplicity with which dependence could be parametrized. Yet, no other model has yet been found that achieves results with regression methodology that are compatible with results where variables are treated as variates. Only the multivariate Normal generalization of $\phi$ exorcises the curse of dimensionality. However, even for the multivariate Normal; $p$, the number of distinct parameters $\theta_1, \theta_2, \ldots, \theta_p$ needed to represent all aspects of distributional structure, increases approximately as the square of the number of variates, $u^2$.

Within linear regression models, the required number of distinct parameters increases approximately as the number of variables, $v$, and not as the square of $v$. Much of the curse of dimensionality is assumed away by the linear regression model. It is no wonder, therefore, that many procedures have been devised to check multivariate Normal and general linear model validity. Whenever such checks fail, the full force of the curse of dimensionality is, at least at the present time, inescapable.

Many curves encountered in vital, demographic and many other branches of applied statistics, are given the picturesque name *bathtub functions*. Linear regression models represent bathtub functions poorly. There is a bad match between: (1) Normal and other symmetric frequency curves and; (2) much social science, demographic, economic, and epidemiological data. This stems, in part, from such issues as the "break-in-period," when the failure of machinery is studied, and "perinatal or neonatal mortality," when the deaths of young human beings are studied.

Many density curves associated with mortality or times-to-failure, assume high values near, and immediately after, the value zero, and then form a wide bathtub-bottom-like shape until what in humans is called *middle age*. Such a curve would be poorly matched by any bell-shaped Normal

## 10.5. Model-fitting considerations

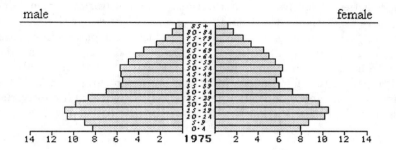

Figure 10.2. An example of a population pyramid.

curve $\phi(x)$ because, among other differences, the Normal is a symmetric curve, while the curve shown in Figure 10.1 is asymmetric.

A population pyramid, such as the curve shown in Figure 10.2, is a stereo snapshot of a population's age density that displays estimates of two, side-by-side density curves; one for males, $f_M$, and one for females, $f_F$ (US Census Bureau, 1998). Age or date variate values are coded along the ordinate $y$-axis, and the frequency proportionate to the chance of randomly selecting an individual at this age is plotted along the abscissa.

As suggested by the population pyramid shown in Figure 10.2, the density curves that population pyramids estimate often have rough edges caused, for example, by: wars; depressions; baby booms; and waves of migration. (These are illustrated in an elegant fashion by the U.S. Census Bureau at www.ac.wwu.edu/~stephan/Animation/pyramid.html, Dickerson and Johnson (1999)). Since age is related to many variates and variables studied by economists, epidemiologists, demographers, and other health scientists, population pyramids cast their shadows on almost all curves encountered in studies that involve natural populations of human beings or animals. For this reason, rough-edged densities encountered by applied statisticians are rarely just artifacts to be explained by the catch-all term "sampling variation." Instead, these edges are natural features that cannot be smoothed away. Again, as in the case of bathtub functions, $f$-type curves with rough edges differ substantially from Normal densities.

The curse of dimensionality is encountered so often in close proximity to population pyramids a warning is in order. In the many circumstances where the outlines of marginal densities of single demographic and vital statistics variates are roughened by the influence of age-distribution irregularities, the roughening process casts an even darker shadow on the joint densities of multivariate distributions. The multivariate Normal density curve is often far too smooth to accurately serve as the choice of function $f$ that prefixes the variates, variables, and parameters within $f(y_1, y_2, \ldots, y_u; \theta_1, \theta_2, \ldots, \theta_p | x_1, x_2, \ldots, x_v; \omega_1, \omega_2, \ldots, \omega_q)$. Even when they are interested in a single function of $x_1, x_2, \ldots, x_v$ that is linked to a single parameter of $f$'s model, biostatisticians, demographers, and other scientists often cannot depend on linear regression model and multivariate Normal model compatibility.

## 10.6 The variance curve property and bathtub functions

Besides the possibility that properties like $E(X)$ and $\text{Var}(X)$ are infinite or ambiguously defined, there is an additional problem with statistical methodology based on either of these two properties. The problem even affects methodology based on the model parameters $\mu$ and $\sigma$ that are usually associated with these properties. Suppose the age-at-death variate density bathtub function graphed in Figure 10.1, were estimated for the 1958-59 population of the USSR (Shryrock, Siegel, et al., 1973 ). Suppose, for the sake of argument, that the curve property $E(X)$ of this $f$-type curve is 30 years. The rectangular solid sketched in Figure 10.3 describes the left-side-bathtub rim's contribution to the property $\text{Var}(X)$. Because of the height

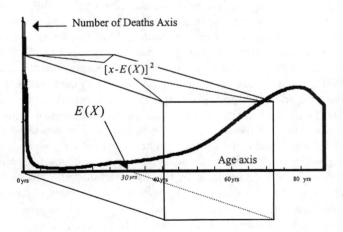

Figure 10.3. Contribution to the variance of a bathtub-shaped function's left edge.

## 10.7. Regression and least squares

of the rim is multiplied by the base of the block shown in Figure 10.3, this contribution dwarfs other density-curve contributions.

Because bathtub and other asymmetric and peculiar-shaped densities are encountered often in demography, survival studies, and vital statistics, it follows that the property $\text{Var}(X)$ is less relevant in these areas than it is elsewhere in applied research. More than the two curve properties $E(X)$ and $\text{Var}(X)$ are often required to provide an accurate description of univariate and multivariate distributions.

Besides illustrating the unfortunate sensitivity of the $\text{Var}(X)$ curve property to the edges of bathtub curves, Figure 10.3 provides a clue concerning a general Var curve property Achilles heel. This weak spot is the counterpart of the absolute deviation Adm' curve property's indifference to changes made to a pdf near the population median, a weakness that was illustrated in Chapter 3. The Var curve property is often swayed disproportionately by curve-tail behavior.

To visualize why curve tails, as compared to a curve's central structure, sway Var's value, consider the volume of the box shown in Figure 10.3. Suppose that a similar box was drawn with one edge extending from the value of $E(X)$ to the age variate value, 10 years. The pdf's height at 10 is much lower than it is within the 0-to-1 year interval. What this bathtub-shaped pdf looks like over the interval from 2 to 30 will change the Var curve property far less than any corresponding change made in the height of the large box shown in Figure 10.3.

## 10.7 Regression and least squares

Before discussing nonparametric procedures in the next chapter, a brief historical outline of parametric regression will be presented in this section. While this book concentrates primarily on curve and parameter selection and not on estimation issues, there are two reasons why an exception to this emphasis is made in this section. The first is that a later chapter will make considerable use of a relationship based on the Gauss-Markoff Theorem that is discussed here. The second is that two general estimation procedures, least squares and the method of averages, illustrate the varied usage of the curve property $E$.

An early exposition in the statistical literature of the distinction between curves based on parameters like $\mu$, on the one hand, and nonparametric regression curves based on curve properties like $E(Y)$ on the other, appears in Marek Fisz's 1963 text, *Probability Theory and Mathematical Statistics*. There Fisz contrasted the direct estimation of the curve property $E(Y|x)$, with procedures for estimating the parameters $\alpha$ and $\beta$ that minimize the curve property $E\{[Y-(\alpha+\beta x)]^2\}$. At $x_0$, for some specific set of parameter $\alpha$ and $\beta$ values, $[Y-(\alpha+\beta x_0)]^2$ is a new variate formed from a response variate $Y$.

Even though $(\alpha+\beta x_0)$ is an expression that is based on the two parameters $\alpha$ and $\beta$ the term $(\alpha+\beta x_0)$ is treated as if it were a constant. Thus $Y-(\alpha+\beta x_0)$ is a translated or relocated version of variate $Y$, and $[Y-(\alpha+\beta x_0)]^2$ is a modified variate, specifically, $Y-(\alpha+\beta x_0)$ squared.

Suppose this modified variate has a density $f_2$ with an expectation $E_2(\alpha,\beta)$, a curve property related to the two quantities, $\alpha$ and $\beta$. Then, since $\alpha$ and $\beta$ of $\alpha+\beta x_0$, are determined by minimizing all possible curve properties $E_2(\alpha,\beta)$, the $\alpha$ and $\beta$ quantities are themselves curve properties. On the other hand, if each candidate $f_2$ is defined in terms of $\alpha$ and $\beta$, it follows that $\alpha$ and $\beta$ are parameters. Thus, regression methodology that concerns the minimization of the expression $E\{[Y-(\alpha+\beta x_0)]^2\}$ is based on the correspondence of parameters $\alpha$ and $\beta$ with the specific curve properties $E_2(\alpha,\beta)$.

Under the assumption of equal $Y$ variate variances despite differences in the value of $x$, in other words, assuming the *homoscedasticity* condition or, alternatively, assuming prior knowledge of relative variability, the mathematical finding called the Gauss-Markoff Theorem concerns $E\{[Y-(\alpha+\beta x_0)]^2\}$. From among a specialized class of parameter estimators, this theorem states that the least squares method proposed by Gauss provides the best alternative. The quality of the estimator is gauged by a curve property that, like $E\{[Y-(\alpha+\beta x_0)]^2\}$, assesses: *mean*, as in $E$; *square*, as in the exponent $^2$ of $E\{[Y-(\alpha+\beta x_0)]^2\}$ and; *error*, as in $|Y-(\alpha+\beta x)|$. Curve, parameter, or property, estimator quality is gauged by *mean-square error* or "mean-square deviation."

The Gauss-Markoff Theorem links an algorithm with a criterion. A curve property, the $E$ of $E\{[Y-(\alpha+\beta x_0)]^2\}$, is not itself a statistical end in the sense of a goal, but rather it is the means by which the end's suitability is gauged in terms of its hopefully small size. This important distinction is illustrated by a comparison with the "method of averages", a general curve construction technique that is also based on the $E$ curve property. Unlike methodology that is based on the Gauss-Markoff Theorem, the method of averages approach takes the curve property $E$ to be a statistical end in itself. It does this by estimating the path taken as the $E(Y|x)$ curve property changes as the variable $x$ changes over a finite sequence of values.

Specifically, suppose $n$ data points of a sample are allocated among a sequence of contiguous strips cut parallel to the ordinate axis. Where these strips are located is determined such that several points fall within all but the leftmost and rightmost strips which, below, are disregarded. Within the $j^{th}$, together with all other strips, the arithmetic averages, the sum of

## 10.7. Regression and least squares

the $j^{th}$ strip-specific $y$-ordinate values divided by the number $n_j$ of points that fall in the strip, are then determined.

Potentially, the points formed by the strip midpoints as $x$-coordinates and strip averages as $y$-coordinates, could then be joined together with line segments to create a nonparametric regression curve. However, in the 1920s, and then until digital computation became available, the method of averages was applied solely to estimate the same sort of parametric regression curves that were then estimated typically by least squares methodology.

Unlike the method of least squares' Gauss-Markoff Theorem theoretical underpinning, the method of averages is justified primarily by empirical investigations. For example, Lipka (1918) calls regression slope parameters "...constants or coefficients appearing in the equation," and for the one parameter special case $y = \beta x$, Lipka indicates that: "...the method of Least Squares gives the best value of the constant but requires the most calculation; the method of averages gives, in general, the next best value of the constant and requires but little calculation..."

As the word constant indicates, although the method of averages is based on the estimation of a sequence of $E(Y|x)$ evaluations at different $x$-values, members of this sequence are envisioned solely as steppingstones to condensed information. This data reduction process, where $n$ data points are abstracted to one or two parameters, is a special case of a procedure called the method of moments, described in Kendall and Buckland's dictionary (1960) as: "A method of curve fitting which proceeds by identifying the lower "moments" of the observed data with those of the particular curve form being fitted."

The method of averages can determine a line that passes through two estimates of the moment $E(Y|x)$. For example, $E$, the first-moment curve property of the $Y$ variate given the value of a variable $x$, can be estimated within each of two wide strips that together enclose all $n$ points. Then a line can be passed through the two points with strip midpoint $x$ coordinates, $x_1$ and $x_2$, and estimated $E(Y|x)$ ordinates $\hat{y}_1$ and $\hat{y}_2$, of $E(Y|x_1)$ and $E(Y|x_2)$, respectively. Two such curve properties are illustrated in Figure 6.1. From these two points, $(x_1, \hat{y}_1)$ and $(x_2, \hat{y}_2)$, the two-point formula for a straight line determines estimators of the $\alpha$ and $\beta$ components of $\alpha + \beta x$. Thus, in this example, the method of averages is based on estimates of the two curve properties, $E(Y|x_1)$ and $E(Y|x_2)$. On the other hand, although Gauss-Markoff-based methodology also calls upon the curve property $E$, when it does so it utilizes the multiple properties $E\{[Y - (\alpha + \beta x_0)]^2\}$, each determined for some particular choice of $\alpha$ and $\beta$ parameter values.

# CHAPTER 11
# Nonparametric Adjustment

## 11.1 Age-adjustment and logistic regression

The previous chapter illustrated the poor correspondence between multivariate Normal models on the one hand, and complex bathtub $f$-type functions as well as rough-edged population pyramids on the other. Age-adjustment procedures that were devised to deal with these complex curves have a long history. However, during much of this period, they have been applied exclusively to deal with a single binary-valued response variate that assumes only one of two values of the sort dealt with by logistic regression.

Age-adjustment procedures are based on designer or artificial curve properties. Unlike the $E$, AD, Var, and Cov curve properties, adjusted rates do not correspond to some particular parameter. Here the word "correspond" is taken to imply that for some special case, like the multivariate Normal distribution, a curve property, such as Cov divided by the square roots of two variances, can be taken to be equal numerically to a parameter like $\rho$.

There is no theoretical reason why a density- or data-generation model cannot be constructed which is expressed in terms of a parameter that corresponds to an adjusted rate. Such rates share a basic characteristic possessed by the partial correlation parameter discussed later in this chapter. However, it is of questionable value to express any of the highly complex curves shown in Figures 10.1 and 10.2 by any elementary function-based model. Consequently, because adjusted rates are based on composites of such curves, there has been little attempt to match these designer properties with any model component.

Today, when computational capacity is substantial, adjustment procedures can be focused on many variate types. Nevertheless, for historical reasons, how this approach works can be explained best in terms of a binary-valued response variate and the nuisance variate, age. Unlike logistic regression, age-adjustment is a nonparametric procedure whose goal is spurious effects reduction. This important tool has been applied in demographic and other statistical studies where a single key variable distinguishes two or more states, cities, census tracts, or other population subgroups; or where some key variable indexes gender, socio-economic, or racial status. The response variate in question usually has designated the morbidity or mortality status of individuals within a population.

When mortality, rather than morbidity, is studied, age-adjustment handles one or more nuisance variables in the following way: Statistical quantities, called *age-specific* death rates estimate the mortality curve $m(y|x)$ of live-dead status $Y = y$ given age $x$. Here $y = 1$ denotes that a subject has died, and the term "specific," designated by the shorthand pair of symbols, $|x$, is applied in the sense of being specific to some value of the nuisance variate or variable. Thus, as in previous chapters, $|x$ denotes the phrase, given the value $x$ of the variate or variable.

To indicate that a conditional $f(y|x)$ describes a mortality response variate's distribution, the letter $f$ is replaced by the letter $m$. Of course, in the mortality context, $Y$ is binary-valued, not the continuous sort of variate that is dealt with by correlation techniques. Age and life-death status variates have a peculiar relationship. Even though a building's death might be taken to occur at the date it becomes unoccupied, the aging process of this building would still continue. Yet, this analogy does not apply to the human aging process. Hence, the special symbol $m(y|x)$, where $x$ is a value of the variate age, represents the specialized mortality curve that relates response variate $Y$'s value $y$ to $m(y|x)$, given $x$.

Consider, for example, the entire population of a large country such as Italy. Suppose on January 1, 1959, a subpopulation was comprised of those particular citizens of Italy who, at some time during the 365 days of the year 1959, were $x$ years old. Some members of this subpopulation died in 1959. The remainder, of course, lived until 12:00.00001 AM, January 1, $1959 + 1$.

When the term "degree of abridgment" appears in demographic and other applied statistical publications, it denotes the interval, in this example year-long, over which data are aggregated. As is true for an abridged dictionary, *abridgment* refers to information that is condensed or lost. Information concerning exactly when—within the year-long interval—any citizen of Italy died, is discarded.

Abridgment processes surmount a problem with the status-curve $m(1|x)$'s interpretation. For instance, suppose a small time interval is selected. If the degree of abridgment decreases still further and approaches zero, then the associated probability that a given citizen of Italy will die during this interval would also decrease—ultimately to zero. To avoid the problem whose formulation goes at least as far back as the Greek philoso-

## 11.1. Age-adjustment and logistic regression

pher Zeno, $m(1|x)$ is taken to be the curve upon which a comparison of the chances of dying at two or more different ages is based. However its values do not assess the actual probability of death at age $x$ exactly.

Suppose the left-hand side, in other words the male side, of a population pyramid describes a distribution of the variate age $X \in g(x)$. (It is assumed in this chapter that the precise ages of so many persons are described by this pyramid that there is no appreciable difference between the pyramid edge coordinate and some density $g$'s evaluation.)

A value $g(x_0)$ is proportionate to the probability that at some specific date $t$, an individual who is picked at random from the population of interest will actually be $x_0$ years of age. The curve $g(x)$ is a snapshot of an age distribution that is taken at a given date $t$. To avoid confusion between the date $t$ and the age-determining-running variable $x$, it should be kept in mind that $g(x)$ represents what otherwise might be symbolized as $g(x|t)$, the distribution of age at the given date $t$ and that, at this same date, the $m(1|x)$, chance-of-dying, curve applies to the population described by $g(x)$.

Suppose the chance-of-dying curve $m(1|x)$ is estimated by the curve $\hat{f}(1|x)$. Then an age-adjusted death rate combines the information conveyed separately by the two curves $\hat{f}(1|x)$ and $g(x)$. The purpose served by this particular curve combination is the estimation of an artificial curve property. This property discloses the death rate if the population from which $m(1|x)$ is estimated was some other standard population, such as that of the United States at date $t$.

As discussed cogently by Gross and Clark (1975), the curve that $\hat{f}(1|x)$ estimates—when taken as a function of $x$ and not as the function that is evaluated at the value 1—is, strictly speaking, a curve that is called conventionally a *hazard function*. In order to be a legitimate candidate for death at $x$, the individual must be alive immediately prior to death.

Even though Gross and Clark use the Greek symbol $\lambda$ to denote the hazard function, in order to both emphasize that it is a conditional density that assumes the value 1, and also to continue reserving Greek letters as symbolic representations of either parameters or the standard Normal density $\phi$, in the remainder of this chapter the function that $\hat{f}(1|x)$ estimates is denoted by $f(1|x)$ in the general case and $m(1|x)$ when the value 1 is the code for death. (In many applications, the curve $m(1|x)$ can be interpreted as the ratio $d(x)/S(x)$ formed by dividing the density $d(x)$ of the age-at-death variate $X$ by the survival curve $S(x)$ that discloses the probability of an individual surviving to reach the age $x$.)

The age-adjusted rate based on $f(1|x)$ or $m(1|x)$ is a standardized quantity that helps cope with a response variate's distribution's relationship to the nuisance variable age. This standardization or adjustment process is designed to assure that the key variables which differentiate different countries, states, or communities can be compared without contamination attributable to age-related spurious association. For example, suppose that when both Italian and USA rates are based on the same standard population pyramid, the age-adjusted mortality rate of Italy is less than that of the USA. Then Italy and USA mortality difference would not be attributed, even in part, to the nuisance variable age.

In the demographic literature, $m(1|x)$ is often denoted by a subscripted symbol like $m_x$, while for a specified date $t$, the sample counterpart of $m(1|x)$ is often defined to equal $_nD_x/_nP_x$. $_nD_x$ designates the number of deaths counted out of an age $x$ "population at risk," a quantity that, in turn, is designated by the three letters $_nP_x$. However, within demographic notation like $_nD_x$ and $_nP_x$, unlike other places where the lower case Latin letter $n$ appears in the statistical literature, the symbol $n$ does not traditionally represent sample size. Rather, it denotes the degree of abridgment. For example, when $x = 5$ and $n = 2$, then $_nP_x$ denotes the number of persons who are from 5 to 6.999999 years of age, while $_nD_x$ denotes the comparable number of deaths among this group of persons. It is also not unusual for the symbol $_nm_x$ to be applied to represent $_nD_x/_nP_x$.

The notation $m(1|x)$ will be retained in this chapter because: (1) The notation $m(1|x)$ resembles the notation $f(1|x)$, in the sense that because both of these expressions describe a distribution, not a sample, degree of abridgment need not be represented; (2) The symbols $m(1|x)$ show more clearly than does the notation $m_x$ that some curve $m(0|x)$—which might, for example, refer to the chance of $x$-year-olds living through some interval—could just as well be used in conjunction with adjustment procedures, as could $m(1|x)$.

There is no theoretical reason why a function $m(y|x)$ of a continuously-valued variate with value $Y = y$ could not be adjusted in the same way that the structure of $m(1|x)$ can be adjusted when $m(1|x) + m(0|x) = 1$ for all $x$. Although there is a lack of literature on this subject, many nuisance variate adjustment problems can be dealt with that also can be studied using partial correlation, $\rho_{12\bullet 3}$-based methodology. However, there is no reason to believe that both parametric $\rho_{12\bullet 3}$-based and nonparametric adjustment will always arrive at exactly the same conclusion, just as there is no reason to believe that the two varieties of adjustment techniques discussed in this chapter will yield identical findings.

Suppose the unlikely situation occurs where death occurs independently of age. Then, for all $x$, the curve $m(1|x)$ satisfies the relationship $m(1|x) = f_m(1)$, where $f_m(1)$ designates the frequency of death irrespective of age, a frequency that is often estimated by a statistic called the *crude death rate*. For the population of Italy at a given date, say $t = 1959$, the crude death rate at $t$ is estimated by the fraction formed by counting the number of deaths and dividing this number by the size of the Italy's population at date $t$. Note that there being no difference between: (1) the crude death rate; and (2) any age-specific death rate, means that ascertaining a person's age yields no information concerning his or her chances of survival. Hence, for the two variates, age and live-dead-status, independence corresponds to an identity between the two curves estimated by: (1) crude and; (2) age-specific death rates, respectively.

## 11.2  Crude and specific rates

When a specific rate $m(1|x)$ equals the crude rate $f_m(1)$, this relationship equates a conditional at variable $x$'s value and the single value assumed by a marginal density, the value $f_m(1)$. Hence the identity over the positive part of the $x$-axis, $m(1|x) \equiv f_m(1)$, defines a horizontal line segment of height $f_m(1)$. A simple geometric interpretation of an expression like $m(1|x) \equiv f_m(1)$ is an exception, not the rule. Its simplicity here stems from the binary structure of the morbidity variate.

Relationships between joint, conditional, and marginal densities are obvious when multiple variates are as similar to each other as the outcomes of independent rolls of a pair of dice. Unfortunately, two variates, such as live-dead (LD) status and age, are often intertwined in complex ways. What, for instance, is the age of a person whose LD status is dead? Because of this semantic snag we will now see that care must be taken when a researcher studying Italian mortality decides to use $m(1|x) \equiv f_m(1)$ to make adjustments that take the age variable $x$ into account. Here the phrase, "take into account," is applied in the same sense that specialized correlation techniques can be called upon to take a nuisance variate into account. On the other hand, most crude death rate estimators, like most estimates of regression $\beta$ parameters, contain a trace or more of nuisance contamination.

Crude death rates correspond to specialized, binary-valued, marginal density curves. The word "crude" does not necessarily mean that a crude rate is less informative than any single specific death rate. However, even though the marginal density curve of a variate $Y$ can be obtained from the joint density curve of $Y$ and any other variate, say $X$, the reverse is not true. Knowledge of the marginal density of $Y$ and the marginal of $X$ only determines the joint density of the variate pair $(X,Y)$ completely in cases where additional information, such as the information that $X$ and $Y$ are independent variates, is provided.

Suppose $Y$ and $X$ variates are actually independent. Consider any variate value $y$ and variable value $x$, together with the conditional $f(y|x)$; for example for $y = 1$, the density $m(1|x)$ of the variate value $Y = 1$ given the value $x$. Then this conditional is equal to the counterpart marginal density curve evaluated at $y$, for instance $f_m(1)$. This statement is a generalization of the relationship between crude and specific death rates where, assuming independence, the uncertainty about the occurrence $Y = 1$, irrespective of the value of $x$, equals the uncertainty about this same occurrence, given the value $x$.

## 11.3 Age-adjustment; marginal, joint, and conditional curves

The age-adjustment process illustrates that the word "curve" can convey a misleading impression about binary-valued variates. For example, a binary-valued variate, such as live-dead status, need not be handled by methodology that is fundamentally different from procedures that deal with continuous variates.

By viewing statistical approaches from the vantage point of curve relationships that apply to many types of variates, issues like the definitions of the terms: event; inference; causality; and probability, can be treated separately from mathematical issues, such as the multiplication of functions or the continuity of a curve. In particular, no matter what definition the concept—probability proportionate to some curve $f$—is given, it is unlikely that this definition will invalidate deductions based on premises such as statistical independence, or on related definitions, such as the definition of a conditional density. As a bonus, from a curve-based perspective these deductions are equally valid for both continuous and binary-valued variates.

The steps by which adjusted rates are constructed are determined by distinctions between marginal, joint, and conditional density curves. As far as actual step details are concerned, the shape assumed by these curves is irrelevant. For instance, the slash that precedes the $x$ within the symbolic representation $m(1|x)$ indicates that $m(1|x)$ is a conditional density evaluated at 1. As a consequence, multiplication of this age-specific rate by the $g(x)$ curve of the corresponding population pyramid yields a product $m(1|x)$ times $g(x)$ that is identical to the joint density $m(1, x)$. Below, expressions like $m(1|x)$ times $g(x)$ will be represented by a multiplication sign, as in $m(1|x) \times g(x)$.

Suppose that when $y$ equals 1 within $m(y, x)$, the function $m(1, x)$ pertains to a person who is freckled. Correspondingly, suppose $y = 0$ indicates that a person is unfreckled. (It is assumed that all members of the

## 11.3. Age-adjustment; marginal, joint, and conditional curves

population can be divided into two categories, freckled or unfreckled.) Were $m(1, x)$ proportionate to the probability that an individual picked at random from a population will be both freckled and be $x$ years old, it then would follow that the two curves, $m(1, x)$ and $m(0, x)$, where $m(0, x) = 1 - m(1, x)$, together describe the joint distribution of the freckled status variate, and the variate age, completely.

Suppose $Y$ is a letter chosen to represent the freckled-unfreckled variate itself, and $y$ represents either of this variate's possible values, 1 or 0 (the two values that designate, freckled or unfreckled status, respectively). In this context the numbers 0 and 1 are the binary counterparts of the one to six uniformly-distributed dots that, as discussed in Section 8.3, are placed on dice. For example, consider the rolls of a pair of dice, one green and one red, where the probability of rolling a two with one die equals the sum of: (1) the probability of rolling a two with the green die and rolling a one with the red die, plus (2) the probability of rolling a two with the green die and a two with the red die, plus (3) ..., plus (4) ... , plus (5) ..., plus (6) the probability of rolling a two with the green die and a six with the red die.

Like the marginal dice density, the crude freckle-rate $f_m(y)$, in other words the marginal density of the variate $Y$, is equal to the definite integral

$$\int_0^\infty m(y|x)g(x)dx.$$

This definite integral sweeps, or combines, $m(y|x)g(x) = m(y,x)$ values. An age-adjusted rate is based on the replacement of the curve $g(x)$ that appears within the integrand of

$$\int_0^\infty m(y|x)g(x)dx$$

with an alternative curve $g_s(x)$. The subscript of the curve $g_s(x)$ discloses that this changeling curve describes some standard population. For example, suppose mortality is studied from the perspective of a binary-valued key variable that distinguishes two different countries, such as Italy and the USA, and their respective $m(y|x)$ curves. Suppose also that the same $g_s$ curve is substituted within the crude rate definite integral, but the two $m$ curves, one for Italy and one for the USA, are both called upon to compute this

$$\int_0^\infty m(y|x)g(x)dx$$

integral.

In this example, the key variable that is associated with the response variate has merely distinguished the Italian population from the American population. By comparing these two different adjusted rates, the connection between this key variable and the response variate, freckled status, can be assessed by the adjusted rates that are based on the same standard population pyramid. Consequently, differences in the age structures of the Italian and USA populations have been equalized which, in turn, assures that when a comparison is based on adjusted rates, this comparison is unaffected by population age-distribution differences. The age variable has, in this way, been partialed, filtered, or adjusted, so that what remains is, in technical terms, the part of the variate $Y$ that is "unconfounded" with age.

## 11.4 Age-adjustment and partial correlation

The age-adjustment process resembles procedures to be discussed in detail in the next chapter that are based on the partial correlation coefficient $\rho_{12 \bullet 3}$. Specifically, let $Y_1$ represent a continuous response variate, $Y_2$ represent a continuous key variate, and $Y_3$ represent a continuous nuisance variate; while $\rho_{12}, \rho_{13}$, and $\rho_{23}$, denote the association parameters corresponding to pairs of these $Y$ variates with matching subscripts. For example, $\rho_{12}$ is the response variate with key variate association parameter. The symbolic expression $\rho_{12 \bullet 3}$ is defined to equal

$$(\rho_{12} - \rho_{13}\rho_{23})/\{\sqrt{1-\rho_{13}^2}\sqrt{1-\rho_{23}^2}\}$$

(Anderson, 1971).

The numerical difference between a crude rate and an age-adjusted rate plays a role similar to that played by the parameter difference $\rho_{12} - \rho_{12 \bullet 3}$. Here $\rho_{12 \bullet 3}$ denotes the association parameter that indexes a response variate's degree of relationship with a single key variate—given the value of a third variate, such as the nuisance variate age. (Conditions under which $\rho_{12 \bullet 3}$ is a single number will be discussed in the next chapter.)

The paucity of literature that concerns adjusted-death-rate validity stems from the binary structure of the live-dead response variate. For example, were it some morbidity response variate, such as the number of days absent from work due to illness, that is age-adjusted, then the counterpart of an adjusted death rate would be a function of a continuous variable. As we will see is true of certain generalizations of $\rho_{12 \bullet 3}$, there is no reason why this function of a continuous variable has to be a horizontal line. Similarly, two adjusted morbidity rates can easily be zero for some levels, and nonzero

## 11.4. Age-adjustment and partial correlation

for others. After adjustment for age, it is quite possible for morbidity frequencies to differ solely when levels exceed some threshold value. (The existence of such a threshold can easily be masked by a nuisance variate or some combination of variates like smoking and alcohol consumption.)

Like $\rho_{12\bullet3}$-based methodology, crude and adjusted rates have limited scope. Yet the many recent publications that discuss high-dimensional model-free curve estimators have, as a consequence, helped extend the applicability of adjustment techniques. In the past, it has primarily been computational difficulties that limited adjustment procedures to a binary-valued response, and a key variate that merely distinguishes two or more groups such as the USA and Citizen of Italy populations on the basis of mortality. Today, the integral

$$\int_0^\infty m(y|x)g(x)dx$$

can be defined in a variety of ways, none of which require that the symbol $y$ designates simply the value of a binary, rather than a continuous or discrete, variate.

The quantity $|\rho_{12} - \rho_{12\bullet3}|$ is a parameter that increases in magnitude as spurious association increases. Similarly, suppose the health of two communities is compared by assessing crude mortality rates that assume different values, yet when adjusted rates are substituted for crude rates—mortality differences vanish. Then this situation resembles a study where $|\rho_{12}|$ is large, but $|\rho_{12\bullet3}|$ differs insignificantly from zero.

On the other hand, suppose adjusted rates differ to a significantly greater extent than crude rates. This suggests that the nuisance variate or variable introduced statistical noise and this noise was filtered out by the adjustment process. The same inference can be made when $\rho_{12\bullet3}$ is estimated to be larger than the corresponding estimate of the ordinary, non-partial association parameter or correlation curve property $\rho_{12}$.

Unless applied in a way that duplicates the partial-correlation-based approach, linear regression $\beta$ parameters cannot be applied to accomplish nuisance variate and variate adjustment tasks. There is a conceptual difference between the independent quantities $Z_1, Z_2$, and $Z_3$, upon which partial correlation models are based, and the regression model terms $x_1, x_2$, and $x_3$ which, were $x_1, x_2$, and $x_3$ variate values, not variable values, would in most cases be dependent upon one another. Because of this conceptual difference, a procedure that leaves a linear regression model term in, and later leaves it out, of a curve's representation, will always be a shaky substitute for: (1) partial correlation; and (2) age-adjustment methodology (unless it produces results that could also have been obtained with the help of methodologies (1) or (2) in the first place).

Among the parameters that help formulate many models of key variable and response variate relationship, the regression coefficient $\beta_3$ of nuisance variable $x_3$ seems to be the sole representative of this variable's contribution. Yet the remaining $x_2$ term is not guaranteed to be a filtered, partialled, or adjusted value. A researcher who turns a blind eye to the distinction between the observed $x_2$, and independent component $Z_2$, in no way assures that Nature will play along. Instead, partialing with regression methodology in order to do what age-adjustment and partial correlation can often do effectively, amounts to what ostriches supposedly do when they encounter a dangerous nuisance.

## 11.5 Direct and indirect adjustment

As the word "direct" suggests, there is a well-known alternative to the direct age-adjusted rate that is referred to as the *indirect age-adjusted rate*. The somewhat artificial statistic referred to below as the *one year abridged age-specific freckled rate*, will help clarify the differences and the similarities between direct and indirect adjustment procedures. This statistic estimates the value attained by a particular conditional $f(1|x)$ where, as before, within $f(1|x)$ the value 1 designates the numerical code that classifies a person as being freckled, while $x$ denotes some particular age at which this classification is made. (Besides its neutral connotation, an advantage of examples that concern freckled rather than dead people is that when a person develops freckles, that person is not removed from the population. For this reason, the two variates, age and being freckled, are simpler conceptually than the two variates, age and death; and freckled-with-age analyses do not require the application of specialized curves such as the hazard function $d(x)/S(x)$.)

Suppose that the subscript $c$ of the symbol $f_c(1|x)$ designates a particular community whose characteristics are being studied, while the subscript $s$ of $f_s(1|x)$ designates a different community called a *standard*. Suppose, for example, that interest is focused solely on individuals who are 85 years old when, for any particular value of $x$, both $f_c(1|x)$ and $f_s(1|x)$ are known curves. Given this narrow focus, the curve properties $f_c(1|85)$ and $f_s(1|85)$ suit the purpose of community $c$ and $s$ comparison. However, when two communities are compared in realistic applications, age rarely can be taken to be some constant value like 85 years. Instead, a nuisance variate or variable assumes, typically, a spectrum of values. Yet even granting that this spectrum is of interest, were it the case that $f_c(1|x) > f_s(1|x)$ for all $x$,

## 11.5. Direct and indirect adjustment

then conclusions would still be obvious, nuisance variates $X$'s distribution notwithstanding.

Inequalities like $f_c(1|x) > f_s(1|x)$ need not be valid for all $x$. Consequently, in order to compare the effect that a key variate (one that takes on the value $c$ or $s$) has on freckled status, morbidity, or mortality—age distinctions notwithstanding—a single numerical value, not a difference between two entire curves, can be determined.

Decisions about communities are often made from an all-or-nothing perspective. For instance, were $f_c(1|x) < f_s(1|x)$ for one value of $x$, and $f_c(1|x) > f_s(1|x)$ for another, then these inequalities can leave a decision maker uncertain whether or not some health intervention is needed in order to equilibrate community $c$ and community $s$.

The direct adjusted rate is a property of a curve that, in turn, is a product of two other curves, one of which is $f_c(1|x)$, where $f_c$ typically describes a particular race, gender, or other subgroup within some community. (Unless otherwise specified, as in the next paragraph, not only $f_c$, but other curves are associated with the same combination of race and gender.)

Suppose $g_s$ is proportionate to a marginal density that specifies the probability that an individual selected at random from community $s$ will be found to be $x$ years old. Given the above definition of $g_s$, it follows that the product, $f_c(1|x) \times g_s(x)$, quantifies the chance that those individuals picked at random from community $s$, (which happens to have the chance-of-being-freckled characteristics that pertain to community $c$) will be both freckled and age $x$.

To see precisely what goal is accomplished by direct and indirect adjusted rates, consider the area beneath the two curves $f_c(1|x) \times g_c(x)$ and $f_s(1|x) \times g_s(x)$, where $g_c(x)$ is the population-pyramid-edge-curve counterpart for community $c$ of community $s$'s curve $g_s(x)$. The values assumed by $f_c(1|x) \times g_c(x)$ and $f_s(1|x) \times g_s(x)$ both equal the values assumed by joint densities. Each value assumed by the curve $f_c(1|x) \times g_c(x) \equiv h_c(1, x)$ is proportionate to the probability of picking someone at random from community $c$ who is both freckled and $x$ years old. Correspondingly, $f_s(1|x) \times g_s(x) \equiv h_s(1, x)$ is the joint density that describes the bivariate distribution of two variates associated with the standard community $s$. Consequently, if the two $h$-curves properties,

$$h_s(1) = \int_0^\infty h_s(1,x)dx \quad \text{and} \quad h_c(1) = \int_0^\infty h_c(1,x)dx$$

were somehow computed with perfect accuracy, then these properties would equal the true "crude" freckled rates of communities $s$ and $c$, respectively.

Theoretical crude rates, such as hs(1) and hc(1), are to some pair of age-specific rates what some pair of marginal densities is to some pair of conditional densities. Therefore, neither $h_s(1)$ nor $h_c(1)$ is specific to some given age or adjusted for age. Instead, were the population pyramids considerably different for communities s and c, then a difference between $h_s(1)$ and $h_c(1)$ could be attributable to community age distribution differences rather than to strictly biological considerations.

Suppose that the two curve properties $h_{s,c}(1)$ and $h_{c,s}(1)$ are defined as

$$h_{c,s}(1) = \int_0^\infty f_c(1|x)g_s(x)dx \quad \text{and} \quad h_{s,c}(1) = \int_0^\infty f_s(1|x)g_c(x)dx.$$

Unlike $h_c(1)$ and $h_s(1)$, the two properties $h_{c,s}(1)$ and $h_{s,c}(1)$ are not merely marginal density curve evaluations. Instead, they are values determined by the direct adjusted rate, and the indirect adjusted rate denominator, respectively.

The quantity $h_{c,s}(1)$ discloses the aggregate effect produced by applying the rates that pertain to community c to modify community s's data. In this way, a community's assessment that is based on this curve property can be adjusted for the nuisance variable age. The fraction $[h_s(1) \times h_c(1)]/h_{s,c}(1)$, the curve property estimated by the indirect age-adjusted rate, accomplishes a similar purpose. While the purpose served by $h_{c,s}(1)$ is fairly obvious, the interpretation of $h_s(1) \times h_c(1)/h_{s,c}(1)$ is more complex.

Consider, for example, the identity

$$h_s(1) \times h_c(1) = \int_0^\infty f_s(1|x)\,g_s(x)dx \times \int_0^\infty f_c(1|x)g_c(x)\,dx.$$

Within the above two integrals, the expression

$$\int_0^\infty f_c(1|x)\,g_s(x)\,dx \int_0^\infty f_s(1|x)\,g_c(x)\,dx$$

is obtained when the four symbols: s; s; c; and c, are permuted to form the sequence: c; s; s; and c. This new expression is the product of the curve property estimated by a direct adjusted rate, times the denominator targeted by the indirect adjusted rate.

Were it permissible to apply

$$\int_0^\infty f_c(1|x)\,g_s(x)\,dx \int_0^\infty f_s(1|x)\,g_c(x)\,dx$$

## 11.5. Direct and indirect adjustment

in place of

$$\int_0^\infty f_s(1|x)\, g_s(x)\, dx \quad \int_0^\infty f_c(1|x)\, g_c(x)\, dx,$$

it would follow that

$$h_s(1) \times h_c(1) = \int_0^\infty f_c(1|x)\, g_s(x)\, dx \quad \int_0^\infty f_s(1|x)\, g_c(x)\, dx.$$

It would then follow that

$$\int_0^\infty f_c(1|x)\, g_s(x)\, dx = h_s(1) \times h_c(1) / \int_0^\infty f_s(1|x)\, g_c(x)\, dx.$$

Unfortunately, the left side (the curve property estimated by the direct age-adjusted freckled rate) is seldom equal to the right side (the curve property estimated by the indirect age-adjusted rate).

Because what is direct, in a mathematical or statistical sense, can run counter to what is practical, which often is the curve pairing, $g_c$ and $f_s$ within

$$\int_0^\infty f_s(1|x)\, g_c(x)\, dx,$$

it is advisable to obtain $g_c$ from the country or community whose rates are being standardized, and $f_s$ from a standard population. Direct and indirect age-adjusted freckled rate comparability depends on the distortion produced by the $s; s; c;$ and $c,$ to $c; s; s;$ and $c,$ permutation. Were: (1) the curves, $g_s$ and $g_c$, similar in shape or were; (2) the curves studied over regions over which they are approximated by horizontal line segments; then these curves could be interchanged with impunity. For reason (1) the United Nations has divided the countries of the world into homogeneous categories. For reason (2) it is advisable to adjust over only a short interval of age values. Within small age intervals the edges of many population pyramids can be fit fairly well by straight, nearly horizontal lines.

The indirect age-adjusted death rate often helps determine expected frequencies obtained by multiplying the size of a community's population by an estimator of

$$h_s(1) \times h_c(1) / \int_0^\infty f_s(1|x)\, g_c(x)\, dx,$$

the quantity which equals the number of deaths that would be expected were standard forces of mortality $f_s(1|x)$ applied to this population. For some community $c$ the symbols $e_c$ denote the product of an estimate of

$$h_s(1) \times h_c(1) / \int_0^\infty f_s(1|x) g_c(x) \, dx,$$

such as the statistic discussed in the next section, and $c$'s size, $\mathcal{P}_c$. Although the capitol letter $E$ represents a particular curve property, the small letter $e$ of $e_c$ actually represents a complex, but useful, specialized variate. Both a community's population size, and the statistic that estimates

$$h_s(1) \times h_c(1) / \int_0^\infty f_s(1|x) g_c(x) \, dx,$$

are themselves random variates and, as a consequence, $e_c$ is a random variate. Examples of this sort of variate will be discussed in the next chapter where, after transformation to yield approximately Normally distributed variates, *expected* morbidity rates are applied to adjust observed morbidity rates.

## 11.6 The computation of adjusted rates

Suppose $m(x : n)$ represents the number 1000 times the ratio $_nD_x/_nP_x$ that was defined in Section 11.1. (By convention, no hat is placed above the letter $m$, even though here $m$ represents a statistic, and not a curve or curve property.) It is handy to represent $_nP_x$ by the six in-line symbols $P(x : n)$, and use the two subscripts, $c$ and $s$, to characterize the sampled populations that $m(x : n)$ and $P(x : n)$ describe. For example, $m_c(x : n)$ and $P_c(x : n)$ describe some country, community, or perhaps county, $c$, while $m_s(x : n)$ and $P_s(x : n)$ describe some standard counterpart of $c$, represented by the letter $s$.

The adaptation of

$$\int_0^\infty m(y|x) g(x) dx$$

to estimate the indirect rate requires the explanation of one important detail. Suppose $\mathcal{D}_s$ and $\mathcal{P}_s$ respectively represent the total number of deaths, and the number of individuals at risk, within a standard population during

## 11.6. The computation of adjusted rates

time $t$, while $\mathcal{D}_c$ and $\mathcal{P}_c$ represent the country, community, or county counterparts of $\mathcal{D}_s$ and $\mathcal{P}_s$. Also suppose that the degree of abridgment of $K$ age intervals is the constant value n, and that the sequence; $x_1, x_2, \ldots, x_K$, designates the left endpoints of each interval. For example, $x_K$ equals $x_{K-1} + \text{n}$. Then the indirect age-adjusted death rate is often estimated by the statistic

$$IR = \left\{\frac{\mathcal{D}_s}{\mathcal{P}_s}\right\} \left\{\frac{\mathcal{D}_c}{\sum_{i=1}^{K} m_s(x_i\colon \text{n}) \mathrm{P}_c(x_i\colon \text{n})}\right\}$$

where, since $m(x : \text{n})$ represents the number 1000 times some ratio of deaths to population at risk, IR now represents 1000 times an adjusted ratio.

There may seem to be an omission from IR's formula. While terms, $\mathcal{D}_c, \mathcal{D}_s$, and $\mathcal{P}_s$ appear in this formula, the statistic $\mathcal{P}_c$ seems to be missing as a divisor of $\mathcal{D}_c$. The statistic $\mathcal{P}_c$ does not appear explicitly because of cancellation illustrated by the relationship

$$\left\{\frac{\mathcal{D}_s}{\mathcal{P}_s}\right\} \left\{\frac{\mathcal{D}_c/\mathcal{P}_c}{\sum_{i=1}^{K} m_s(x_i\colon \text{n}) \mathrm{P}_c(x_i\colon \text{n})/\mathcal{P}_c}\right\} = \left\{\frac{\mathcal{D}_s}{\mathcal{P}_s}\right\} \left\{\frac{\mathcal{D}_c}{\sum_{i=1}^{K} m_s(x_i\colon \text{n}) \mathrm{P}_c(x_i\colon \text{n})}\right\}.$$

However, the adjusted expected number of cases, $e_c$, is estimated by the indirect rate times $\mathcal{P}_c$, an estimator that equals

$$\left\{\frac{\mathcal{D}_s}{\mathcal{P}_s}\right\} \left\{\frac{\mathcal{D}_s}{\sum_{i=1}^{K} m_s(x_i\colon \text{n}) \mathrm{P}_c(x_i\colon \text{n})/\mathcal{P}_c}\right\}.$$

# CHAPTER 12
# Continuous Variate Adjustment

## 12.1 Observed and expected rates

Three's a crowd but, as far as a few types of continuous variates are concerned, there is a jostle-free passageway. However, before discussing the multivariate Normal route to partialling and adjustment, a method will be discussed that seems, often deceptively, to be an easy way to deal with nuisance variates.

In a study of environmental contamination, suppose $o_j$ cancer cases are diagnosed in the $j^{\text{th}}$ of 426 census tracts during some specific period of time. (The lowercase letter $o$ distinguishes this letter from the number 0.) To deal with one or more nuisance variates, it seems natural to divide each $o_j$ by its counterpart number of expected cases, $e_j$, where each $e_j$ is calculated by the indirect age-adjustment methods outlined in Section 11.5. Traditionally the logarithm of ratios like this are calculated. However, in many realistic applications, some ratio numerator, for example the component $o_j$ of $\log(o_j/e_j)$, assumes the value zero. This often happens when no cases of a disease occur during the period being studied.

For any $C > 0$, as $u$ approaches 0 from above, $\log(u/C)$ approaches $-\infty$. This motivates the choice of some positive constant $\hat{\tau}$ that is added to each ratio $o_j/e_j$ prior to logarithm computation. The addition of $\hat{\tau}$ seems to be merely an obvious computational expedient that does not merit the status of a well-thought-out threshold parameter estimator. However, to see why caution is definitely called for, consider that $\log([o_j/e_j] + \hat{\tau})$ equals $\log([o_j + \hat{\tau}e_j]/e_j)$ which, in turn, equals $\log(o_j + \hat{\tau}e_j) - \log(e_j)$. Yet applications of $\log(o_j + \hat{\tau}e_j) - \log(e_j)$ are motivated by a transformed estimate of the variate $o_j$ provided by $\log(e_j)$, not of the variate $o_j + \hat{\tau}e_j$. As a consequence, the inclusion of a constant within a log transform argument turns out to not be as simple as might be supposed at first glance.

175

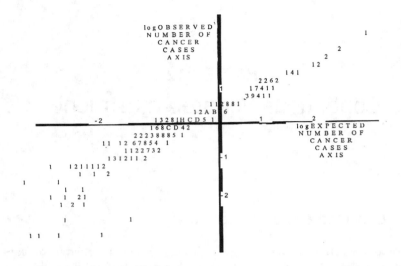

**Figure 12.1.** Log-observed and log-expected cancer frequencies.

Figure 12.1 illustrates the damage that can be done when sample values are over transformed by the selection of too small a $\hat{\tau}$. This *frequency diagram* describes $n = 426$ $Y_j = \log(o_j + \hat{\tau})$ variates plotted on the ordinate axis against the associated counts of $X_j = \log(e_j + \hat{\tau})$ on the abscissa (Tarter, Cooper, and Freeman, 1983). For example, the number 2 at the top right designates that two tracts had log-observed as well as log-expected number of cases both greater than to 2.

The choice $\hat{\tau} = 0.5$ yields a comet-shaped pattern with a broad tail preceded by a narrow head that points towards the upper right side. The more $\log(o_j + \hat{\tau})$ differs from $\log(e_j + \hat{\tau})$, the greater the distance between, on the one hand, the corresponding data point and, on the other, the diagonal, $y = x, 45°$ line that passes through the origin. Points at the lower left of Figure 12.1 help describe census tracts smaller than tracts described by points at the upper right. Thus the choice of $\hat{\tau}$ as too small a constant has overemphasized small tracts. Obviously, large tracts are more informative than small tracts and, too small a choice of $\hat{\tau}$ could have led to disastrous data-analysis consequences.

As described by Tarter and Hill (1997) and many other statistical papers, research on this topic has concentrated solely on the lognormal model, based on the assumption that variate $Y = \log(X - \tau)$ has a $N(Y; \mu, \sigma)$ density. Yet, as far as adjustment by the division of an observed count or rate by an expected count or rate is concerned, the problem is that ratios involve two quantities which, like $o_j$ and $e_j$, each require a separate value of the constant $\tau$. The next section will discuss partial correlation methods that can often circumvent difficult nuisance-variate-control issues that cannot be dealt with by methodology based on ratios and log ratios. However, as is true of most powerful tools, partial correlation is not applicable to every problem. Consequently Section 12.5 describes a curve-estimation-based alternative.

## 12.2 Trivariate data-generation and additive regression models

Even when a data transformation like the logarithmic function is unnecessary, in the trivariate simplest case there are alternative ways to interpret the individual symbols within the relationship

$$\rho_{12\bullet 3} = (\rho_{12} - \rho_{13}\rho_{23})/\{\sqrt{1-\rho_{13}^2}\sqrt{1-\rho_{23}^2}\}$$

that define the partial correlation $\rho_{12\bullet 3}$ in terms of unconditional correlations $\rho_{12}, \rho_{13}$, and $\rho_{23}$. For instance, each rho symbol has a geometric curve property interpretation that was illustrated by Figures 8.1 and 8.1. Alternatively, each can represent a parameter within a data-generation model, a ratio of a covariance to the square root of the product of two variance curve properties, or a parameter within a bivariate density model.

For example, the equations below transform independent continuous variates $Z_1, Z_2$, and $Z_3$ to possibly dependent variates $C_1, C_2$, and $C_3$. The letter $C$ is chosen to place emphasis on the assumption that each member of the $C_1, C_2$, and $C_3$ trio is a simulated continuous variate. These, and the $Z_1, Z_2$, and $Z_3$ variates, do not merely assume a "denumerable" sequence of values, as does a Poisson or any other discrete variate. Instead, $C_1, C_2$, and $C_3$ are variates with continuous cdf's where it is assumed that:

$$\begin{aligned} C_3 &= \mu_3 + \sigma_3 Z_3; \\ C_2 &= \mu_2 + \sigma_2\{\rho_{23}Z_3 + \sqrt{1-\rho_{23}^2}Z_2\}; \quad \text{and} \\ C_1 &= \mu_1 + \sigma_1\{\rho_{13}Z_3 + \sqrt{1-\rho_{13}^2}[\rho_{12\bullet 3}Z_2 + \sqrt{1-\rho_{12\bullet 3}^2}Z_1]\}. \end{aligned}$$

Dependent variates can be simulated in many ways. However, the advantage of the three equations above is that they demonstrate important features of partial correlation. Regarding correlation notation, since a comma symbol can be confused easily with the small font subscript one, 1, even though a comma is omitted, the symbol $\rho_{13}$ still designates the association parameter that relates the $1^{st}$ and the $3^{rd}$ variate, not some $13^{th}$

parameter. Two subscripted indices that designate the same sort of quantities, for example variates that are partialled, are often expressed without any separating symbol like a comma. However this convention cannot be adopted when subscripts refer to both partialled, and partialing, variates. As a consequence, the indices of partialing variates are set out after a period or a bullet, as in the notation $\rho_{12.3}$ or $\rho_{12\bullet 3}$.

The period of $\rho_{12.3}$ and the bullet of $\rho_{12\bullet 3}$ are often chosen as substitutes for the slash sign $|$. However, like a comma, a slash sign within a small font subscript can be mistaken for the number 1. Hence in this and later chapters, a large or small bullet that appears within a subscript is applied in place of the slash $|$ which denotes the word "given," as in $Y|x$, the variate $Y$ given the variable $x$. In some statistical writings, $Y|x$ is expressed as $Y_x$ or even $Y(x)$. As discussed in Section 2.2, a time series can be described by a sequence comprised of many variates whose values are usually equally-spaced evaluations of some function $Y(t)$, where $t$ designates time. Consequently, the notation $Y|x$ helps distinguish the left-hand side of a data-generation model from a function like $Y(t)$.

The variate $Y|x$'s value is related to the variable value $x$ while, similarly, here the variate $C_1$ is expressed in terms of other quantities, specifically $C_2$ and $C_3$. It is convenient to assume that $C_2$ represents a key variate that corresponds to the variable $x$, which follows the slash sign between the $Y$ and the $x$ of $Y|x$. On the other hand, $C_3$ is assumed to be a nuisance variate that is unmatched by any

$$Y|x = \alpha + \sum_{k=1}^{m} \beta_k (x - \bar{X})^k + \sigma Z$$

polynomial model counterpart.

The single variable, $m = 1$, special case of model component

$$\alpha + \sum_{k=1}^{m} \beta_k (x - \bar{X})^k,$$

often is generalized to form an additive multiple regression data-generating model like $Y = \alpha + \beta_1 x_1 + \beta_2 x_2 + \sigma Z$. For this generalization to provide a valid description of $Y$-variate generation, $\sigma Z$ must be shared by the two components $\beta_1 x_1$ and $\beta_2 x_2$. If $x_1$ and $x_2$ are researcher-selected variables, the sharing issue is moot; but it often is an issue when $x_1$ and $x_2$ are *variate* values. For example, suppose that for a particular choice of the sequence $\alpha_1, \beta_1, x_1, \alpha_2, \beta_2$, and $x_2$, the variate pair $(Y_1, Y_2)$ is defined in terms of the pair of standard independent Normal variates $(Z_1, Z_2)$ by the expressions $Y_1 = \alpha_1 + \beta_1 x_1 + \sigma_1 Z_1$ and $Y_2 = \alpha_2 + \beta_2 x_2 + \sigma_2 Z_2$. Then this suggests the question: Is the response variate $Y = Y_1 + Y_2$ a variate that could also have been simulated by calling a single standard Normal variate $Z$ into service within the data-generation model $Y = \alpha + \beta_1 x_1 + \beta_2 x_2 + \sigma Z$?

Any sum of Normal variates is a Normal variate. Thus, for $\alpha$ set equal to $\alpha_1 + \alpha_2$, and $\sigma = \sqrt{\sigma_1^2 + \sigma_2^2}$ the answer to the question above is yes. When "additive" data-generation model $Y = \alpha + \beta_1 x_1 + \beta_2 x_2 + \sigma Z$ is valid, for Normally-distributed residuals like $\sigma_1 Z_1$ and $\sigma_2 Z_2$, it makes no difference whether residuals from components like $\beta_1 x_1$ and $\beta_2 x_2$ are modeled individually, say as $\sigma_1 Z_1$ and $\sigma_2 Z_2$, or modeled collectively, say as $\sigma Z$. Yet when $Z_1$ or $Z_2$, or both $Z_1$ and $Z_2$, happen to be nonNormal variates, it can make a considerable difference whether residuals are modeled individually or collectively. For example, if $\beta_1 \neq 0$ and $\beta_2 \neq 0$, then when the variates $\beta_1' x_1 + \sigma_1 Z_1$ and $\beta_2' x_2 + \sigma_2 Z_2$ are Laplace, instead of Normally, distributed, it is not the case that $Y = \alpha + \beta_1 x_1 + \beta_2 x_2 + \sigma Z$, where $Y$ and $Z$ are Laplace variates. Similarly, were $Y_1 = \beta_1 x_1 + \sigma_1 Z_1$ a Normal variate and $Y_2 = \beta_2 x_2 + \sigma_2 Z_2$ a Cauchy variate, then no weighted sum of $Y_1$ and $Y_2$ with nonzero weights is either a Normal or a Cauchy variate. Hence it is wise to check on the Normality of any variates that supply the values of terms within often-assumed models like $Y = \alpha + \beta_1 x_1 + \beta_2 x_2 + \sigma Z$.

## 12.3  Regression and data generation

To compare the simple additive linear regression model to the second member of the data-generation model pair,

$$C_3 = \mu_1 + \sigma_1 Z_1, \quad C_2 = \mu_2 + \sigma_2(\rho Z_1 + \sqrt{1-\rho^2} Z_2),$$

suppose the upper summation limit $m$ of

$$\alpha + \sum_{k=1}^{m} \beta_k (x - \bar{X})^k$$

equals 1, and $(x - \bar{X})$ equals $C_3$. Then $Y|x$ equals

$$\alpha + \beta C_3 + \sigma Z = \alpha + \beta(\mu_1 + \sigma_1 Z_1) + \sigma Z = [\alpha + \beta \mu_1] + \beta \sigma_1 Z_1 + \sigma Z.$$

It follows that were $C_2 = Y$, then $\sigma_2 \rho = \beta \sigma_1$ and for any $\sigma_1 \neq 0$, the parameter $\beta$ equals $\rho \sigma_2 / \sigma_1$. The $\beta = \rho \sigma_2 / \sigma_1$ formula expresses the linear regression parameter (sometimes called a coefficient), $\beta$ in terms of the association parameter $\rho$.

Despite the existence of the formula $\beta = \rho \sigma_2 / \sigma_1$ that links regression with correlation in the above simple special case, models that define $Y|x$

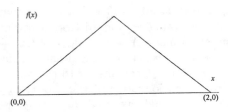

Figure 12.2. Triangular density.

and models that define $C_2$ can differ substantially. One reason why they differ was discussed in the previous section in terms of questions concerning the term $(\rho Z_1 + \sqrt{1-\rho^2} Z_2)$ of

$$C_2 = \mu_2 + \sigma_2(\rho Z_1 + \sqrt{1-\rho^2} Z_2).$$

Suppose, for example, that two variates are independently distributed and both have a uniform distribution with density $I_{[0,1)}(x)$. Their sum will have the isosceles triangular density shown in Figure 12.2 with vertices located at $(0, 0)$, $(2, 0)$ and $(1, 1/2)$. Thus, for the discrete outcomes of dice and for the triangular variate-sum density of Figure 12.2, when the $Z_1$ and $Z_2$ components of the composite variate $\sigma_2(\rho Z_1 + \sqrt{1-\rho^2} Z_2)$ are uniformly distributed and when, in addition, $\rho \neq 0$, the composite variate $\sigma_2(\rho Z_1 + \sqrt{1-\rho^2} Z_2)$ has a density that differs substantially from the densities of the components $Z_1$ and $Z_2$.

The seemingly simple weighted sum $\sigma_2(\rho Z_1 + \sqrt{1-\rho^2} Z_2)$ of nonNormal $Z_1$ and $Z_2$ usually has a distribution described by a density that must be modeled as an elementary function that differs in structure from any elementary function model for the densities of the $Z_1$ and $Z_2$. It is therefore not surprising that regression approaches are often based exclusively on variables treated as constants and, as a consequence, on quantities that are assumed to differ from the sort of variate values which complicate non-Normal $\sigma_2(\rho Z_1 + \sqrt{1-\rho^2} Z_2)$.

## 12.4 Correlation, regression, and nuisance variables

In many applications it is neither the $C_1$ response variate, the $C_2$ key variate nor the $C_3$ nuisance variate that can correctly be assumed to be generated by equations in the component $Z_1, Z_2, Z_3$ variates (Kowalski, 1970). Instead it may be some functions of the response, key, and/or nuisance

## 12.4. Correlation, regression, and nuisance variables

variates that can be determined by these equations. Perhaps, for example, it is the logarithms of the observed variates, and not their untransformed values, that can be taken to be generated from $Z_1, Z_2,$ and $Z_3$.

Suppose however, for the sake of simplicity, that a preliminary data transformation is not required to assure that the observed $C_1, C_2,$ and $C_3$ are multivariate Normal variates. This Normality assumption will allow us to focus on the correlation-based and regression-based simulation models without the distraction of side issues such as transformations to Normality.

Kendall and Stuart (1979) call parameters such as $\rho$ "..coefficients of interdependence..". They selected the term "interdependence" to distinguish these quantities from $\beta$ parameters, quantities that determine dependence. Unfortunately, Kendall and Stuart's choice of terminology notwithstanding, the distinction between interdependence and dependence can often be blurred.

Consider the parameter $\rho_{12\bullet 3}$ within data generating model component

$$\rho_{12\bullet 3} Z_2 + \sqrt{1 - \rho_{12\bullet 3}^2} Z_1,$$

where $Z_2$ is the key variate component, lacking nuisance variate $Z_3$'s influence on the response variate's value. Consequently, $\rho_{12\bullet 3}$ assesses the part of the key variate's influence upon the response variate $C_1$, attributable solely to the key variate independent component $Z_2$. Component $Z_2$'s important conceptual advantage is that it is statistically independent of the nuisance variate. On the other hand, $\sqrt{1 - \rho_{12\bullet 3}^2}$ weights the contribution of component $Z_1$. The $Z_1$ variate is taken to have nothing whatsoever to do either with key variate $Z_2$ or with nuisance variate $Z_3$. In some applications, $Z_2$ is thought of as a variate that is free of nuisance variate contamination, while $Z_1$ is thought of as the statistical equivalent of extraneous noise. How much information $Z_2$ conveys about $C_1$ is measured by $\rho_{12\bullet 3}$ without $Z_3$'s or, equivalently, nuisance variate $C_3$'s, interference.

The component $Z_1$ within the expression

$$\rho_{12\bullet 3} Z_2 + \sqrt{1 - \rho_{12\bullet 3}^2} Z_1,$$

is similar to the term $Z$ of $\alpha + \beta_1 x_1 + \beta_2 x_2 + \sigma Z$. However, $Z_2$ and $Z_3$ differ from the pair of variables $x_1$ and $x_2$ in a way that affects their interpretation greatly. This difference is not strictly due to $x_1$ and $x_2$'s roles as values of variables, not of variates. Instead, it arises from an overall distinction between the three quantities; $Z_1, Z_2,$ and $Z_3$; on the one hand, and the three quantities; $C_1, C_2,$ and $C_3$; on the other.

In order to focus upon association parameter issues, suppose scale parameters are set equal to 1, and the $x_1$ and $x_2$ variables' values are obtained by first setting $x_1$ equal to a $C_2$ key variate value, and then setting $x_2$ equal to a corresponding $C_3$ nuisance variate. For example, $C_1, C_2$, and $C_3$ could be taken, collectively, to be trivariate-Normal distributed with a density that is represented in an abbreviated form as $N(c_1, c_2, c_3)$ where, when expressed explicitly, the formula for $N(c_1, c_2, c_3)$ contains among its parameters the two parameters $\rho_{13}$ and $\rho_{23}$.

Here the subscripts of $\rho_{13}$ and $\rho_{23}$ correspond to the indices of the variates $C_1, C_2$, and $C_3$. Also it is worth pointing out that here, two trivariate curve properties, $\tilde{\rho}_{13}$ and $\tilde{\rho}_{23}$, called *product moment-based correlation coefficients* or *Pearson correlations*, are known to take on values equal to the parameters $\rho_{13}$ and $\rho_{23}$, respectively, within the formula for $N(c_1, c_2, c_3)$.

The curve properties $\tilde{\rho}_{13}$ and $\tilde{\rho}_{23}$, and the corresponding data-generation model components $\rho_{13}$ and $\rho_{23}$, quantify the extent to which those variates whose indices appear in the subscripts $_{12}$ and $_{23}$, are associated. In addition, $\tilde{\rho}_{12}$ designates the particular correlation curve-property of the bivariate Normal density that corresponds to the association parameter $\rho_{12}$ within both the data-generation model term $\rho_{12} Z_2 + \sqrt{1 - \rho_{12}^2} Z_1$, and within symbolic representations of the bivariate Normal pdf. The parameters $\rho_{13}$ and $\rho_{23}$ fulfill similar roles in the trivariate case for the first and third, and the second and third, variates, respectively.

There is a perfect match between the data-generation model parameters, like $\rho_{12}$, and Normal curve properties, like $\tilde{\rho}_{12}$. However, when variates are generated from nonNormal $Z_2$ and $Z_1$, it is certainly not obvious that some correlation curve property corresponds to parameter $\rho_{12}$ of $\rho_{12} Z_2 + \sqrt{1 - \rho_{12}^2} Z_1$. For example, for $\rho_{12} \neq 0$ consider a pair of independent Cauchy variates $Z_1$ and $Z_2$, where $C_3 = Z_1$ and

$$C_2 = \rho_{12} Z_2 + \sqrt{1 - \rho_{12}^2} Z_1.$$

Then even if the covariance curve property of $(C_3, C_2)$'s joint density equals some finite constant, since the variances of $C_3$ and $C_2$ are infinite it follows that $C_3$ and $C_2$'s correlation would equal zero even though $\rho_{12}$ is defined to be nonzero. This mismatch between parameters and properties is not surprising. After all, why should it always be possible to abstract all the information contained in a bivariate density curve to one number, for example a correlation coefficient, that encapsulates variate association or relationship?

Besides the perfect match-up between both the data-generation as well as the density model parameter $\rho_{12}$ on the one hand, and curve property $\tilde{\rho}_{12}$ on the other, the Normal $f$-type model has several other useful characteristics. For instance, the model for the conditional density $N(c_1, c_2|c_3)$ can be expressed in terms of an association parameter constant $\rho_{12 \bullet 3}$ where

## 12.4. Correlation, regression, and nuisance variables

here, irrespective of the coordinate $c_3$ at which the trivariate is sliced, the parameter $\rho_{12\bullet 3}$ assumes the value also assumed by the correlation-type curve property of density $N(c_1, c_2|c_3)$. In addition, for independent standard Normal $Z_1, Z_2$, and $Z_3$ variates, the parameter $\rho_{12\bullet 3}$ that appears within the explicit model for $N(c_1, c_2|c_3)$ equals the data-generation model component $\rho_{12\bullet 3}$ within

$$C_1 = \mu_1 + \sigma_1\{\rho_{13}Z_3 + \sqrt{1 - \rho_{13}^2}[\rho_{12\bullet 3}Z_2 + \sqrt{1 - \rho_{12\bullet 3}^2}Z_1]\}.$$

The conditional density $N(c_1, c_2|c_3)$ is identical to the trivariate density $N(c_1, c_2, c_3)$ divided by the marginal density $N(c_3)$ of the nuisance variate $C_3$. In the trivariate Normal case, the numerical value assumed by $\rho_{12\bullet 3}$ is unrelated to the actual value $c_3$ assumed by $C_3$. This convenient property of the Normal underlies the potential for correlation parameters to encapsulate information concerning the relationships between variates. Since the value assumed by the $c_3$ of $N(c_1, c_2|c_3)$ determines where the distribution's density is sliced, the parameter $\rho_{12\bullet 3}$'s lack of relationship to $C_3$'s contribution can be viewed in terms of density slices. Specifically, when $\rho_{12\bullet 3}$ does not change when $C_3$ changes, then it follows that no matter where—perpendicular to the $c_3$-axis—the density $N(c_1, c_2, c_3)$ is sliced, the parameter $\rho_{12\bullet 3}$, within the explicit conditional Normal model for the density $N(c_1, c_2|c_3)$, remains the same. Thus, as was true of its parametric interpretation in terms of data generation model component $\rho_{12\bullet 3}Z_2 + \sqrt{1 - \rho_{12\bullet 3}^2}Z_1$, when $\rho_{12\bullet 3}$ designates a property of the conditional Normal density model it measures the response-with-key-variate relationship as a constant that is unaffected by the influence of the nuisance variate $C_3$.

Even when Normal $N(c_1, c_2, c_3)$ scale parameters are set equal to 1 and all scale parameters within the three equations that define $C_1, C_2$, and $C_3$ are also set equal to 1, the parameter $\rho_{12\bullet 3}$ still does not usually correspond to the parameter $\beta_1$ within $\alpha + \beta_1 x_1 + \beta_2 x_2 + \sigma Z$. For example, suppose a pair of Normal variates, say $C_2$ and $C_3$, is chosen to simulate the values of $X_1$ and $X_2$ within $Y = \alpha + \beta_1 X_1 + \beta_2 X_2 + \sigma Z$. (Since $X_1$ and $X_2$ are taken to be variates, they are denoted by uppercase letters.) Suppose also that $Z, X_1$, and $X_2$ are generated in a way that assures that $Y, X_1$, and $X_2$ have a trivariate Normal distribution, and let $\rho_{13}$ represent the Pearson correlation of the variates $Y$ and $C_3$, while $\rho_{23}$ denotes the correlation between $C_2$ and $C_3$. Then regression coefficient $\beta_1$ equals $\rho_{12\bullet 3}\sqrt{1 - \rho_{13}^2}/\sqrt{1 - \rho_{23}^2}$.

The above relationship between one regression coefficient and the partial correlation parameter shows that the identity $\beta_1 \equiv \rho_{12\bullet 3}$ is valid only if $|\rho_{13}| = |\rho_{23}|$. When the absolute value of the correlation between the response and the nuisance variates equals the absolute value of the correlation between the key and the nuisance variates, then and only then is the regression parameter $\beta_1$ equal to the association parameter $\rho_{12\bullet 3}$.

The different roles played by $\beta_1$ and $\rho_{12\bullet 3}$ can be interpreted in terms of the three data generation equations from which $C_1, C_2$, and $C_3$ are obtained from simulated variates $Z_1, Z_2$, and $Z_3$. The crucial distinction between the $C$-type and the $Z$-type variates is that the former are observed, and usually interrelated variates, while the latter are taken to have been generated as independent, and hence unrelated variates. It follows that whenever it is important to assess the portion of the key variate's contribution that is unattributable to the nuisance variate's effect on the response variate, for nuisance variate partialling purposes the parameter $\rho_{12\bullet 3}$ is preferable to the regression parameter $\beta_1$. Indeed, one parameter will be zero when the other is zero. Yet the value assumed by a nonzero regression coefficient is affected by association parameters other than $\rho_{12\bullet 3}$.

One way of viewing the distinction between the partial correlation and $\beta_1$, is to view $\beta_1$ as the coefficient of the possibly contaminated key variate $X_1$. Unlike its partial correlation counterpart, the parameter $\beta_1$ provides doubtful protection against possible spurious association. The response variate is related to (1) the key or/and, (2) the nuisance variate's contribution. Hence, as pointed out by Kendall and Stuart (1979), a single regression coefficient like $\beta_1$ discloses the contribution of one member of a sequence of variables to the dependence of this response variate on the sequence as a whole. The coefficient $\beta_1$ does not disclose the interdependence between response, key, and nuisance variates.

When interest is focused on the influence exerted by multiple key variates on a response variate, two or more association parameters, such as $\rho_{12\bullet 34}$ and $\rho_{13\bullet 24}$, are often compared. For example, $\rho_{12\bullet 34}$ might assess the response variate's association with key variate $C_2$, free from the influence of key variate $C_3$ and nuisance variate $C_4$. Alternatively, $\rho_{13\bullet 24}$ might assess the response variate's association with key variate $C_3$, free from, the influence of key variate $C_2$ and nuisance variate $C_4$. Each of two key variates $C_2$ and $C_3$ is taken, in turn, as a nuisance variate alongside nuisance variate $C_4$.

When more than three indices characterize some parameter $\rho$, interdependence patterns become highly complex. The trivariate relationship $\beta_1 = \rho_{12\bullet 3}\sqrt{1-\rho_{13}^2}/\sqrt{1-\rho_{23}^2}$, after all, concerns merely one key variate and one nuisance variate. Consequently, caution is called for when the two parameters $\beta_1$ and $\beta_2$ of $\alpha + \beta_1 x_1 + \beta_2 x_2 + \sigma Z$ are compared in order to assess which, among candidate key variables, has the most effect on a response variate. Regression limitations do not stem entirely from the estimation procedures that determine what the values assumed by parameters such as $\beta_1$ and $\beta_2$, are, nor from differences in key and other variate scale parameter values. There simply is a right way and a wrong way to compare key variates or variables. Partial correlation approaches are right and regression approaches based solely on a comparison of two or more regression slope $\beta$ parameters are wrong unless, of course, they are adapted to yield results that are identical to those that could be obtained by partial correlation methodology.

## 12.5 Trivariate Normality graphics

There are so many ways to check on bivariate Normality that it is reckless to apply correlation techniques without first applying at least one of the techniques surveyed by Kowalski (1970). In addition, by using modern nonparametric curve estimation techniques, the trivariate Normality assumption upon which partial correlation's validity is based can be checked graphically. For example, the simulated data experiment described in this section illustrates the implications of full trivariate interrelationships that cannot be dealt with completely by $e_j$-type adjustment. Based on parameter choices $\rho_{12\bullet 3} = \mu_1 = \mu_2 = \mu_3 = 0$, $\sigma_1 = \sigma_2 = \sigma_3 = 1$, $\rho_{23} = 0.75$, and $\rho_{13} = -0.75$, suppose 1000 trivariate Normal variates are generated. Since

$$\rho_{12\bullet 3} = (\rho_{12} - \rho_{13}\rho_{23})/\{\sqrt{1 - \rho_{13}^2}\sqrt{1 - \rho_{23}^2}\},$$

the assumption $\rho_{12\bullet 3} = 0$ implies that $\rho_{12} - \rho_{13}\rho_{23} = 0$ and therefore $\rho_{12} = \rho_{13}\rho_{23} = -0.5625$.

This nonzero value of the parameter $\rho_{12}$ can be conceptualized as a twist placed in a trivariate distribution by variate interrelationships. On the other hand, the partial correlation $\rho_{12\bullet 3} = 0$ can be thought of as the untwisted parameter that discloses key with response relationship freed from the distortion produced by the nuisance variate.

A modern counterpart of the method of averages (MA) that was discussed in Section 10.7 can help untwist variate interrelationships without requiring trivariate Normality. Towards this end, Tarter and Lock (1993, Chapter 8) show that the general statistical approach known as maximum likelihood estimation can be blended with the MA to yield a wide variety of regression procedures. An advantage of this approach is that unlike Gauss-Markoff-methodology based on

$$E\{[Y - (\alpha + \beta x)]^2\}$$

(where $\mu_x$ is constrained to equal $\alpha + \beta x$ or to equal some user-supplied generalization of $\alpha + \beta x$), in the sense that it can conform to any path traced by $\mu_x$ as $x$ changes, the MA counterpart of $\mu_x$ is nonparametric.

The solid curve of Figure 12.3(a) was based on an estimator of $N(c_1, c_2|1)$, while the corresponding response-variate-with-key-variate least squares determined line is shown as the broken curve. These curves, and corresponding curves based on estimators of $N(c_1, c_2|-1)$ and $N(c_1, c_2|0)$ that are shown in Figures 12.3(b) and 12.3(c) were obtained from the data samples above (the $\rho_{12\bullet 3} = \mu_1 = \mu_2 = \mu_3 = 0$, $\sigma_1 = \sigma_2 = \sigma_3 = 1$, $\rho_{23} = 0.75$, and $\rho_{13} = -0.75$ sample) superimposed on a scatter diagram that depicts response variate $C_1$'s and key variate $C_2$'s coordinates. The three solid curves were obtained from the four-dimensional surface formed by plotting the estimated frequency as a function of variate values $c_1, c_2$, and $c_3$, and then slicing this surface perpendicular to the $c_3$ axis, through the three values; $c_3 = 1, -1$, and $0$; respectively.

Because of its flexibility, a nonparametric curve lacks the stiffness of a simple model like $\mu_x = \alpha + \beta x$. Consequently, it tends to be particularly accurate where data are tightly packed. This explains why, unlike the solid curve shown in Figure 12.3(b), for the particular sample under study, the curves shown in Figures 12.3(a) and 12.3(c) are almost perfectly horizontal near their centers but show pronounced slopes at their edges.

Figures 12.3(d) and 12.4 show the same conventional least-squares-fitted straight line also shown in the previous 3 figures, together with the nonparametric MA-fitted curve now determined from bivariate, as compared to the previous trivariate, data. When the nuisance variate values play no role in the estimation of the MA curve, the new and the conventional procedures provide similar estimators. They are so similar, in fact, that to see the difference between the parametric and nonparametric curves one must enlarge the image shown in Figure 12.3(d) between plus and minus one on the $x$- and $y$-axes to obtain Figure 12.4. As was true in the three conditional cases, the nonparametric curve that is in accord with the curve predicted by parameter values provides a validity check.

## 12.5. Trivariate Normality graphics

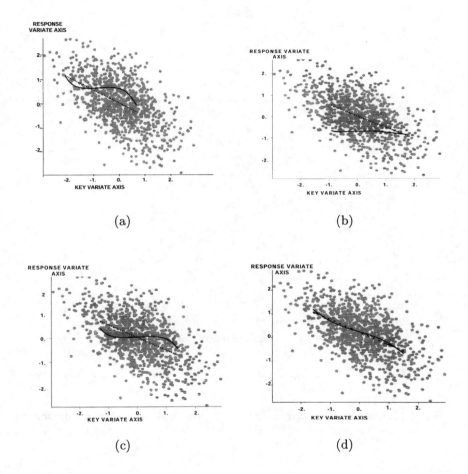

**Figure 12.3.** Scatter diagram and parametric as well as nonparametric curves based on: (a) $N(c_1, c_2|1)$, (b) $N(c_1, c_2|-1)$, (c) $N(c_1, c_2|0)$, (d) $N(c_1, c_2)$.

188                                                    12. Continuous Variate Adjustment

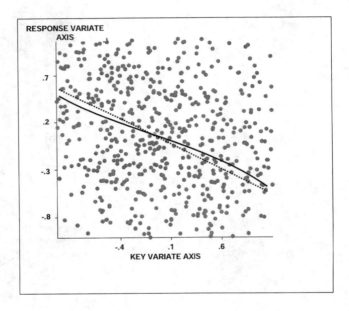

**Figure 12.4.** Zoomed Scatter diagram and parametric as well as nonparametric curves based on $N(c_1, c_2)$.

# CHAPTER 13
# Procedural Road Maps

## 13.1 The organization of statistical data and statistical methods

Whether data are Normal or Cauchy-distributed, and whether a regression curve is a line or some other curve, can be referred to as high-order uncertainties. Beneath this level there are other underlying decision layers that are basic to data-analysis success. In general, these have more to do with the cdf curve type than the shape of some $f$-type curve. This section and the next trace the historical trend away from specific models and towards more basic structures. Sections 13.3 and 13.4 then proceed to suggestions about these structures.

If there has been any formal attempt to spell out where the borders between different statistical uncertainties are, word of this attempt has not reached this author. However, one thing is certain. Where these borders presently are was determined during the pre-computer era. Hence many key concepts were formulated at a time when there was no practical way to try out and evaluate the scope of pivotal yet simple ideas.

For example, before the 1960s many statistics texts recommended a procedure called *Yates' correction to a chi-square table*. The theoretical reasoning that underlies this modification of the basic chi-square statistical tool was universally accepted until, thanks to modern computation and simulation methodology, careful computer-based investigations by Grizzle (1967) showed that Yates' correction often does more harm than good.

In the sense that they dealt with variates that are attribute-valued, Grizzle's studies dealt with a specialized, although commonly applied, statistical procedure. While a binary or dichotomously-valued variate indexes two subgroups, an attribute-valued variate such as race or country of origin indexes two or more subgroups. As often encountered as binary with binary associations are, Grizzle's work was merely a scratch on a surface whose extensive scope has yet to be surveyed comprehensively.

As is the case for any form of new media, alongside the benefits of enhanced computation, there is a downside. The downside has affected the division of statistical subject matter. When a new way to repackage old ideas, old music, or any form of literature is invented, a backlog is typically ready to be dusted off and re-presented. However, it is not always the case, as it was with Yates' correction, that new media and machinery are applied in a timely fashion to reassess and reaffirm validity. Quite the contrary, old war horses are trotted out of their stables no matter what their capacity to carry modern burdens.

The box plot and other graphical and computational tools that are discussed in later chapters are accepted currently as vital components of any statistical investigation. They can now help assess the weight of modern burdens in the light of older techniques' carrying capacity. Yet these easy-to-apply tools did not become readily available overnight. Instead, in its early forms the digital computer placed heavy demands on the mathematical and logical skills of its users. For example, pull-down menus only became available at least thirty years after the first commercial statistical program package appeared on the market.

Hence, even granting the questionable premise that during the early computer era the number of available quantitatively-adept individuals increased, a considerable proportion of these individuals' efforts was absorbed by computational as opposed to statistical details. So much time has been spent implementing and then applying pre-computer era statistics by means of post-computer machinery, that there has been little residual time left to carry out studies such as those which showed that Yates' correction is of questionable value. Hence the validity checking capacities of computer-based graphical procedures like the box plot have not kept pace with the pouring of old statistical wine into new and convenient to pour bottles. It is because of this imbalance that the hypothesis-test-sampling-distribution concepts that are taken to be introductory material in many pre-computer texts are placed as later chapters in this monograph. This author believes that sampling distribution-based considerations become relevant only after more basic uncertainties are addressed fully.

Statistical foundation excavation has been accelerated by representational generality. With the help of tools and props provided by new computational devices, incomplete guesses about curve structure and data-generation processes were dug out and replaced gradually by metrics and criteria posed from the subjective viewpoint of the researcher. What nature did was left for the data to decide. Where the researcher wanted the most precision, and what he or she stood to lose by a wrong or an inaccurate answer, were spelled out in detail.

One example of the above change of emphasis is a book by Devroye and

Gyorfi (1985) entitled *Nonparametric Density Estimation: the $L_1$ View*. The word "view" designates the *viewpoint* of the researcher that the absolute value of curve $f$'s estimation error, as in $|f - \hat{f}|$, should be considered alongside some alternative measure of goodness of fit, such as $|f - \hat{f}|^2$ which, in turn, is called the "squared error metric." Here $|f - \hat{f}|^2$ is associated closely with the basic curve estimation metric, the mean integrated square error (MISE), whose modifications and interpretations were elaborated upon in Section 5.2.

The criterion $|f - \hat{f}|^2$ would be designated "the $L_2$ view." Neither $L_1$, nor $L_2$, metrical views are descriptions—with blanks left to be filled in—of what a curve's expression is like. Rather, they describe how inaccuracies affect the curve estimation process. If the estimation process were to be compared to a race, the selection of $L_1$ resembles the rules under which some contestant, for example the contestant represented by the symbol $\hat{f}$, is to be judged. On the other hand, the selection of a parametric model is analogous to a head start in the form of some assumed elementary function model within which unspecified parameters are to be determined from data.

The phase of the problem-solving process called *estimation* by Fisher was concerned exclusively with these unspecified parameters. Fisher admonished statisticians to restrict their efforts to this problem solving phase, and leave the choice of head start to investigators with specialized knowledge of the subject matter under study. By what could hardly be treated as a coincidence he was the major proponent of the highly effective estimation tool known as the method of *maximum likelihood* (ML). Hence, according to Tapia and Thompson (1978), Fisher was really saying that his colleagues should leave to other, non-statistically oriented scientists, problems that were then beyond the scope of Fisher's own newly devised methodology.

It is not so much that Fisher was wrong. In the era that stretched from the 1920s until the end of the Second World War, computational and conceptual machinery was not sufficiently developed to distinguish alternative opening gambits to what often seemed to be some lengthy game. Only recently, thanks to the contributions of Gaskins, Good, Tapia, Thompson, and many others, has ML been extended in the form of penalized likelihood to deal with representational forms that do not require the head start granted by Fisher. What is specified is not so much a limited guess of what nature has or has not done. Instead, it is the assumption of a roughness penalty, metric, or decision criterion that balances representational complexity on the one side against data availability on the other.

A survey of parametric approaches resembles the critical evaluation of a game such as chess. The chess endgame—where few pieces are left on the board—and the beginning moves called opening gambits, are easier to evaluate than are the intermediate states of play. When Fisher admonished his statistical contemporaries to leave specification issues to subject matter specialists, he restricted attention to the statistical endgame. What are sometimes called *infinite parameter space* methodologies are therefore not mere extensions of finite parameter space methods. Rather, they serve to shift attention back to the statistical game's starting positions or opening gambits, if you will.

Whether or not one particular shift or extension is better or worse than any alternative modern approach is less relevant than is the circumvention of the requirement that statistical investigations be divided between estimation and specification phases. Only in the pre- computer era was it necessary to rely on the algebraic manipulation of the handful of symbols that were bases for the specification of statistical curve's representation. A key component of the specification problem—the part that ascertained which particular handful of symbols should be used—was rendered moot once the power of modern forms of computation was paired with forms of curve representation that were not limited guesses.

Generalized representation did not, itself, free statistics from the cloistering effects of model-based exclusivity. The direct application of Fisher's maximum likelihood methodology to the general forms of curve representation, such as Fourier series that are elaborated upon in the next chapter, was found in the 1950s and 1960s to overfit data. For example, when a density curve represented as a Fourier series is estimated by means of Fisher's unpenalized maximum likelihood methodology, the resulting curve is zero between any two adjacent ordered sample members. In other words, this asserts that it is impossible for a variate to attain any value different from already observed values.

When unpenalized likelihood is applied to an iid sample of discrete data points, and an unweighted Fourier series represents each individual point's density, all true knowledge is treated as being within the sample. The trouble is, this seemingly reasonable assertion suggests that the occurrence of anything between and beyond sample values is impossible.

When constants, called *multipliers*, that modify maximum likelihood estimators of individual Fourier coefficients, are assessed on the basis of the $L_2$ metric, the estimators that result (as well as their wavelet counterparts) have many applications. Both penalized likelihood, as well as ordinary likelihood called upon in conjunction with the $L_2$ metric, work well with representational forms that bypass the problem of specification. Yet, the development of these nonparametric approaches had to await the

## 13.2. Log and log(-log) transformations

computational machinery of the 1960s and 1970s. Among nonparametric techniques, only those that were graphical and rank-based were practical prior to the 1960s.

Because the computer is a new factor in the totality of statistical history, the model-based literature is vastly more comprehensive than is the nonparametric literature. This is true despite pre-computer era statistical investigations of a curve representation, the Gram Charlier Edgeworth (GCE) series described briefly in Appendix (I.4), that bears a striking resemblance to today's kernel and Fourier series approaches. However, GCE did not make use of penalized likelihood, an $L_1, L_2$, or indeed, any metric. Because it applied the same estimation procedures that had been applied to simple parametric models, and consequently could not take a metric-based point of view, its attempts at generality failed.

Despite the GCE dead-end, many elementary function models of the sort studied by Fisher are still appreciated greatly today. Once the question of model selection was disentangled from the problem of computational tractability, it was not long before many different models were tried as alternative curve-based descriptions of data samples.

When tractability ceased to be an insurmountable data-analysis barrier, many novel statistical approaches became practical. Yet, with a forty year methodological backlog, it is not surprising that much sorting out was required. Such sorting has been streamlined by an object oriented trend that grouped statistical procedures in terms of modules. For example, simulation methodology was devised by means of which it was easy to interchange files of simulated data that have different distributional characteristics. This interchange capability facilitates performance checks of techniques that are based on one set of assumptions when applied with data that satisfies a different set of assumptions.

As far as representational tools are concerned, thanks in large part to the digital computer, different alternative expressions have been proposed which describe variate structure, and the general shape of the curves that illustrate the distribution of variates. Studies conducted in the mid 1960s began an era in which the following question could be answered: How is one to pick from among a variety of alternative models, expressions, and procedures whose validity depends on these expressions?

## 13.2 Log and log(-log) transformations

Many examples illustrate how hard it is to select the proper elementary function curve representation. Consider the following three alternatives: (1) the simple assumption that a variate has a Normal density $\phi[(x - \mu)/\sigma]/\sigma$; (2) the assumption that data are Normally distributed after points

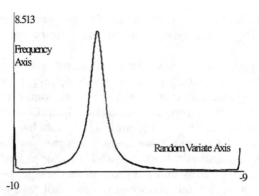

Figure 13.1. High central mode loglog Cauchy.

are transformed by the application of a log transformation; and (3) the assumption that data are Normally distributed after some log(-log) transformation. Alternative (1) implies that the iid variates in question have cdf $\Phi[(x-\mu)/\sigma]$; alternative (2) implies that for any value $x > \tau$ the variate's cdf is $\Phi[(\log\{x-\tau\}-\mu)/\sigma]$ ; while alternative (3) implies that within the open interval $(\gamma_1, \exp[-\gamma_2])$, this cdf equals

$$\Phi[(\log\{-\log(x-\gamma_1)-\gamma_2\}-\mu)/\sigma],$$

where $\exp[-\gamma_2] > \gamma_1$. (The first negative sign of $-\log(x-\gamma_1)-\gamma_2$ is required because for the log function to be real-valued, it is necessary that its arguments be positive.)

Alternative (1) concerns a two-parameter model; alternative (2) concerns a three-parameter model within which the value of the threshold parameter $\tau$ is sometimes assumed to be known; similarly, alternative (3) is based on a four-parameter model within which the values of either or both the $\gamma_1$ and $\gamma_2$ parameters are sometimes assumed to be known.

In many applications, the support region (the region over which a variate's density is nonzero) of the variate being studied must be bordered on the left by the value zero. Thus the superiority of models (2) and (3) over model (1) can be due either to the better fit of these models over all support points or, alternatively, due to models (2) and (3) being in accord with prior knowledge concerning the zero-bordered support region. Because model (3) is a more complex curve representation than model (2) which, in turn, is more complex than model (1), were either the parameter $\tau$ of model (2) or the parameters $\gamma_1$ and $\gamma_2$ of model (3)—estimated—rather than constrained to equal zero, then model (3) would be found typically to fit data better than model (2), while model (2) would almost certainly fit better than model (1).

## 13.2. Log and log(-log) transformations

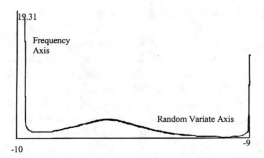

**Figure 13.2.** Low central mode loglog Cauchy.

A generalization of the Cauchy model illustrates how difficult it is to choose between curve representation alternatives to curves (2) and (3). Rather than the loglog Normal, which is based on a variate

$$Y = \log\{-\log(X - \gamma_1) - \gamma_2\}$$

that is Normally distributed, suppose the function $\phi$ of density $(\phi[(\log\{x\} - \mu)/\sigma])/(\sigma x)$ is replaced by the standard Cauchy model $\varphi$ where

$$\varphi(z) = 1/\{\pi(1 + z^2)\}$$

(equivalently, $\varphi$ is a Student-t density with one degree of freedom).

The author and Lock (1991) discuss a finding of Tukey (1958) that for some values of the parameter $\sigma$ the curve represented by the model $(\varphi[(\log\{x\} - \mu)/\sigma])/(\sigma x)$ will be multimodal. A log transformation will transform certain iid log Cauchy data samples that have a common multimodal density and yield a transformed variate with a unimodal symmetric density. Yet because of its multimodality, the multimodal log Cauchy density can be modeled mistakenly as a mixture.

The loglog Cauchy density formed by differentiating

$$F_C[(\log\{-\log(x - \gamma_1) - \gamma_2\} - \mu)/\sigma],$$

where $F_C(y) = (1/2) + \pi^{-1}\tan^{-1}(y)$ is a curve with properties that are even stranger than those of the log Cauchy itself. For example, as is shown by Figure 13.1, when $\gamma_1 = -10, \gamma_2 = \mu = 0$, and $\sigma = 0.1$, the loglog Cauchy is trimodel with a pronounced middle mode. However if $\gamma_1 = -10, \gamma_2 = \mu = 0$, but $\sigma = 0.5$, then the trimodal density shown in Figure 13.2 shows a low central mode and yet, when $\gamma_1 = -10, \gamma_2 = \mu = 0, \sigma = 10$, the curve's central mode disappears from view.

## 13.3  Methodological alternatives.

The loglog Cauchy illustrates that multiple modes and other structures of a univariate density cannot always be relied upon to classify various types of curve uncertainties. Multiple modes suggest the use of a mixture model representation. Yet the loglog Cauchy is a curve that, after the appropriate sequence of data transformations are executed, can be dealt with by using methodology designed with an eye to the unimodal and symmetric Cauchy model.

Statisticians like J. Tukey who have discussed the misleading shape of the log Cauchy density have also been major proponents of computationally intensive data analysis (especially Tukey, 1965 and 1977). This is not an accident. The confusing intersection of the class of curves containing both the log Cauchy and the loglog Cauchy on the one hand, and the class densities with more than a single mode on the other, is only one of many examples that illustrate why computationally intensive modeling methods are used.

Beyond density shape—as in a bell-shaped $f$-type curve—and the particular elementary function representation of a variate's density, there are alternative ways that statistical techniques can be categorized in terms of fundamental concepts such as the continuity of an $F$-type curve. For example, the following characteristics of statistical data can help provide clear-cut boundaries:

1. The consistency of the case-by-case data file structure;

2. The choice between conditional density representation and representation in terms of a joint or marginal density;

3. cumulative distribution function structure; (often corresponds to distinctions between continuous, discrete, ordered, and attribute-valued variates.)

4. The distinction between response, key, and nuisance, variates or variables.

Off-the-shelf statistical programs usually are designed to help study data *casewise*. The term "casewise" means that the multiple variate values obtained from a subject are arranged in neat columns that are usually separated by space or tab delimiters. Package programs often deal with these columns as wholes, in the sense that the name of each variate designates an entire column of data as a single entity called a *data column vector*.

Casewise categorization is convenient, particularly when the assumption that *all* elements of each individual column vector assume values that

## 13.3. Methodological alternatives.

are either uniformly attribute, uniformly ordered, uniformly discrete, or uniformly continuous, is valid. This is the case even though an attribute-valued variate's distribution is described by some density, but not unambiguously by any cdf function, and an ordered-valued, non-discrete and non-continuous variate is known to have a value greater or less than some other value. A common example of an ordered variate will be discussed later in this section.

In alternative data case categorizations some elements are of one type, for example, attribute-valued, and some are another, say continuous. The word "continuous" implies that the variate's cdf, $F$, is continuous in the mathematical sense of this term. Obviously, for example, Poisson variates are not continuous but are discrete-valued because their cdfs are step functions, while any Normal, Cauchy, Laplace or logistic cdf describes the distribution of a continuous variate.

All measuring instruments have a limited range of precision. Consequently, although exact size, temperature, or gestational age are usually taken to be continuous variates, there are often instances when it is known only that a measurement is less than, or greater than, some limiting value or detection limit. Suppose the exact value assumed by a variate is unknown, but it is known that this value must be less than, or greater than, some known value $y$. Then $y$ is often said to be the value assumed by an ordered variate $Y$, referred to as either a *censored* or *incomplete* measurement.

Details that distinguish different types of censored and incomplete measurements will be discussed in several later chapters. However, for the purposes of this introductory section suffice it to say that few, if any, samples are comprised of incomplete and/or censored measurements exclusively. For instance, suppose that in order to study bulb survival, at a certain time many new light bulbs are lit simultaneously. Then only an unusual study would have a period of observation that was so brief that no bulbs were observed long enough to yield a single complete measurement because the study stopped before the first bulb burnout occurred. The true time of bulb burnout is a continuous variate. Hence almost all data vectors that contain a value of an ordered-valued incomplete measurement (bulb still burning), also contain some other value that corresponds to a continuously-valued complete measurement (from some bulb that is known to have burned out and, therefore, with exact time-to-burnout known). Speaking generally, such a vector records the values of multiple types of measurement.

Unfortunately, no term exists which designates that all measurements within a given data column vector are of the same type. Hence below for purposes of clarity this situation will be characterized by the term "unitary." The opposite situation, where data that bears on the value of a given variate is comprised of more than a single type of measurement, will be designated by using the term "hybrid." (Obviously it is easier to categorize, analyze and simulate unitary-type data files than their hybrid counterparts.)

## 13.4 Conditional and joint density models.

The distinctions discussed in the previous section suggest that a first cut at dividing statistical methodology can be made along the border that separates approaches that deal with unitary, from those that deal with hybrid, data files. It also suggests that since whether or not a data column vector is unitary is an important consideration, data type should serve as a categorization principle by which statistical methodology is linked with forms of data. Yet, such a categorization would be inadequate without several other basic distinctions. For instance, somewhere along the line, all statistical problems that concern two or more variates and variables are always divided into two basic classifications.

All statistical problems that deal with more than one variate, or else concern one or more variates and variables, apply either representation based on some conditional density $f(y|x)$ or representation based on an expression for a joint density $f(y,x)$. Thus, whether a slash, |, or a comma is used to separate the symbols for variates and other variates or variables, provides an important clue concerning the referents of the term "statistical uncertainty."

The distinction between $f(y|x)$ and $f(y,x)$, and the distinction between unitary and hybrid data files, are related. For example, the notation $Y \in f(y|x)$ designates that a scalar or vector-valued variate $Y$'s distribution is described by a density $f$, where the word "density" is singular, and yet this single density's structure changes with values of the variable $x$. Were measurements listed within a $Y$ data vector not unitary types that are all continuous or all attribute-valued, this usage of the term density would be stretched beyond the breaking point. In hybrid cases it is stated explicitly that some $Y \in f(y|x)$, while the distribution of some other data vector values are described by a density represented by some symbol other than the single letter $f$.

For instance, since the symbol $F$ of the notation $F(y_o|x)$ refers to the probability of being less than or equal to $y_o$, when the symbol $Y$ designates that, given $x$, a variate's value is less than some value $y_o$, then, strictly speaking, the probability density of $Y$ at $y_o$ is $F(y_o|x)$. As this example illustrates, for some hybrid types of data the symbol $f$ of the expressions $Y \in f(y|x)$ and $(Y, X) \in f(y, x)$ is not serviceable.

Even though the left-hand side of $(Y, X) \in f(y, x)$ lists the pair of

## 13.4. Conditional and joint density models.

symbols $Y$ and $X$, it is typical for the symbol placed before the $\in$ symbol within $(Y, X) \in f(y|x)$ to be the single capital letter symbol $Y$. Hence, both to help simplify notation and because the difference between notation $Y \in f(y|x)$ and $(Y, X) \in f(y, x)$ can be confusing, when unitary data are described, it is handy to drop the $f$ and distinguish two basic forms of statistical data cases symbolically by calling upon the comma and slash symbols within expressions like $(Y, X)$ and $(Y|X)$.

Subscripted forms of non-italicized symbols that call upon the a, o, d, and c of $Y_a, Y_o, Y_d,$ and $Y_c$, could denote which particular data type: attribute; ordered; discrete; or continuous, is assumed by $Y$ variate data case components. Unfortunately, small font subscripts do not label procedural road maps as clearly as do uppercase letters. For this reason, attribute-, ordered-, discrete-, and continuous-valued variates and variables will be denoted by the capital letters $\mathcal{A}, \mathcal{O}, \mathcal{D},$ and $\mathcal{C}$, respectively.

For example, the two expressions, $(\mathcal{C}, \mathcal{A})$ and $(\mathcal{C}|\mathcal{A})$, distinguish two forms of data. The first form corresponds to some model or more general form of curve representation that describes the joint density of a continuous variate and an attribute variate. The second corresponds to the counterpart expression for a conditional density of a continuous variate, given some attribute-valued *variable*. When $(\mathcal{C}, \mathcal{A})$ and $(\mathcal{C}|\mathcal{A})$ data notation helps distinguish between variates and variables, there is no need to distinguish referents by case, such as $X$ or $x$. When a symbol is placed before a slash or within parentheses that do not contain any slash symbol, it designates some variate. On the other hand, when any symbol is placed after a slash, it designates a variable value that can be thought of as distinguishing one probability density function "slice" from another.

The $(\mathcal{C}, \mathcal{A})$ and $(\mathcal{C}|\mathcal{A})$ notational distinction can be augmented by subscripts that help designate whether the value is assumed by a response, a key, or a nuisance variate. For instance one can assign the subscript 1 so that the particular variate referred to is a response variate, and then follow this with symbols subscripted consecutively until all response variates are denoted. Key variate subscripts can then be assigned so that no key variate has a subscript whose value is larger than the subscript of any nuisance variate. A large subscript value suggests that the subscripted entity has been pushed to the background. Nuisance variables and variates cannot be swept under some representational rug. Yet despite the need to pay attention to them, there is little reason for them to take representational preference over any of their key, or response, counterparts.

We read English and many other languages from left to right, and therefore, the ordered assignment of subscripts to variates and variables can serve as a handy mnemonic device. Nuisance variates are, after all, poor relations who cannot be turned away, but with which we would prefer to have minimal contact. Since their values are basic to goal assessment, response variates take their place of honor at the opposite end of the scientific pecking order. Not knowing just what variate value we would like to predict and/or change implies that we are in the dark about our scientific goal. Midway between response and nuisance variates are the key variates, variates that correspond to our areas of expertise and whose purpose is to affect, in the sense that turning a key affects a door, or to predict, one or more response variate values.

The above notational devices are summarized as follows:

1. Symbols like $f$ or $F$ or symbolic representations of specific curves like the Poisson or Normal, lognormal, or loglog Cauchy, densities, are omitted, although the parentheses that had once enclosed the arguments of curves like $f(y|x)$ or $f(y,x)$ are retained;

2. When some form of representation like $f(y|x)$ or $f(y,x)$ is modeled or when a distribution is expressed by model-free representational devices, the presence or absence of a slash symbol designates whether or not this representation describes a conditional density. For example, the notation $(C_1, C_2, C_3|C_4)$ indicates that a data file is represented statistically by some expression for a conditional density of the three variates $C_1, C_2$ and $C_3$ *given the* $C_4$ *variable*;

3. Subscript ordering proceeds from response-to-key-to-nuisance. For example, if measurements are taken of two response variates while one key variate as well as one nuisance variable are assigned, then the symbol pair $C_1, C_2$, designates dual response variates, $C_3$ represents a key variate, and $C_4$ represents a nuisance variable.

4. The letter of the alphabet that denotes a variate and/or variable indicates whether it is continuous or some other type, such as attribute, ordered, or discrete. For many general statistical classifications (such as multivariate as opposed to univariate regression methodologies) the variety of approach; (for example as logit versus conventional forms of regression, or analysis of variance (ANOVA) as opposed to analysis of covariance (ANCOVA)), corresponds to a particular letter-sequence choice.

When the number of "data case" configurations like $(C_1, C_2, C_3|C_4)$ is compared to the number of topics covered by statistical texts, then in the sense that most texts and off-the-shelf packaged programs do not cover these configurations, it is apparent that there are many methodological gaps in the statistical literature. For example, in comparison to methods that deal with other case structures, a disproportionate number of available

## 13.4. Conditional and joint density models. 201

approaches concern ANOVA data, in the sense that for every program that deals with $(\mathcal{C}_1|\mathcal{O}_2,\mathcal{C}_3)$, a multiplicity of packaged programs can be bought that are designed to help study ANOVA data of the form $(\mathcal{C}_1|\mathcal{A}_2,\mathcal{A}_3)$.

One useful characteristic of $(\mathcal{C}_1|\mathcal{A}_2,\mathcal{A}_3)$ or $(\mathcal{C}_1|\mathcal{O}_2,\mathcal{C}_3)$ notation is that it allows a researcher to stand back and obtain a wide angle view of statistical alternatives. From this vantage point it is clear that there are a limited but large number of: (1) letter; (2) comma or slash symbol and; (3) subscript alternatives. For example, $(\mathcal{C}_1,\mathcal{C}_2,\mathcal{C}_3)$, $(\mathcal{C}_1|\mathcal{C}_2,\mathcal{C}_3)$, $(\mathcal{C}_1,\mathcal{C}_2|\mathcal{C}_3)$, $(\mathcal{O}_1,\mathcal{O}_2,\mathcal{O}_3)$, $(\mathcal{O}_1|\mathcal{O}_2,\mathcal{O}_3)$ and $(\mathcal{O}_1,\mathcal{O}_2|\mathcal{O}_3)$ are only 6 of the many possible permutations that involve three continuous and ordered variates and variables. The fact that methodological divisions involve regions of disproportionate size or scope, raises issues of scientific and historical importance. One such data-analysis issue is: Why are there so many practical procedures designed to deal with either continuous or attribute forms of information, while so few off-the-shelf program packages have been constructed with discrete and ordered data in mind? This and other related questions will be considered in the next chapter.

# CHAPTER 14
# Model-based and Generalized Representation

## 14.1 Multiple properties and parameters

Parameter types, such as those characterized by the words "location" and "scale," are context dependent. For example, the words "location" and "scale" together with the symbols $\phi(, [, x, -, ]$, and $/$, could mean the same to us as they do to the inhabitants of some distant planet and yet it would still be possible that on this distant planet the Normal curve is represented as $\phi([x - \sigma]/\mu)/\mu$ in terms of a location parameter $\sigma$ and scale parameter $\mu$. The role that symbols placed within an expressions like $[x - \mu]/\sigma$ or $[x - \sigma]/\mu$ play, characterizes location and scale parameters universally, in the sense that it is not the particular Greek letter chosen to represent it but the interpretation of surrounding symbols that gives the parameter type its identity.

A major advantage of elementary function-based statistical modeling is that parameter universality helps to divide knowledge into categories. For instance, saying that a variate $X$ has a Normal distribution is tantamount to asserting that the curve designated by the symbol $\phi$ of $\phi([x-\mu]/\sigma)/\sigma$ has a known formula. Thus, the assumption $X \in \phi([x-\mu]/\sigma)/\sigma$, is equivalent to the assumption that two numerical values, $\mu$ and $\sigma$, suffice to encapsulate all that can be learned about the distribution of $X$. In addition, the knowledge categorization process proceeds a step further; in the sense that when both $\mu$ and $\sigma$ are unknown, then neither the location, in the sense of position, nor the scale, in the sense of spread, of the curve in question are known.

In summary, in our statistical world the three symbols $\phi, \mu$ and $\sigma$, provide a shorthand way to categorize assumptions about which a researcher is uncertain. This process is extended still further when association, mixing, and threshold parameters are added to the list that begins with the three Greek letters $\phi, \mu$ and $\sigma$. Moreover, for mixtures as well as in the bivariate and multivariate case, there can be multiple location, scale, and even association parameters such as $\rho_1$ and $\rho_2$. A decided strong point of the parameterization process is that referring to two parameters as $\rho_1$ and $\rho_2$, and not either $\mu_1$ and $\mu_2$, or $\sigma_1$ and $\sigma_2$, provides a code which asserts that what these quantities have in common is that both these values help determine the association between two variates.

The neat compartmentalization provided by statistical parameter systems does not usually apply to curve properties. Unlike the typical usage of a location parameter, it would be strange for anyone to assert that he or she knows what the curve $f$ is within $\int_{-\infty}^{\infty} xf(x)dx$, but does not know either the value assumed by the curve property $E(X)$ or whether this value does or does not exist. Consequently, since curve properties like $E(X)$ and $\text{Var}(X)$ do not help much to categorize statistical knowledge, exactly what is their role? One answer often is that when the full information content of a population is reduced to what, in all but a few instances, is a single number, individual curve properties provide helpful hints about the sample distillation process.

The statistician's job is somewhat more complicated than what has often been expected of the mathematicians who made use of Descartes' brilliant linkage of parameters as algebraic entities with curve properties as geometric entities. Besides algebraic and geometric contingencies, statisticians need to consider data. Perhaps because data as well as some statistically relevant curve property like $E(X)$ are both more tangible than the typical statistical model's parameters, statisticians once assumed that it was easier to associate curve properties with data than to link parameters to data. However, research published by R. A. Fisher in the 1920s and 1930s directed emphasis away from curve properties (like the moments discussed in this section), and towards approaches for which individual parameter-with-individual-curve-property linkage was irrelevant.

In the sense that many curve properties, like the mode, can provide information that cannot be extracted easily from any mixture or other model, Fisher's victory over the proponents of property-based approach is now considered by some statisticians to have been a Pyrrhic victory. For different standard models $g$ that are assumed within the density model $f(x) = g([x-\mu]/\sigma)/\sigma$, likelihood approaches devised by Fisher yield typically different estimators of the parameter $\mu$ or the parameter $\sigma$. Yet, in order to check on $g$'s validity by fitting $f$ to a histogram or other nonparametric density estimator, candidate values of the two parameters $\mu$ and $\sigma$ are required.

Section 2.3 introduced the important idea that two cdf curves will often appear to be similar even though their density derivatives appear dissimilar. For the data analyst, curve choice can play a subtle role. Curves that are

## 14.1. Multiple properties and parameters

themselves functions of some density and its associated cdf can be devised that can help tease apart uncertainties that pertain individually to $g$, $\mu$ and $\sigma$. Hence a major advantage of methodology based, at least in part, on curve property estimation, is that these approaches can help tease apart the overall concept, *goodness of fit*, so that when fit is found to be poor, follow-up steps are spelled out. On the other hand when, for example, some curve estimator $\hat{f}(x) = \phi([x - \hat{\mu}]/\hat{\sigma})/\hat{\sigma}$ and some histogram are determined from the same sample of univariate data, it may not be obvious whether it is the Normality assumption that concerns $\phi$, the estimation of location parameter $\mu$ by $\hat{\mu}$ or, alternatively, the estimation of scale parameter $\sigma$ by $\hat{\sigma}$, or some combination of these three, that is at fault when poor fit is indicated.

In part to deal with goodness of fit follow-up issues, during the 1950s and '60s the histogram was promoted from its job as a check against some estimator like $\hat{f}(x) = [x - \hat{\mu}]/\hat{\sigma})/\hat{\sigma}$, based on estimators, $\hat{\mu}$ and $\hat{\sigma}$, and began to be assigned other more demanding responsibilities. One such responsibility was as a substitute for the curve $f(x)$ that appears within an expression that defines a particular curve property.

When $f$ appears within an expression like $\int_{-\infty}^{\infty} x f(x) dx$, it seems roundabout to initially ascertain the density's structure by estimating this curve by a histogram or by any alternative nonparametric methods and then, as a second step, to estimate a curve property like $E(X)$ by substituting the nonparametric curve estimator, say $\tilde{f}$. So much attention has been paid to the sample mean and its robust modifications during the 1970s and 1980s that, at least in the simple univariate case, when the single curve property $E(X)$ is estimated, the substitution of nonparametric curve estimator $\tilde{f}$ is, indeed, of questionable value. However, the curve substitution approach does makes considerable sense when the complexity of curve models based on multiple parameters forces validity issues to the fore. For example, validity questions often concern curve-truncation issues that complicate conventional parametric approaches such as maximum likelihood or, alternatively, these questions concern the details of mixture-model representation.

In an early attempt to deal with a complex mixture decomposition problem, the pioneer statistician Karl Pearson (1894) studied particular subsequences of curve properties known as *cumulants*. The sequence of cumulants has $E(X)$ and $\text{Var}(X)$ as its first two members, and goes on to include simple functions of the higher-order ordinary moments. For the Normal density, these functions of higher-order ordinary moments assume a value equal to zero. Hence, during the period when most goodness of fit problems concerned the fit of some Normal curve, it was natural to turn to the third and fourth cumulants to deal with skewness and kurtosis issues, such as those introduced in Section 4.2.

When $g$ is the standard Normal or any other standard symmetric density, it is obvious why the term $\mu$ within $g(x-\mu)$ is called a location parameter, and why $\sigma$ of $g([x-\mu]/\sigma)/\sigma$ is called a *scale parameter*. In light of its connection with center of gravity, since the definite integral $\int_{-\infty}^{\infty} xf(x)dx$ is known to equal the coordinate at which curve $f$ balances, it is also reasonable to define $E(X)$ as $\int_{-\infty}^{\infty} xf(x)dx$. However, why the word "variance," which sounds so similar to the important statistical terms variability and variate, is defined as $\int_{-\infty}^{\infty} [z - E(X)]^2 f(z)\,dz$ is far from obvious.

The variance is also called either the *second cumulant* or the *second ordinary moment about the mean*. As discussed in Chapter 3, the phrase, "ordinary moment about the mean," is based on an answer to the question: Within the squared term, why are deviations measured from $E(X)$ rather than from the population MEDIAN curve property that is defined by the equation

$$\int_{-\infty}^{\text{MEDIAN}} f(z)dz = 0.5?$$

As far as other alternatives are concerned, why not substitute the absolute deviation $|z-\text{MEDIAN}|$, the $1.5^{\text{th}}$ or the third power of the term $|z-E(X)|$, as in $|z - E(X)|^{1.5}$ or $|z - E(X)|^3$ to quantify variability as

$$\int_{-\infty}^{\infty} |z - E(X)|^3 f(z)dz, \int_{-\infty}^{\infty} |z - E(X)|^{1.5} f(z)dz,$$

or even

$$\int_{-\infty}^{\infty} \left[(z - E(X))^2 + |z - E(X)|\right] f(z)dz?$$

None of these choices stands out clearly as better or worse than any other and, even granted that the term $|z - E(X)|$ should be squared, $\text{Var}(X)$ itself is not even the curve property counterpart of the Normal's scale parameter $\sigma$.

In elementary statistics texts, the symbol $\phi$ is defined for the general Normal variate, $X \in \phi([x-\mu]/\sigma)/\sigma$ by a choice of formula selected so that it is the square root of the second cumulant (referred to as the *standard deviation*), not $\text{Var}(X)$ itself, which equals the scale parameter $\sigma$. Despite the Var property's indirect relationship to the Normal scale parameter, for historical and computational reasons the variance's square root is the default scale-associated curve property. More papers were written about the squared deviation from $E(X)$ integrand component than about the *absolute deviation*, $\int_{-\infty}^{\infty} |z - \text{MEDIAN}| f(z)dz$, and almost nothing is known about

$$\int_{-\infty}^{\infty} |z - E(X)|^{1.5} f(z)dz$$

## 14.1. Multiple properties and parameters

or
$$\int_{-\infty}^{\infty} \left[(z - E(X))^2 + |z - E(X)|\right] f(z)dz.$$

Emphasis was placed on the second cumulant because multiple curve properties are grouped together typically to form families, and interest in one particular family predated interest in all others. The first four population moments about the mean: $E(X)$; $\text{Var}(X)$;

$$\int_{-\infty}^{\infty} [z - E(X)]^3 f(z)dz; \quad \text{and} \quad \int_{-\infty}^{\infty} [z - E(X)]^4 f(z)dz,$$

have been studied for approximately 200 years. Tradition aside, today there is no particular reason why this family of four properties is preferable to many other candidate sequences. For example, beginning with the value 1, the smaller the positive integer-valued exponent of a deviation is, the less problem there will generally be, both computationally and with regard to *outliers*. Consequently, when it is known that f is symmetric, there is much that can be said in favor of the sequence

$$\text{MEDIAN}, \quad \int_{-\infty}^{\infty} |z - \text{MEDIAN}| f(z)dz,$$

$$\int_{-\infty}^{\infty} |z - \text{MEDIAN}|^2 f(z)dz, \quad \int_{-\infty}^{\infty} [|z - \text{MEDIAN}|]^3 f(z)dz.$$

(The word "outliers" refers to uncharacteristic measurements that will be discussed at length in later sections.)

Elderton (1953) applied a simple procedure, based on integration by parts, to members of the Pearson curve family discussed in Appendix I.4. Extending the idea, that by including the 1/2 term as part of the standard Normal model's exponent, the scale parameter $\sigma$ can be assumed to equal the curve property, standard deviation; Elderton associates parameters of the Pearson curve family on the one hand, to members of the moments-about-the-mean family of curve properties on the other. Thanks to this linkage, were some $f$-type Pearson curve family member specified as a valid model, then the match-up of this Pearson model's parameters to moment-curve properties can be made by performing simple computations.

It is much easier to perform repeated multiplication than it is to raise a number to a fractional power like 1.5. Hence, during the long era when practicing statisticians lacked an easy way to do simple computations, sample moments seemed preferable to statistics that were the sample counterparts of

$$\int_{-\infty}^{\infty} |z - E(X)|^{1.5} f(z)dz \quad \text{or} \quad \int_{-\infty}^{\infty} \left[(z - E(X))^2 + |z - E(X)|\right] f(z)dz.$$

From a statistical point of view, a variate $X$ whose distribution is described by a general Normal density model $\phi([x-\mu]/\sigma)/\sigma$ is characterized by the values of the two parameters $\sigma$ and $\mu$. Consequently, it was not at all surprising that when baby brothers and sisters were added to the family whose two initial siblings were $E(X)$ and $\sqrt{\text{Var}(X)}$, the eventual roles that these newcomers were expected to play concerned the generalizations of $\phi([x-\mu]/\sigma)/\sigma$ which are described in Appendix I.4.

During the early years of the twentieth century, higher-order moments were often applied to either the problem of determining skewness and kurtosis or to what has come to be called the *mixture decomposition problem*. It took a long time to realize that $E(X)$'s and $\sqrt{\text{Var}(X)}$'s convenient match with $\mu$ and $\sigma$ did not necessarily imply that higher-order moments can be happily mated with threshold and mixing parameters.

## 14.2 Specification and generalized representation

By the 1960s, improved histograms and other curve estimators made it easier to discern differences between actual $f$-type curves and any one particular simple density model like the Normal. Also, by this time the ordinary population moments that fitted the Normal and other Pearson family curves so well had lost their computational justification. This raised the basic uncertainty issue: if the data are not Normally-distributed, then how are they distributed?

To circumvent many basic uncertainty issues, the finite number of explicit symbols of a model-based typographical expression are often replaced by some *generalized* representation. These are based on a variety of notational devices, including, for example, the dots at the end of $k = 0, 1, 2, \ldots$. Sometimes the term "closed-form" designates the antonym of generalized representation. Within the statistical literature, generalized representation is divided into the three categories: (1) "orthogonal series" and "wavelets;" (2) "splines;" or (3) "generalized histograms" or "kernels."

For example, when the letter i denotes $\sqrt{-1}$, type (3) generalized representation is often based on the sequence whose terms are $e^{-2\pi ikx}$ for $k = 0, \pm 1, \pm 2, \pm 3, \ldots$; and from this sequence of curve building-blocks, many densities and other functions can be expressed as the Fourier series

$$\sum B_k e^{2\pi ikx}.$$

Here $\{B_k\}$ denotes the sequence of $f$'s "Fourier coefficients."

## 14.2. Specification and generalized representation

Suppose a density $f$ has been scaled so that the unit interval (0,1) can be taken for all practical purposes to be its support region. Then the term "Fourier coefficient" refers either to the parameter $B_k$ of $\sum B_k e^{2\pi i k x}$ or the curve property

$$B_k = \int_0^1 f(x) e^{-2\pi i k x} dx.$$

This important curve property can be thought of as what the $k^{\text{th}}$ moment about the mean would become if the term $\{I_{[0,1]}(x)e^{-2\pi i k x}\}$ replaced the term $[z - E(X)]^k$ within

$$\int_{-\infty}^{\infty} [z - E(X)]^k f(z) dz.$$

To indicate that $k$ takes on all negative and positive integer values, as well as the value 0, the limits of the summation are unspecified above the capital sigma symbol of the expression $\sum B_k e^{2\pi i k x}$. This openness and absence of restriction is characteristic of wavelet and Fourier series as well as all other forms of generalized representation.

For the price of a potentially unlimited number of terms, the following representational advantages are purchased: (1) with a few exceptions, such as the parameters of the Pearson family and the sequence of moments about the mean, parameters and properties are associated more clearly with each other than they are typically when elementary-function-based models describe curves like $f$; (2) the linear assemblage $\sum B_k e^{2\pi i k x}$ facilitates many computational and mathematical operations, for instance integration and differentiation. (3) histograms, like the curve estimator shown in Figure 1.1 of Chapter 1, can be described easily when building blocks $e^{2\pi i k x}$ serve as the basis of Fourier representation.

The raising of a variate value $x$ to a large $k$ power often results in values that are more unwieldy than values attained when $\cos(2\pi k x)$ or its sine counterpart are evaluated. The numerical advantages of $\{I_{[0,1]}(x)e^{-2\pi i k x}\}$ counterbalance whatever advantages stem from Pearson family relationships between curve parameters and the moment property sequence. For example, the usual estimators of cumulants are referred to as k-*statistics*. However, in comparison to recently studied alternatives, such as those based on Fourier/wavelet expressions, k-statistics are difficult to work with while, with respect to simplicity and elegance, no representational system is as good as the system based on $\{I_{[0,1]}(x)e^{-2\pi i k x}\}$. For a given number of terms, as a match for complex regression and $f$-type curves, no system can vie with $\{I_{[0,1]}(x)e^{-2\pi i k x}\}$'s wavelet counterparts (Chui, 1992; Tarter, 1998).

Generalized statistical curve representation is a hallmark of nonparametric or model-free curve estimation. However, strictly speaking, the symbolic representation $f(x)e^{-2\pi ikx}$ is semiparametric, since the $e^{-2\pi ikx}$ component of $f(x)e^{-2\pi ikx}$ is a specific elementary function, while the four symbols that designate the curve $f(x)$ of the product $f(x)e^{-2\pi ikx}$ designate this curve implicitly. They do not spell out the detailed formula by means of which this curve can be evaluated numerically.

Because $\{I_{[0,1]}(x)e^{-2\pi ikx}\}$ can be expressed in terms of sine and cosine components, the term "trigonometric moments" is sometimes used to denote the sequence of Fourier coefficients $\{B_k\}$. Even though the word "trigonometric" calls to mind wiggly curves such as time series, the component $I_{[0,1]}(x)$ of $\{I_{[0,1]}(x)e^{-2\pi ikx}\}$ can serve as a mask that delimits *noncyclic* curves. The need to delimit representation to intervals shorter than $(-\infty, \infty)$ had, prior to the development of wavelet techniques, been once thought to be the price required for representational generality.

For functions that cannot be modeled, either because no particular choice of model can be specified validly or because no model for the curve in question exists, Fourier series and wavelet generality allow curve properties like $B_k$ to be dealt with effectively, in the sense that a property of a curve like $f$ can be defined without reference to any particular model family like that of Pearson. Hence Fourier coefficients $B_k; k = 0, \pm 1, \pm 2, \pm 3, \ldots$ play dual roles as properties or parameters, and for this reason have a versatility that is not encountered typically when elementary-function-based model parameters are associated with curve properties. The price exacted by the three dots at the end of $k = 0, \pm 1, \pm 2, \pm 3, \ldots$ and the implicit upper and lower infinite summation limits of $\sum B_k e^{-2\pi ikx}$ pay for many representational luxuries.

$\text{Var}_f(Y)$ is defined in terms of the deviations from the curve property $E_f(Y)$, and consequently, without reference to this second curve property, the property $\text{Var}_f(Y)$ is ambiguous. Unlike individual cumulants like $E_f(Y)$ and $\text{Var}_f(Y)$, $B_k$ can be clearly distinguished from a different Fourier coefficient, such as $B_{k+1}$, in the sense that it can be specified without any reference to any other Fourier coefficient. Its neighbor's role notwithstanding, each $B_k$ has its own role to play.

The neat division of curve properties and roles stems from a mathematical characteristic of $\{I_{[0,1]}(x)e^{-2\pi ikx}\}$ and counterpart wavelet functions known as *orthogonality*. Orthogonal functions bear a close resemblance to the statistically independent $Z_1, Z_2$, ,and $Z_3$ variates discussed in previous chapters. On the other hand, the individual elementary functions within a model serve roles similar to those played by dependent $Y_1, Y_2$, ,and $Y_3$ variates. Therefore it is not surprising that model-based representation is prone to confounding.

## 14.3 Identifiability of generalized versus extended model representation

Mixture models can be expressed in terms of parameter combinations that are indistinguishable from each other. For example, consider the identifiability characteristics of the five parameter model,

$$f(y; \Theta_1, \Theta_2, \ldots, \Theta_5) = P f_1(y) + (1-P) f_2(y),$$

where $f_1(y)$ and $f_2(y)$ are the densities $(\sigma_j)^{-1} g((y-\mu_j)/\sigma_j)$;; $j = 1, 2$; respectively; mixing parameter $P$ satisfies the inequalities $0 \leq P \leq 1$; and g is the Normal $\phi$ or some other standard model. For the remainder of this chapter, $\Theta_1, \Theta_2, \ldots, \Theta_5$ designate the five parameters $\Theta_1 = P, \Theta_2 = \mu_1, \Theta_3 = \mu_2, \Theta_4 = \sigma_1, \Theta_5 = \sigma_2$.

For any constants $a, b, c,$ and $d$, such that $d > 0$, the three different sets of parameter values: $\{\Theta_1 = 0, \Theta_2 = a, \Theta_3 = b, \Theta_4 = d, \Theta_5 = d\}$; $\{\Theta_1 = 1, \Theta_2 = b, \Theta_3 = a, \Theta_4 = d, \Theta_5 = d\}$ and; $\{\Theta_1 = c, \Theta_2 = b, \Theta_3 = b, \Theta_4 = d, \Theta_5 = d\}$, are interchangeable. For example, $\{\Theta_1, \Theta_2, \ldots, \Theta_5\}$ that equal $\{0, 100, 2, 1, 1\}$, $\{1, 2, 100, 1, 1\}$, $\{0.5, 2, 2, 1, 1\}$, and $\{0.05, 2, 2, 1, 1\}$ describe the same curve. Differences between parameter values do not themselves imply necessarily that the curves represented with the help of these parameters are different.

Unlike the mixture model, Fourier/wavelet representation relates curves to symbols unambiguously. If any single Fourier coefficient differs within two Fourier series expressions, these two series *can* always be treated as representations of two distinct curves. Conversely, two distinct curves that differ over an interval of any length $C$, where $C > 0$, must have two different sequences of Fourier coefficients. A great deal of trouble is taken in all branches of statistics to determine whether or not there is a difference between two curves. Hence the clear association between curve difference and curve representational difference, is a major advantage of Fourier series nonparametric methodology.

The curve known as a *characteristic function* which has equally spaced evaluations $B_k; k = 0, \pm 1, \pm 2, \pm 3, \ldots$; determines a unique curve. On the other hand, neither the curve known as the *moment-generating function*, nor the infinite sequence of all ordinary moments, determines a density curve $f$ uniquely. For example, a lognormal density and a second, distinct $f$-type curve can both have the same sequence of moments. Thus, while even perfect knowledge of all moments does not pin down the identity of the curve that has these moments, in the following sense the Fourier coefficients $B_K$; where $k = 0, \pm 1, \pm 2, \ldots$, do determine a unique curve.

Two different curves, $f_1$ and $f_2$, can have the same Fourier expansion only in regions over which these curves enclose area zero. This is not an important proviso, because when $f_1$ is a density, it follows that no regions which enclose an area equal to zero can lie within $f_1$'s support subregion. Properties of curves, in lieu of curves in their entirety, are supposed to simplify statistical processes. Thus the uniqueness of the properties, Fourier coefficients, implies that simplification is not purchased at the price of ambiguity.

Many popular curve-uncertainty-reduction methods are based on a match-up of one or more estimated curve properties, such as a sequence of moments, with the corresponding properties of fitted curves. Nevertheless, basic statistical formulas often depend solely on the existence of finite and well-defined curve properties, and do not require that curves belong to specific families such as the family studied by Karl Pearson. Hence the connection between particular models like the Normal or Cauchy, on the one hand, and the existence of certain curve properties, on the other, often is the actual motivation for the model selection. For example, the Cauchy's Var property and the $E$ property of the "folded Cauchy" density $g(x) = 2I_{[0,\infty)}(x)/\{\pi(1+x^2)\}$ are both infinite, and thus many statistical relationships cannot be called upon to assist in the analysis of Cauchy and folded Cauchy data.

In general, there are two crucial assumptions: The first assumption, which is usually met in practice, is that the properties $E(Y)$ and $\text{Var}(Y)$ are finite-valued. The second assumption, which is the source of much statistical difficulty, is that the properties $E(Y)$ and $\text{Var}(Y)$ have some simple connection with location and squared scale parameters. The irony is that it is the same truncation processes that limit the support of what otherwise would be curves like the ordinary and folded Cauchy, and hence assure the existence of $E(Y)$ and $\text{Var}(Y)$ that loosen the connections between these curve properties on the one hand, and $\mu$ and $\sigma$ parameters on the other.

Truncation is certainly not the sole source of property-with-parameter mismatch. For example, suppose a model, such as the mixture of two Normal densities, validly describes the distribution of some variate within an iid data set. Then $E(Y)$ and $\text{Var}(Y)$ both exist and are well-defined. But, if a mixing parameter $P$ has a value between zero and one, a mixture of two Normals must be expressed in terms of dual sets of both location and scale parameters. In practice, what parametric duality means is that, with one notable exception, convenient and simple methods provided by an off-the-shelf statistical program package seldom can be used for mixture data analysis.

The methodology called collectively *cluster analysis* does sometimes provide a convenient substitute for true mixture decomposition procedures.

## 14.3. Identifiability of generalized versus extended model representation

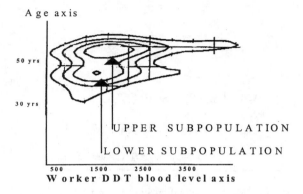

Figure 14.1. Mixture specification ambiguity.

However, as illustrated by the following example, it is unlikely that this approach can be sharpened sufficiently to cut to the core of many basic duality issues.

Suppose that as described by Tarter and Silvers (1975), blood levels of DDT are depicted along the $x$-axis and ages are shown along the $y$-axis. Specifically Figure 14.1 shows estimated equiprobability contours that describe the estimated joint density of the DDT and age variates. (In the sense that they enclose regions whose borders correspond to equal estimated probability density, the contours shown in this picture resemble the isobars or isotherms of a weather map.)

Suppose there are two distinct groups of fieldworkers, parents and offspring, and within each group some log transformation of the DDT variate will produce two individually multivariate Normal subpopulations, each with the same logDDT variate location parameters. (In other words, after transformation, both sets of Normal contours will be centered at points that lie on the same vertical line.) Suppose also that the parents' logDDT variate scale parameter assumes a value that is larger than the value assumed by the scale parameter that helps describe their offspring's logDDT variate's distribution. (Were population subcomponent Normal contours sections of watermelons, then the parents' watermelon would be longer than the offsprings' watermelon.)

In their untransformed, lognormal state, in comparison to the lower contour, the upper contour will look as if it were pulled the right and, of course, be less symmetric than the idealized upper contour shown in Figure 14.1. However, even when their shape is elliptical, the two contours

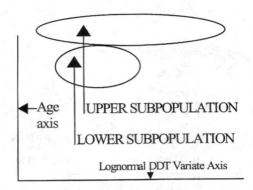

Figure 14.2. Mixture specification ambiguous ellipses.

shown in this figure illustrate that a large value of Pearson's r can be expected even though what has happened in reality is that scale $\sigma_1$ versus $\sigma_2$ difference has been confounded with a nonzero-valued $\rho$ parameter. Due to component scale parameter differences, the observed age/DDT contours shown in Figures 14.1 and 14.2 indicate a pronounced relationship between age and blood DDT level. There is, indeed, a relationship here, possibly due to second-generation-induced homogeneity, but it is not an association that has any health implications.

In this example, because of mixture structure, the properties $E(X)$ and $\text{Var}(X)$ do not have some simple connection with location and squared scale parameters. The $E(X)$ curve property of the upper older subpopulation is, indeed, much larger than the $E(X)$ curve property of the lower younger subpopulation. Yet this difference is due solely to scale parameter difference. Again, parametric multiplicity is the root cause of scientific uncertainty.

## 14.4 The E(X) curve property's relationship to location and scale

The above example is realistic but complex. That the curve property $E(X)$ can be related to a scale as well as to a location parameter is also illustrated by the general rectangular or uniform density curve $\mathcal{I}(x; \mu, \sigma) = \sigma^{-1} I_{(0,\ 1]}\left((x-\mu)/\sigma\right)$ where, as before, $I_{(0,\ 1]}(z)$ equals one if $0 < z \leq 1$, and zero elsewhere. A variate $X$ whose density is $\mathcal{I}(x; \mu, \sigma)$ has the interval $(\mu, \mu + \sigma]$ as its support region. Therefore, expectation $E(X)$ equals $\mu + \sigma/2$, not $\mu$.

## 14.4. The E(X) curve property's relationship to location and scale

As was done by inserting the term one-half within the exponent of the Normal density model, in order to match a property with a parameter, the expression $\sigma^{-1} I_{(0,1]}((x-\mu)/\sigma)$ can be exchanged for

$$\sigma^{-1} I_{[-1/2, 1/2]}((x-\mu)/\sigma),$$

which is a model that, when valid, is defined so that location parameter $\mu$ equals the curve property $E(X)$. While in the rectangular density case, identifiablity problems stem from the selection of a standard model, the age/logDDT example is prone to problems that cannot be resolved merely through representational choice. Difficulties can be traced to the interpretation of location, scale, and association, parameter values. When a single bivariate Normal model provides valid representation, the inequality $\rho > 0$ indicates variate association clearly. But when mixtures of lognormal variates are encountered, scale parameter differences can appear in a $\rho > 0$ disguise.

The reason why it is easy to mistakenly assert that $\rho > 0$ for data sets that resemble those of the age/logDDT example, is that an estimator of the parameter $\rho$ will, in this situation, typically indicate that the association parameter is positive. In the sense that the age and logDDT variates are dependent, this is a correct inference. However, the question why are these variates dependent? is answered correctly only by paying careful attention to identifiablity issues. Cluster analysis, as its name implies, aims at the subdivision of data sets to form homogeneous subsets called *clusters*. It does not target either parameters or curve properties that concern any cluster individually. Similarly, the problem illustrated by the fieldworker example is unlikely to be corrected by applying distribution-free methods. Quite the contrary, it is likely that data structure complexities would be swept under the rug when these robust procedures are applied. Robustness connotes getting the right answer for the wrong reasons. Problems with the fieldworker study stem from not asking the right question: Why are these variates dependent?

The major strength of the model-based approach is that when applied carefully, it can uncover subtle distributional clues. There are nonparametric approaches that can also pick up these clues. But there are also nonparametric approaches that resemble the incautious assumption of a single bivariate Normal distribution of DDT-blood-level/age variates, and thus tender misleading clues. It is really not the case that nonparametric and model-based approaches vie with each other. Instead, they can serve as crosschecks on clues to validity, identifiability, and even curve-property existence.

Crosschecks are important because questions pondered today differ substantially from those that occupied statisticians' time prior to the modern computer. Data analysis has become assumption analysis. What should happen after a given data set is matched to valid assumptions is far less an issue than assumption validity.

CHAPTER 15

# Parameters, Transformations, and Quantiles

## 15.1 Location and scale parameter representation of continuous variates

When uncertainty issues concern curve types $f, F$, and $F^{-1}$, what the word "uncertainty" refers to is clear. On the other hand, what the curve types discussed in this chapter, most notably: confidence bands; prediction bands and; continuous conditional box plots, actually are often misunderstood. However, before turning to these complex issues, one important related concept requires some explanation.

The roles played by location and scale parameters within density and cdf curve notation suggest why continuous data types, in comparison to their discrete or ordered counterparts, can be studied conveniently. Suppose the expressions, $f(x) = g[(x - \mu)/\sigma]/\sigma$ and $F(x) = G[(x - \mu)/\sigma]$ represent the density and cdf, respectively, of a continuous variate in terms of the standard curves $g$ and $G$, respectively. Then, almost invariably, either $g$ or $G$ is taken to be a simple elementary function, or, with the notable exception of the Normal cdf $\Phi$ both $g$ and $G$ are expressed as elementary functions. The elementary function *forms* discussed in Section 14.2 that were so important to Fisher and his contemporaries are often now taken to be computational conveniences. Today, besides models like the Normal, Fisher's likelihood procedures are applied even though neither $g$, nor $G$ is an elementary function.

Previous chapters have explained why a symbol $\mu$ within argument $(x - \mu)/\sigma$ of some function $G$ is called a as a *location* parameter and why $\sigma$ is called a *scale* parameter. Location and scale parameter notation implies that when the cdf $F$ is expressed in terms of some elementary function model, it follows that representational structure is separable into two portions, the standard model or expression $G$ and the term $(x - \mu)/\sigma$. Of course, unlike the cdf model, to assure that $f$ integrates to one over the $X$ variate's support region, any density's model is a composite function $g[(x-\mu)/\sigma]$ that must be divided by the parameter $\sigma$.

These notational conventions are straightforward when $X$ is some continuous variate. On the other hand, a discrete-, ordered-, or attribute-valued variate's density and cdf model can rarely be parametrized, either as $g[(x-\mu)/\sigma]/\sigma$ or, in the case of a cdf, as $G[(x-\mu)/\sigma]$. For example, most discrete measurements take on integer values exclusively. This implies that for $f(x) = g[(x-\mu)/\sigma]/\sigma$ or $F(x) = G[(x-\mu)/\sigma]$ to be models that describe one of these discrete variate's distributions, it cannot also be true that the parameters $\mu$ and $\sigma$ assume a full range of values. For instance, if the model $f_1(x) = g(x)$ represents a discrete variate's density whose support is integer-valued, it follows that the density $f_2(x) = g[(x-0.5)]$ will not have an integer-valued support region. It can assume nonzero values only over integer—plus one-half—values.

Because of the limitations of discrete variate modeling processes, models for the Poisson, binomial, and many other discrete variate density models are not expressed in terms of some expression $g$ as $f(x) = g[(x-\mu)/\sigma]/\sigma$. Instead, the representations of these curves describe complex relationships between variate and parameter values. This, in turn, renders the algebraic treatment of these curve representations more cumbersome than the treatment of typical continuous variate densities.

An additional advantage of continuous variate curve representation $F(x) = G[(x-\mu)/\sigma]$ is that these expressions can be easily extended by replacing the variate value x by a simple function like

$$\log(x-\tau),\ \log[\log\{x-\gamma_1\}-\gamma_2],$$

or even a fractional power transformation function $(x-\gamma)^{1/(K+1)}$ where here $K$ is some specified shape-determining parameter whose value differs from -1. As is the case of their location and scale parameter representational mismatch—discrete-, ordered-, or attribute-valued variates match poorly with expressions involving the data transformation functions $\log(x-\tau)$, $\log[\log\{x-\gamma_1\}-\gamma_2]$, or $(x-\gamma)^{1/(K+1)}$.

There are two notable exceptions to the mismatch between discrete variates and variate transformations. There is usually little reason for a fractional power transformation like $(y-\gamma)^{1/(K+1)}$ or a log transform to be applied to a variate $Y$ that usually assumes one of only a handful of distinct values. For example, the number of daughters born to an American middle-class family would not be the sort of variate whose log would be modeled in a simple fashion. However, the two transformations $\sqrt{Y} + \sqrt{Y+1}$ and

$$\sin^{-1}\sqrt{Y/(N+1)} + \sin^{-1}\sqrt{(Y+1)/(N+1)}$$

## 15.1. Location and scale parameter representation of continuous variates

are sometimes applied in order to modify discrete-valued Y variates.

Suppose a large number $m$ of separate replicate independent samples of a Poisson distributed variate $Y|x_k; 1, 2, ..., k, ..., m$ are gathered where, for example, $x_k$ provides an index that identifies the $k^{th}$ replicate subset. (The $x$ that follows the slash sign serves the same role played by a subscript $x$ and is placed in line with the symbol $Y$ to allow room for its own subscript.) The sample variances determined for each of such samples will be found to be roughly proportionate to the corresponding sample means, where variance-to-mean correspondence implies that since the Poisson is a one parameter, $\lambda$, model, all of its curve properties that can vary from context to context, must be determined by the value of $\lambda$. In a similar way, since the general Normal model is parameterized in terms of the parameters $\mu$ and $\sigma$, for this model any curve property that (unlike some measures of skewness and kurtosis) varies from context to context, must be related to one or both of the parameters $\mu$ and $\sigma$.

Suppose most of the $m$ means of $Y|x_k; 1, 2, ..., k, ..., m$; replicated subsets are less than 0.8. Then the tranformation $\sqrt{Y} + \sqrt{Y+1}$ is said to yield an "approximately stabilized variance" (Dixon and Massey, 1983). Specifically, where the original $Y$ variates might have an assortment of variances whose values depended on the value of $\lambda$, for each value of $x_k; 1, 2, ..., k, ..., m$; the $m$,

$$\sqrt{Y|x_k} + \sqrt{(Y|x_k)+1}$$

variates, are taken to have the same variances.

In an analogous fashion, suppose discrete $(Y_k/N)|x_k; 1, 2, ..., k, ..., m$; variates are proportions of subsamples of size $N$. Given some particular $k$, for example, suppose $Y_k$ daughters are born among a number of births $N > 10$. If $Y_k$ is a binomial variate then, if $E(Y) = Np$ and if the inequality $Np(1-p) > 0.8$ is valid, the arc sin transformed variate

$$\sin^{-1}\sqrt{Y_k/(N+1)} + \sin^{-1}\sqrt{(Y_k+1)/(N+1)}$$

has a variance approximately equal to the constant value $1/(N+\frac{1}{2})$. A well-written outline of the reasoning that underlies transformations like the arc sine and the $\sqrt{Y} + \sqrt{Y+1}$ variance stabilizing transformation is provided in Section 7.45 of Bennett and Franklin's book, *Statistical Analysis in Chemistry and the Chemical Industry* (1954). Both the square root and arc sine transformations are particularly valuable as preliminaries for the ANOVA procedures that are discussed in the next chapter.

## 15.2 $\rho$-focused transformations and $\sigma$-focused transformations

The values attained by a variate are often ranked so that the variate which has the smallest value among sample size $n$ counterparts is renamed $X_{(1)}$, the second is named $X_{(2)}$, the $j^{th}$, $X_{(j)}$, and finally the variate with the largest value is represented by $X_{(n)}$. Later sections will describe parameter-specific estimation procedures based on weighted sums of the order statistics $X_{(1)}, X_{(2)}, ..., X_{(j)}, ..., X_{(n)}$. The value of the ordered variate $X_{(j)}$ can be thought of as a pair of numbers, $x_{j'}$ and $(j)$ where $j'$ (which usually assumes an integer value different from the value assumed by $j$) designates the index of this variate prior to ranking, and $(j)$ designates $X_{(j)}$'s rank.

Why $x_{j'}$ and $(j)$ are difficult to conceptualize in bivariate and multivariate contexts is illustrated by a comparison to many athletic contests. Consider the borders that keep runners in their own lanes. From a general perspective, these borders constrain the athlete ranking process to pathways that are, roughly speaking, one-dimensional. Thus many disputed athletic and other winner selection processes can be thought of as the effect that the curse of dimensionality has on ranking procedures.

The ranking process is the equivalent for a sample of what a specific application of the cdf curve $F$ is for a population. In the same way that the inverses of many $F$-type curves are pressed into service within data-generation models, a $F$-type curve itself often is thought of as a variate transformation that yields standard uniform variates. Specifically, suppose the continuous variate $X$ has a density $f(x)$. Then, if $f$ is the derivative of cdf $F$, it follows that ($F$ of $X$-variate)-variate, $Y = F(X)$, has a uniform density with support over the closed unit interval.

An evaluation of the curve $F$ discloses values of the probability that a variate will attain some value or less. Consequently, the enumeration of $F$ is seldom generalized to bivariate and higher-dimensional applications. On the other hand, ranking methodology can be based on univariate adaptations of the conditional cdf $F(y|x)$.

Consider the actual process by means of which n members of an iid sample can be ordered to obtain the sequence $\{(j)\}$ within $\{x_{j'}, (j)\}$. The function $F^*$, called the sample cumulative distribution function (scdf) is defined, in part, by the relationships $F^*(x_{j'}) = j/n; j = 1, 2, ..., n$. This step function has risers at each variate value and contains horizontal line segment stair-like steps that connect, for example, the tops of one riser to the bottom of the next riser. Multiplication of the scdf by the sample size $n$ provides a transformation or mapping which takes in an unranked variate value argument, $x_{j'}$, and yields the argument's rank $j$. (In keep-

## 15.2. ρ-focused transformations and σ-focused transformations

ing with the name, sample cumulative distribution function, as $n \to \infty$ the scdf, constructed from an iid sample of variates with common cdf $F$, approaches $F$.)

Ranks like $1, 2, 3, ..., n$ are, of course, evenly-spaced, and thus can be thought of as uniformly-distributed discrete variates. As a consequence, because it maps variate values onto the corresponding ranks divided by $n$, the scdf often serves as an approximation of what could theoretically be achieved by the population cdf, say $F$, were it possible to press $F$ into service as a variate transformation that yields standard uniform continuous variates. For instance, suppose each iid unranked variate $X_j; j = 1, 2, ..., n$; with cdf $F$, is mapped into a transformed value equal to $Z_j = F(X_j)$. Then each of the $n$ new $Z_j$ variates will have a rectangular, in other words uniform, density.

As discussed in Section 7.3, if $Z_j; j = 1, 2, ..., n$; are $n$ iid variates, each with density $I_{(0, 1]}(z) = 1$, if $0 < z \leq 1$ and zero otherwise, then it follows that $\Phi^{-1}(Z_j); j = 1, 2, ..., n$; are $n$ iid standard Normal variates. (As before, $\Phi^{-1}$ designates the inverse standard Normal cumulative.) Consequently, if the two transformations, $\Phi^{-1}$ and $F$ are combined to form the curve $\Phi^{-1}$ of $F$, then composite curve $\Phi^{-1}F$ will always transform any variate whose cdf is $F$ to yield a normal variate.

Correspondingly, the inverse Normal scores transformation $\Phi^{-1}F^*$, a composite of some approximation of $\Phi^{-1}$ with $F^*$, Normalizes variates approximately. In the days before this process was accomplished by digital computation, sample values were first ranked. Then tables of the function $\Phi$ were consulted; each rank, in turn, was divided by $(n + 1)$ and these quotients' values located in the table body. Scores were then read from the table margin. (A useful nonparametric inferential procedure called "Van der Waerden's test" is based on inverse Normal scores. For this reason SPSS (1998) states that the "Van der Waerden (Probability Plot)...calculates the expected Normal distribution using the formula $r/(n + 1)$, where $n$ is the number of observations, and $r$ is the rank, ranging from 1 to $n$.")

The availability of $\Phi^{-1}F^*$ as a universal transformation to Normality seems, at least at first glance, to imply that we can throw out logarithmic, arc sine, square root, and other transformations. After all, if there exists a universal transformation, why bother with alternative functions like the log? Why not just apply a curve estimator of $\Phi^{-1}F$, such as $\Phi^{-1}F^*$, to produce Normal variates?

The scope of $\Phi^{-1}F^*$ may be galactic, but it is not really universal. Despite the ease with which $\Phi^{-1}F^*$, and estimates based on generalized representations of $F$ other than the scdf, can be determined, the log and many other elementary-function-based transformations all have their own niches. We will now see that doubts concerning the applicability of inverse

Normal scores can be answered easily in terms of distinctions between parameter types, and that these distinctions are related closely to distinctions between curve representation of conditional-versus-joint density functions.

The answer to transformation niche questions is disclosed by the answer to the related question: What particular Normal model describes the density of Normal-scores-transformed variates? The answer is—the standard Normal. The Normal-scores transformation can throw out the baby with the bath water. It removes information that discloses the values of $E(X)$ and $Var(X)$, the properties targeted in investigations of model location and/or scale parameters. For instance, most methods that are applied to analyze $(C|\mathcal{A})$ data, call upon data-generation models based on multiple $\mu$ parameters. (The subscripts of $C$ and $\mathcal{A}$ variates are omitted to allow a single letter like $C$ or $\mathcal{A}$ to denote one or more variates or variables of a given type.)

Despite its limitations, Normalized scores can be a powerful tool for the many statistical applications where neither the properties $E(X)$ and $Var(X)$, nor location and scale parameters $\mu$ and $\sigma$, are of interest. For example, most multivariate analyses of $(C,C)$, or more complex forms of multivariate data, are conducted with an eye to one or more association parameters such as $\rho$. For such problems, $\mu$ and especially $\sigma$ can be taken to be nuisance parameters.

Applications and the power of tests based on multivariate transformed data are discussed by Kowalski and Tarter (1969), where several examples are provided in which findings that are in accord with biomedical thinking are matched with actual data analyses only after $\Phi^{-1}F^*$ and modern variations of this function are applied.

## 15.3 Quantiles, quartiles, and box-and-whisker plots

The fiftieth percentile is the particular quantile, $F^{-1}(0.5)=$ MEDIAN, and is also called the second quartile. In general, the curve property *quantile* is an evaluation of some inverse cdf at an argument between zero and one. Both the Kendall and Buckland (1960) as well as the Marriott (1960) dictionary define the term "quantile" in terms of the *partition value of a variate*. However, neither dictionary defines the term "partition value." The curve property $F^{-1}(0.25)$ is called the *first quartile* and the property $F^{-1}(0.75)$ is called the *third quartile*. Since $|F^{-1}(0)|$ and $F^{-1}(1)$ are often infinite, the terms "zero$^{th}$ quartile" and "fourth quartile" are seldom applied.

The Normal, Cauchy, Laplace, and logistic density functions are all symmetric. Consequently, if the estimator $F^*$ is obtained from a size $n$ sam-

## 15.3. Quantiles, quartiles, and box-and-whisker plots

ple of iid Cauchy data, and then each sample member is transformed by $G^{-1}F^*$, where $G^{-1}$ is the Cauchy standard inverse cdf

$$G^{-1}(z) = 0.6745\ tan[\pi(z - 0.5)],$$

or this inverse function's Normal, Laplace, or logistic counterpart, then as $n \to \infty$ the new variates' expected value and median will both approach zero. Here the popultation median is equal to the population mean. In addition, for any $0 < p < 1$, the value of this variate's $p^{th}$ quantile will equal 0 minus its $(1-p)^{th}$ quantile.

In contrast to the inverse Normal scores transformation, for any $p$ other than 0.5, if some transformation like the log or square root does not work well, in the sense that it does not yield some symmetrically-distributed Normal or other desired variate, then it will often be the case that the transformed variate's median is *not* at an equal distance from both $p^{th}$ and $(1-p)^{th}$ quantile. Consequently suppose, when $p < 0.5$, three short horizontal lines of equal length are constructed, each above the other, at heights equal to the $p^{th}$, the median $(p = 0.5)^{th}$ quantile, and the $(1-p)^{th}$ quantile. If two line segments perpendicular to the $y$-axis are added to the picture so that the three horizontal line segments are enclosed and the $p^{th}$ and $(1-p)^{th}$ quantile lines serve as a rectangle's border, this figure is referred to as a *box plot*.

A box plot resembles a cigarette placed on end or, in this chapter, on its side, whose filter extends from the $p^{th}$ quantile, where $p < 0.5$, to the median. Like wisps of smoke coming from each end, most box plots are supplemented by appendages called whiskers. Among procedures that are based on estimated quantiles, box-and-whisker plots stand out, both in popularity and with regard to the variety of their applications. These graphical techniques are best described by an example based on actual data. Table 15.1 summarizes a file of California transfusion-associated AIDS cases. Within the file there were twice as many men as women who contracted AIDS through blood transfusion, and approximately twice as many white cases as all other races combined.

As of mid-January, 1992, almost three-quarters of the 648 patients had died. The time from transfusion to diagnosis of AIDS is referred to as *progression*. Over 13% of patients progressed to a diagnosis of AIDS in less than 500 days (about 1.4 years), while 84% developed the disease in the interval beginning 500 days and ending 5,000 days after the first transfusion date, $x_0$, noted in their file. The remaining 3% of patients did not develop AIDS until 5,000 or more days (about 13.67 years) from $x_0$.

Box-and-whisker plots help compare $(\mathcal{C}|\mathcal{A})$-type files indexed by the attribute-, nominal- or qualitative-valued, $\mathcal{A}$-type variable of $(\mathcal{C}|\mathcal{A})$. For

| Race & Gender | Status | | | Progression | | | Transfused after Diagnosis |
|---|---|---|---|---|---|---|---|
| | Total | Alive | Dead | < 500 | 500 - 5000 | > 5000 | |
| Total | 712 | 191 | 521 | 95 | 597 | 20 | 41 |
| Men | 481 | 122 | 359 | 77 | 390 | 14 | 36 |
| Women | 231 | 69 | 162 | 18 | 207 | 6 | 5 |
| White | | | | | | | |
| Men | 341 | 79 | 262 | 55 | 277 | 9 | 29 |
| Women | 136 | 33 | [103] | 9 | 123 | 4 | 2 |
| African Am. | | | | | | | |
| Men | 53 | 17 | 36 | 11 | 39 | 3 | 2 |
| Women | 48 | 19 | 29 | 4 | 44 | - | 1 |
| Hispanic | | | | | | | |
| Men | 58 | 16 | 42 | 10 | 46 | 2 | 5 |
| Women | 48 | 19 | 29 | 4 | 44 | - | 1 |
| Asian | | | | | | | |
| Men | 27 | 9 | 18 | 1 | 26 | - | - |
| Women | 13 | 7 | 6 | - | 13 | - | - |
| Other | | | | | | | |
| Men | 2 | 1 | 1 | - | 2 | - | - |
| Women | 1 | - | 1 | - | 1 | - | - |

Table 15.1. Characteristics of California Transfusion-Associated AIDS, 1981-91.

example, Figure 15.1 describes the $\mathcal{C}$ variate, White female progression. The subsample's data corresponds to the $n = 103$ deaths entry bracketed in Table 15.1. Box-and-whisker plots provide a simple way to compare sample constituents, such as indicators of variation called hinges, that are based on quantile estimators.

A size $n$ sample, where $n_{\text{low}} = (n+3)/4$ and $n_{\text{high}} = (3n+1)/4$ are both integers, yields the box hinges that extend from $X_{(n_{\text{low}})}$ to the value of $X_{(n_{\text{high}})}$. An example of an upper hinge $X_{(n_{\text{high}})}$ is shown in Figure 15.2. Chambers, Cleveland, Kleiner, and Tukey (1983), as well as Tukey (1977), provide comprehensive descriptions of a box plot. The box plot's inventor, J. Tukey, indicates that a hinge can be crudely defined to be a sample quartile. Here the lower box plot hinge estimates $F^{-1}(0.25)$ while the upper hinge estimates the quartile $F^{-1}(0.75)$. The pair of estimates of the quantile curve properties $F^{-1}(0.1)$ and $F^{-1}(0.9)$ often furnish whisker endpoints. For example, the whiskers in Figure 15.1 extend from about 30 days to 3,970 days.

It is common practice to display the positions of individual values larger than an estimate of $F^{-1}(0.9)$ or smaller than an estimate of $F^{-1}(0.1)$ by some particular plotting character. Cleveland shows small-circle-extreme-

## 15.3. Quantiles, quartiles, and box-and-whisker plots

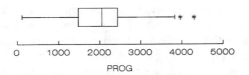

Figure 15.1. White female progression.

value-box-plot *exceedances* on the cover of his (1985) book, *The Elements of Graphing Data*. The default outside value chosen in the SYSTAT (1994) computer package is shown as an asterisk, while a small circle denotes a far outside value in this package. According to the authors of SPSS, box plots are "...based on the median, quartiles, and extreme values. The box represents the interquartile range which contains 50% of values. The whiskers are lines that extend from the box to the highest and lowest values, excluding outliers. A line across the box indicates the median."

The term "exceedance" refers to the number of observations, $m$, in a sample which exceed in magnitude at least $k$ observations in some other sample (Marriott, 1960). It is somewhat unclear whether it is the number $m$ referred to by Marriott and other authors that is itself the exceedance or whether it is each of the $m$ large or small variate values themselves that are defined to be exceedances. Were exceedances defined to be variate values, then the variates with these values can often be thought of as the counterparts of naturally occurring outliers.

Later chapters will review a concept known as censoring that is the root cause of much uncertainty about statistical curves. David (1970, Section 6.3) discusses what he calls the *censoring at the right* that occurs "...when it is decided to stop an experiment as soon as $N(< n)$ times under test have failed." Unlike exceedances, for which it unclear whether it is the number of variates or the variates themselves that are referred to, when censoring occurs the entity that is censored is a variate or measurement.

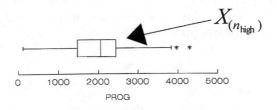

Figure 15.2. Box hinge.

Unfortunately, to make matters still more complicated, a censored measurement can either be the result of *Type I censoring* that results from curtailing a sample below or above a fixed point, or from curtailing a sample both below and above two distinct points. Censoring can also occur when sampling stops as soon as a preselected number of measurements, in David's notation $N$, are taken. This is known as *Type II censoring*. Type II censoring is encountered much more commonly than its Type I counterpart and, hence, the word "censoring" will refer to the Type II variety below.

Censoring often refers to the gathering of data in ways that resemble, for a sample, what the term "truncation" refers to in the context of a population. In actuality what is truncated is a model, not a population. Like censoring, the edges A and B of a truncated model like

$$\{I_{[A,B]}(x)\}\{\phi([y-\mu]/\sigma)\}/\{\sigma K(\mu,\sigma)\}$$

can be determined either by some known interval [A, B] or to some known proportion $K(\mu,\sigma)$. Here $K(\mu,\sigma)$ assures that the area beneath the model equals one. Because both truncation and censoring have so much to do with uncertainties about statistical curves, each is discussed in a separate chapter below, and the remainder of this chapter is devoted to topics that have a close relationship to quantiles and transformations, yet have only peripheral effects on curve specification.

## 15.4 Normal ranges and box sizes

A study by Heath (1946) lists data for students who had what Heath called: "general all-round normal reactions." For example, among 265 individuals, the mean blood pressures recumbent mm. Hg is given as systolic 114.9 and diastolic 71.7, with corresponding ranges, 98-146 and 40-92. Benjamin (1959) describes the Heath study and also summarizes a second study in which, according to Benjamin, "Blood pressures fitted a Normal curve and the dispersion, as indicated by the mean ± twice the standard deviation, was (mm. of Hg): Males, Systolic 107.3-236.9, Diastolic 55.4-127.4." The intervals 107.3-236.9 and 55.4-127.4 described in this second study, the ranges presented by Heath, and "box sizes" such as

$$X_{(n_{high})} - X_{(n_{low})},$$

are not related clearly to each other.

Most packaged statistical programs provide box-and-whisker plotting capability. Thus, since box-and-whisker plots are now so easy to compute,

## 15.4. Normal ranges and box sizes

in contrast to the era in which Benjamin and Heath conducted their investigations, there is less justification today for the presentation of mean ± twice the standard deviation intervals or, even worse—unweighted sample ranges. The word *range* can denote both an interval $[X_{(1)}, X_{(n)}]$ that is bordered by $X_{(1)}$ and $X_{(n)}$, and the single statistic, $X_{(n)} - X_{(1)}$. However, most statisticians today prefer to call the single scalar-valued statistic, $w = X_{(n)} - X_{(1)}$, the *range*.

Interval $[X_{(1)}, X_{(n)}]$ versus scalar $X_{(n)} - X_{(1)}$ ambiguity is not the only usage of the term "range" that is confusing. For instance, if variate values are added to an already existing sample, as data piles up because $n$ increases, the statistic $X_{(n)} - X_{(1)}$ cannot decrease in magnitude. Consequently, for Normal, lognormal, and many other data sets, publications that convey spread-related information solely by disclosing the endpoints of an interval $[X_{(1)}, X_{(n)}]$, as sample size $n$ increases shoot at a moving—in the sense of ever expanding—target. This target is different from that aimed at either by: box-plot hinges; or mean ± twice the standard deviation intervals, both of which estimate particular properties of a single curve that describes a distribution, not a sample.

The concept *normal range* can be misleading. Kendall and Buckland's, Marriott's, and the earlier (1939) Kurtz and Edgerton dictionaries do not legitimize the term "normal range" by defining it. It is ignored by Marriott even though this work prepared for the International Statistical Institute as an update of the Kendall and Buckland dictionary contains entries that define the terms: Normal deviate; normal dispersion; Normal distribution; normal equations; Normal equivalent deviate; normal inspection; Normal probability paper; Normal scores tests; Normalization of frequency function; Normalization of scores; Normalising transform; Normit and; Normix. (So many uncertainties are just unpleasant facts of statistical life that it seems a pity to call up the concept, normal range, when box plot and many other modern alternatives can now be utilized conveniently.)

In the light of the importance of box plots as modern replacements for normal range presentations, it is unfortunate that what *fences* are exactly, is difficult to define. For example, J. Tukey, the developer of the box plot, defines a *fence* as follows: "fence: inner fences are one step outboard of the hinges: outer fences are two steps outboard;" while, as far as outside values are concerned: "outside value: a value outside the inner fences (and if "far out" is also mentioned, inside the outer fences)" (1977).

The definition of a graphical tool like the box-and-whisker plot is complicated by a subtle effect of sample size $n$. For example, when $n$ is between 20 and 200, estimates of quantiles $F^{-1}(0.1)$ and $F^{-1}(0.9)$ serve as reasonable choices of whisker endpoints. For larger samples, endpoints that estimate $F^{-1}(0.05)$ and $F^{-1}(0.95)$, or even $F^{-1}(0.01)$ and $F^{-1}(0.99)$, seem more in accord with hypothesis-testing procedures that, in later chapters, are discussed in terms of statistical parameters and curves. For this reason, whisker endpoints and critical region boundaries differ considerably in a conceptual sense. The latter are based on the distribution of a test statistic like the sample mean. On the other hand, $F^{-1}(0.05)$ and $F^{-1}(0.95)$ are properties of a curve that pertains to a single sample element—in essence a small piece that might, perhaps, later be inserted into the completed test statistic.

We will now consider procedures that can help deal with curve specification issues that are more complex than those that pertain to a single variate's distribution's variability. One of these procedures extends the idea behind what is called an *interval estimate* to deal with uncertainty concerning a specialized regression curve. The other can be thought of as a composite of variability and uncertainty that is of particular value when an individual population member is targeted for prediction purposes.

## 15.5 Confidence bands and prediction bands

Because the singular counterpart of the word "data"—datum—sounds so odd, it is a useful term. When it appears, it provides a clue that what is referred to is, in effect, a sample of size one. A single box-and-whisker plot is designed to summarize what is known about a single datum's distribution. But where does the sample, taken as a whole, come in?

The terms "confidence band" and "prediction band" illustrate one answer. Parametric regression curves are often enclosed within either of two distinct types of bands, *confidence bands* or *prediction bands*. The former are pairs of curves that resemble, in two dimensions, certain one-dimensional hypothesis tests' critical regions. (The curves and parameters upon which hypothesis test construction is based will be discussed in later chapters.) A confidence band encloses, or, with a known probability, does not enclose, a parametric curve like $\mu_x$, or a curve-property-trace like $E(Y|x)$. On the other hand, a prediction band is constructed in reference to a variate $Y$'s value $y$ per se.

Because it concerns $Y$'s distribution, one variety of prediction band can be thought of as a two-dimensional generalization of an interval that begins at the $0.05^{th}$ quantile, $F^{-1}(0.05)$, and extends to the $0.95^{th}$ population quantile, $F^{-1}(0.95)$. It generalizes this interval in two ways: (1) it surrounds a curve whose evaluations change as some variable $x$'s value changes and; (2) like its confidence band counterpart, it encloses this curve with a known probability. It differs from a confidence band because the curve in question is $Y|x$; not $\mu_x, E(Y|x)$, or some other trace of parameter or curve property evaluations.

A confidence band provides information that concerns the credibility of a regression curve's estimator. For this reason, as $n \to \infty$ such bands converge to a single curve. Prediction bands also, like confidence bands, move toward each other as $n$ increases but, unlike confidence bands, rarely collapse to form a single, perfectly predictive, limiting curve. Their job is to describe credibility up to, but not beyond, the limits placed by a distribution's dispersion, as is often quantified by a curve property like Var or a parameter like $\sigma$.

A prediction band conveys information that pertains to an individual population member, while a confidence band assesses what either a parametric regression curve like $\mu_x$, or a property-based curve such as $E_f(Y|x)$, looks like. Therefore, what will occur to a given subject can only be known exactly in the unrealistic situation where $Var(Y|x) = 0$ or $\sigma_x = 0$. However, as $n \to \infty$, any consistent regression curve estimator of $\mu_x$ or $E_f(Y|x)$ will converge to $\mu_x$ or $E_f(Y|x)$.

Tarter, Lock, and Ray (1995) as well as Tarter and Lock (1993, Section 7.6) discuss examples of box plots that are the simplified nonparametric counterparts of prediction bands and nonparametric confidence bands. Box plot simplifications of prediction bands are based on estimates of $F^{-1}(0.75|x)$ and $F^{-1}(0.25|x)$ that, as $x$ sweeps through a sequence of values, together form the upper and lower edges of an idealized box plot that corresponds to a $F^{-1}(0.50|x)$ population median-based nonparametric regression curve. Generally speaking, the box plots of Section (15.4) describe $(C|A)$-type data, while the examination of $(C|C)$-type data can be based on estimates of curves like $F^{-1}(0.75|x)$ and $F^{-1}(0.25|x)$. (On the one hand, like any box plot, the curves that estimate $F^{-1}(0.75|x)$ and $F^{-1}(0.25|x)$, lack the credibility assessment that is provided by a typical pair of prediction bands. On the other hand, since they are nonparametric, these estimators of $F^{-1}(0.75|x)$ and $F^{-1}(0.25|x)$ can assume shapes that are not constrained by the assumption of a particular linear or other parametric model.)

## 15.6 Notches, stems, and leaves

Besides $F^{-1}(0.75|x)$ and $F^{-1}(0.25|x)$ where $x$ varies continuously, box plots are generalized in a second way. *Notched box plots* assess subgroup medians graphically. Consider, for example, the two hinges which specify the distance that separates the narrow edges of a box. These hinges determine box length. On the one hand, both of these hinges are estimators of a specific pair of curve properties while, on the other, box width can either be determined on the basis of aesthetic considerations or, alternatively, can be modified to convey other specialized information.

For example, box-plot bands describe the sweep of a sequence of hinges that change as the variable $x$, within $F^{-1}(0.75|x)$ and $F^{-1}(0.25|x)$, changes. Each individual hinge, for a particular value of the variable $x$, corresponds to the associated conditional density $f(y|x)$. When the marginal density $f(y)$ is described or estimated by a box plot, the edges of the plot can be indented to disclose pertinent information. When two adjoining box-plots are situated so that there is no horizontal line which intersects a pair of box-plot notches, they provide a faint graphical clue that a test of median curve-property difference will, when calculated, prove to be significant at the 0.05 level.

Box-and-whisker plots provide an excellent way to review a single sample and to compare multiple samples. Suppose, for example, as described in Figure 15.3 a single variate such as age is studied in terms of its association with an attribute-valued key variable, such as race, based on Table 15.1 data. For this sort of variate-with-variable analysis, the box-and-whisker plot has few rivals.

To describe the female components of the third column of Table 15.1, in Figure 15.3, $x = 1$ denotes Caucasian American (Am.); $x = 2$, African Am.; $x = 3$, Hispanic Am.; $x = 4$, Am. of Asian ancestry; and $x = 5$, other. Since only one woman corresponded to the bottom row of this table's third column, there was not enough information to widen the line that appears at the right side of Figure 15.3 in order to form a box.

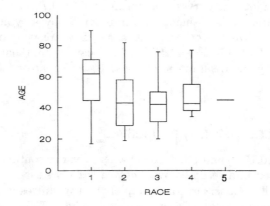

Figure 15.3. Box plots for key variable race.

## 15.6. Notches, stems, and leaves

| | |
|---|---|
| 1  | 779 |
| 2  | 0244 |
| 2  | 56666777788899999 |
| 3  | 001111223444 |
| 3H | 677788899 |
| 4  | 0022222233344 |
| 4  | 555556667777789 |
| 5M | 00011334 |
| 5  | 5667777888889999 |
| 6  | 011112223333334444 |
| 6H | 5555666777778899 |
| 7  | 011222334 |
| 7  | 55566667778999 |
| 8  | 000223 |
| 8  | 8 |
| 9  | 0 |

Figure 15.4. Stem-andleaf plot of AIDS group $x = 1$ data.

Figure 15.4 shows a graphical alternative to a box-and whisker plot called a *stem-and-leaf diagram*. To construct this variety of diagram, first extract the leading digit (here the digit that falls in the ten's place) of each data-value to form a common stem. Then assemble digits horizontally to flesh out the leaf. In addition, since stem-and-leaf plots are paired

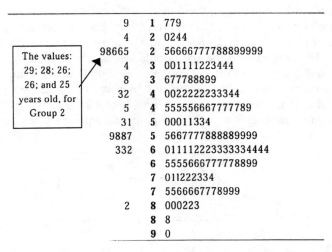

Figure 15.5. Stem-and-leaf, left side Group 2, right side Group1.

traditionally with box plots, box plot hinges and the median are marked by the letters H and M, as shown in Figure 15.4.

Stem-and-leaf displays can be used to compare subsample pairs as shown by Figure 15.5. However, unlike box plots, they are not ideally suited for the comparison of three or more different subsamples. It is hard to separate sample size differences from distributional differences when data from samples of different sizes are described by back to back stem-and-leaf plots. However, for samples of equal size, back-to-back stem-and-leaf plots can play the role for analyses of many varieties of data sets, that population pyramids can play within analyses of census data gathered for men and for women. The left side of Figure 15.5 describes the subsample of the 29 women, listed in the $7^{th}$ row and 3rd column of Table 15.1, whose progression is greater than 500 days and who were not transfused after diagnosis. (One woman's progression equaled 500 days exactly.)

## 15.7 The log transformation and skewness

Box plots can provide a simple way to check on the effectiveness of a transformation like $Y = \log(X + \hat{\tau})$. Box plots describing over-transformed or hyper-transformed data, characteristically resemble the plot shown in Figure 15.6. This pattern occurs because rather than a sample median that bisects the box, previously small values are spread out inordinately, while the spacings that separate upper-estimated quantiles are compressed.

Because today the box plot and many other modern graphical and curve estimation procedures can be computed easily, the pair of curve properties, skewness and kurtosis, seem exotic, in the sense that they are strikingly unusual. In particular, while the population mean, median, standard deviation, and covariance all have one or more simple relationships with the $\mu, \sigma$, or $\rho$ parameters within one or more density, cdf, and data-generation models, this is not true of skewness and kurtosis. For $k = 2, 3, ...$; suppose the curve property, $k^{th}$ ordinary population moment about the mean,

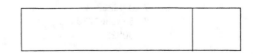

Figure 15.6. Box plot from overtransformed data.

## 15.7. The log transformation and skewness

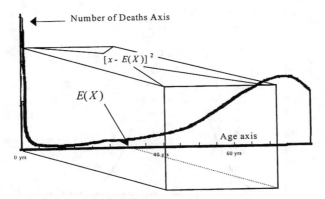

**Figure 15.7.** Contribution to the variance that is enhanced for higher order moments.

defined as the integral

$$E\{[(X - E(X)]^k\} = \int_{-\infty}^{\infty} (x - E(X))^k f(x)dx$$

exists. Then for a unimodal curve with mode MODE, skewness has been defined as

$$(E(X) - MODE)/\sqrt{E\{[(X - E(X)]^2\}},$$

as well as

$$\Theta = \left[\sqrt{\beta_1}(\beta_2 + 3)\right]/[2(5\beta_2 - 6\beta_1 - 9)],$$

where

$$\beta_1 = (E\{[(X - E(X)]^3\})^2/(E\{[(X - E(X)]^2\})^3,$$

$$\beta_2 = E\{[(X - E(X)]^4\}/(E\{[(X - E(X)]^2\})^2,$$

and even as

$$\gamma_1 = E\{[(X - E(X)]^3\}/(E\{[(X - E(X)]^2\})^{3/2}.$$

Besides the problems that stem from truncation which were illustrated in Section 4.5, for $k > 3$, higher-order moments are beset with other computational and conceptual difficulties. When even the standard deviation is sensitive to edge-effects, one can imagine what would happen, for example, were the rectangular block of Figure 10.3 (that is reproduced as Figure 15.7) replaced by a hypercube of $k > 3$ dimensions.

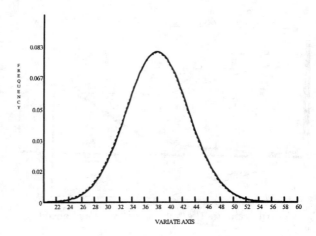

Figure 15.8. Normal (solid) and Lognormal (dotted) curves.

The threshold parameter $\tau$ that appears within the formula for the lognormal and other transformation-based models, can often be interpreted in terms of the concept skewness. The role that $\tau$ can play is illustrated by a study based on the 1936 paper published by Fisher, in which he laid the foundation for the field of discriminant analysis. To illustrate his ideas, Fisher applied what he called a *discriminating linear function* to three species of Iris: Iris setosa; versicolor; and virginica.

Figure 15.8 shows the application of two parametric curve estimators to the 50 virginica scores graphed by Fisher at the end of his paper. The parameter $\tau$ can be estimated efficiently by calling upon a combination of likelihood and graphical techniques (Tarter and Hill, 1997). For example, Figure 15.8 shows two estimated parametric curves that are nearly indistinguishable. One of these curves was estimated from the $n = 50$ virginica sample by the application of graphical procedures. It estimated $\tau$ to be -457.43, and the mean and standard deviation of the $Y = ln(X + 457.43)$ variate to be 6.2055 and 0.0100, respectively.

The other curve applied Normal model-based methodology to this same data to determine the $X$ variate's mean and standard deviation to be 37.9371, and 4.9727, respectively. Yet even though the curves that described the virginica variate's distribution were obtained by these two very different procedures, when these curves were graphed, Figure 15.8 was obtained. Besides demonstrating how accurate the $\tau$ estimation procedure can be, these two almost identical curves illustrate the near identity of certain Normal and lognormal curves.

## 15.7. The log transformation and skewness

The connection between the parameter $\tau$ and skewness is illustrated by the virginica curves. The virginica score support region can be taken to be within the interval that begins at 20 and ends at 60. Nevertheless, $\tau$ was estimated to be -457.43, a number that is far smaller than 20. In general, as shown by the author and Lock (1993), when a threshold is estimated to be far smaller than a support region's endpoint, this suggests that the curve being estimated can be taken to be symmetric. Conversely, were it the case that actually $\tau$ is close to the endpoint at 19, for example, then the resulting curve would look the way lognormal densities are usually thought to look, appreciably skewed.

When skewness can be interpreted in terms of a three-parameter lognormal model, kurtosis can be interpreted in terms of mixtures of identically scaled subpopulations. Each subpopulation has identically-scaled but not identically situated Normal densities. Pronounced kurtosis often results when two mixture components are separated with location parameters whose difference divided by $\sigma, |\mu_1 - \mu_2|/\sigma$ is larger than 0.2. Within the two component mixture model described in Appendix I.4, when the value of the mixing parameter equals 0.5, and when mixture components have a common scale parameter $\sigma$, the density curve that describes this mixture will be symmetric. Yet when $|\mu_1 - \mu_2|/\sigma > 0.2$, this density will either be bimodal or, if $|\mu_1 - \mu_2|/\sigma$ is sufficiently small, it will show a flat plateau-like top that is a distinguishing feature of *kurtotic* curves.

The Cauchy density is a curve with the sort of heavy tails that characterize positive kurtosis. If a packaged computer program estimates kurtosis to be pronounced, then a researcher should be on guard against either the presence of lurking variates that underlie mixture structure or the heavy tailedness characteristic of Cauchy distributions.

# CHAPTER 16
# Noncentrality Parameters and Degrees of Freedom

## 16.1 The $(C_1|A_2)$ case and variate-variable relationships

The box-plot techniques discussed in the previous chapter provide a convenient way to compare continuous response variate densities graphically. These plotting techniques correspond to a class of analytic techniques for the study of experimental and observational $(C_1|A_2)$ data known as analysis of variance (ANOVA) that has a comprehensive literature made up of many fine books and important papers. The analysis of this type of information is facilitated by a quantity called a noncentrality parameter $\delta$. However, before explaining what $\delta$ actually is—in terms of the $(C_1|A_2)$ data case where $A_2$ assumes three or more values—the $t$-test for group differences, which concerns the special case of $(C_1|A_2)$ where $A_2$ assumes two distinct values, will provide relevant background. Assume that the $C_1$ variate has a Normal density and the binary, in other words dichotomous-valued variable $A_2$ determines the index $j$ of the location parameter $\mu_j$ of this Normal density, an index that takes on the value 1 or 2.

Because $A_2$ is binary-valued, a difference between the two curves, for example

$$e^{-(1/2)[(z-\mu_1)/\sigma]^2}/(\sigma\sqrt{2\pi}) \quad \text{and} \quad e^{-(1/2)[(z-\mu_2)/\sigma]^2}/(\sigma\sqrt{2\pi}),$$

is tantamount to a relationship between $A_2$ and $C_1$. As far as tests and estimates of $A_2$-with-$C_1$ relationship are concerned, these are usually based on two individually iid data sets,

$$\{[(C_1|A_2=1)]_{n(1)}\} \quad \text{and} \quad \{[(C_1|A_2=2)]_{n(2)}\}.$$

The subscripts $n(1)$ and $n(2)$ designate the exact numbers of measurements within each data set, where each set is assumed to contain at least one measurement.

For hypothesis-testing purposes, the three $H_o$ null hypotheses: $\mu_1 = \mu_2$; $\mu_1 \leq \mu_2$ and; $\mu_1 \geq \mu_2$, are equivalent, respectively, to the $H_o$ hypotheses: $\mu_1 - \mu_2 = 0$; $\mu_1 - \mu_2 \leq 0$ and; $\mu_1 - \mu_2 \geq 0$. The left-hand side of these three inequalities illustrates why the statistic $\bar{y}_1 - \bar{y}_2$ often serves as part of a statistic's numerator. Here $\bar{y}_1$ denotes the sample mean of the $n(1)$ values of $C_1$ that are contained in the

$$\{[(C_1|A_2 = 1)]_{n(1)}\}$$

subsample. Correspondingly, $\bar{y}_2$ represents the sample mean of the $n(2)$ values of $C_1$ that are contained in the $\{[(C_1|A_2 = 2)]_{n(2)}\}$ subsample.

Before reviewing standard ways to generalize the statistic $\bar{y}_1 - \bar{y}_2$, several important specification issues that concern this statistic should be mentioned. The measured values that contribute to the two sample means whose difference determines $\bar{y}_1 - \bar{y}_2$ are said to be allocated to two cells. Besides the large literature that concerns homoscedasticity of within-cell subsample values, much is now known that concerns other aspects of ANOVA validity. In particular, the next chapter will discuss techniques that can adjust for the poor quality of outlying observations. (Problems with cell-variance differences are often due to these observations.)

Later chapters of this book constitute a single unit that concerns hypothesis-testing issues. While Chapter 17 deals with outliers and validity issues that are related to sample quality, Chapters 18, 19, and 20 concern the specification of population structure. In particular, the last two chapters illustrate the limitations of a basic relationship known as the Central Limit Theorem (CLT), and why ANOVA uncertainties are not necessarily cleared up as sample sizes increase.

The random variate $(\bar{y}_1 - \bar{y}_2) - (\mu_1 - \mu_2)$ measures the difference between the estimator $(\bar{y}_1 - \bar{y}_2)$ of the composite parameter $(\mu_1 - \mu_2)$, and the value of $(\mu_1 - \mu_2)$ itself which, under the null hypothesis, is almost always equal to zero. However, before a $(\bar{y}_1 - \bar{y}_2) - (\mu_1 - \mu_2)$ test-statistic's value can be looked up in a table, the variate $(\bar{y}_1 - \bar{y}_2) - (\mu_1 - \mu_2)$ must be standardized to allow for scale parameter differences. For this reason,

$$t = \frac{\bar{y}_1 - \bar{y}_2 - (\mu_1 - \mu_2)}{s_p \sqrt{\frac{1}{n(1)} + \frac{1}{n(2)}}}$$

is often calculated, where the statistic $s_p^2$, called the *pooled estimate of variance*, is expressed as

$$s_p^2 = \frac{\sum_{i=1}^{n(1)}(Y_{1i} - \bar{y}_1)^2 + \sum_{i=1}^{n(2)}(Y_{2i} - \bar{y}_2)^2}{n(1) + n(2) - 2}.$$

## 16.1. The $(\mathcal{C}_1|\mathcal{A}_2)$ case and variate-variable relationships

The comma is omitted between the number 1 or 2 and the index $i$, and in addition, each individual $Y_{1i}$ represents the $i^{\text{th}}$ $\mathcal{C}_1$ variate of the data case

$$\{[(\mathcal{C}_1|\mathcal{A}_2 = 1)]_{n(1)}\},$$

while each individual $Y_{2i}$ represents the $i^{\text{th}}$ $\mathcal{C}_1$ variate of

$$\{[(\mathcal{C}_1|\mathcal{A}_2 = 2)]_{n(2)}\}.$$

When all $\mathcal{C}_1$ observations are iid Normal (which among other things, implies that the two scale parameters of the two group-specific density models are equal) it follows that the statistic $t$ will be Student-$t$ distributed with $n(1)+n(2)-2$ degrees of freedom. In the special case where $n(1)+n(2)-2 = 1$, this Student-$t$-statistic is a Cauchy variate. (The sampling distribution parameter $\nu$ that represents degrees of freedom is discussed later in this chapter.)

Often $\mathcal{A}_2$ of $(\mathcal{C}_1|\mathcal{A}_2)$ takes on more than two values, where these values index categorizations called *levels* that distinguish distinct cells. In all but one respect, analyses of studies concerning three or more levels are similar to studies of $(\mathcal{C}_1|\mathcal{A}_2)$ two-level data. It is due to this distinction that the statistical technique often applied to study this form of data is called *analysis of variance*.

A new approach that differs from that required by the two-level special case is necessary because of a simple mathematical problem. While the difference between two parameters, for example, $\mu_1$ and $\mu_2$, can be summarized by the single parameter $\mu_1 - \mu_2$, differences among three or more parameters cannot be dealt with in this way. For example, the parameter $\mu_1 - (\mu_2 + \mu_3)/2$ does indicate how much difference there is between the first location parameter on the one hand, and the average of the second and third location parameters on the other. Yet location parameters $\mu_2$ and $\mu_3$ could be different, and $\mu_1 - (\mu_2 + \mu_3)/2$ could still equal zero.

Although the single equality $\mu_1 - (\mu_2 + \mu_3)/2 = 0$ does not necessarily imply that all location parameters are equal, this single example does not preclude the existence of some other function of all location parameters that does imply that all location parameters are equal. Specifically, suppose that three location parameters are averaged to yield a grand mean or, alternatively, a parameter $\mu_{\cdot}$. For this three level case, a dot or bullet follows the symbol $\mu$ which designates that $\mu_{\cdot}$ a quantity called "mu-dot" or the *population grand mean*, equals the location parameter average $(\mu_1 + \mu_2 + \mu_3)/3$.

Now consider the composite parameter, $\delta^2 = (\mu_1 - \mu_{\cdot})^2 + (\mu_2 - \mu_{\cdot})^2 + (\mu_3 - \mu_{\cdot})^2$. This new composite equals zero if and only if all three location parameters $\mu_1, \mu_2,$ and $\mu_3$ have the same value which, of course, would then be equal to the $\mu_{.}$ (parameter's value. The exponent 2 in equation $\delta^2 = (\mu_1 - \mu_{.})^2 + (\mu_2 - \mu_{.})^2 + (\mu_3 - \mu_{.})^2$ plays an important role. Without squaring, or some alternative operation like the absolute value | |, a positive value of $(\mu_1 - \mu_{.})$ could be counterbalanced by a negative value of $(\mu_2 - \mu_{.})$.

The squaring of each difference within the sum of the squared differences $(\mu_j - \mu_{.}^2)$ where $j$ equals 1, 2, or 3, assures that $\delta^2$ equals zero only if every squared difference equals zero. Also, notice that the definition of the parameter $\delta^2$ can be extended to include any number of location parameters, for example, $(\mu_1, \mu_2, \mu_3, \ldots, \mu_m)$, where $m > 3$. When

$$\mu_{.} = (\sum_{j=1}^{m} \mu_j)/m,$$

it follows that $\delta^2$ equals zero when, and only when, $\mu_1 = \mu_2 = \mu_3 \ldots = \mu_m$.

Given that the parameter $\delta^2$ is chosen to be the $m > 2$ level analog of the difference parameter $\mu_1 - \mu_2$, the next important question is: What statistic should be used in place of $\bar{y}_1 - \bar{y}_2$ to estimate $\delta^2$? To answer this question, similarly to the construction of $\mu_{.}$, let $\bar{y}_{.}$ equal the mean of all observations from the combined subsamples of sizes $n(1), n(2), ..., n(m)$. The statistic $SS_m$ is based on the $m$ sample means, each represented by the subscripted letter $y$ topped by the traditional bar, $\bar{y}_i$; $i = 1, 2, .., m$. Here,

$$SS_m = \sum_{i=1}^{m} ((\bar{y}_i - \bar{y}_{.})^2)$$

would seem to be a reasonable estimator of $\delta^2$.

One modification of $SS_m$ is required before the $SS_m$ statistic can estimate a wide variety of $\delta^2$ parameters. When $n(1) = n(2) = n(3) = \ldots = n(m)$, the usual average of the subsample means, $\bar{y}_1, \bar{y}_2, \ldots, \bar{y}_m$, in other words, the average defined as

$$(1/m) \sum_{i=1}^{m} \bar{y}_i,$$

is identical to the $\bar{y}_{\bullet}$ statistic. Yet when any one $n(i), 1 < i \leq m$, does not equal another sample size, $n(k), 1 < k \leq m$, then there is no guarantee that

$$(1/m) \sum_{i=1}^{m} \bar{y}_i$$

equals the $\bar{y}_{.}$ statistic.

## 16.1. The $(C_1|A_2)$ case and variate-variable relationships

To show this, let the $j^{\text{th}}$ measurement of the $i^{\text{th}}$ level be represented by $Y_{ij}$, where $Y$ is chosen to help distinguish this variate from all sample means. For $i \leq m$, it follows that

$$\sum_{j=1}^{n(i)} Y_{ij}$$

equals $n(i)\bar{y}_{i.}$. Setting $n$ equal to $n(1)+n(2)+n(3)+\ldots+n(m)$, the grand mean of all $n$ measurements, it follows that $\bar{y}_{..}$ equals

$$(1/n) \sum_{i=1}^{m} \sum_{j=1}^{n(i)} Y_{ij}.$$

Consequently the grand mean $\bar{y}_{..}$ equals

$$(1/n) \sum_{i=1}^{m} n(i)\bar{y}_{i.}.$$

In other words, the grand mean of the observations equals the weighted average of the subsample, sometimes called *cell means* $\bar{y}_i$; $i = 1, 2, \ldots$, where cell means are weighted by the corresponding sample sizes.

For each index $i$ that is less than or equal to $m$, each of the $m$ individual $(\bar{y}_i - \bar{y}_{.})^2$ components of $SS_m$ is computed from a separate subsample. Consequently, since these subsamples may differ in size, a weighted sum of $(\bar{y}_i - \bar{y}_{.})^2$ is often applied in place of the unweighted sum

$$SS_m = \sum_{i=1}^{m} (\bar{y}_i - \bar{y}_{.})^2.$$

Division of the weighted sum by the quantity $m-1$, yields the mean square for (or between) cells (or categories); an expression that is central to analyses of variance.

The mean square for cells is represented as

$$s_m^2 = \{\sum_{i=1}^{m} n(i)(\bar{y}_i - \bar{y}_{.})^2\}/(m-1).$$

Note that the population analog of $s_m^2$, which is commonly represented as

$$\{(\sigma\delta)^2/(m-1)\},$$

equals
$$\{\sum_{i=1}^{m} n(i)(\mu_i - \mu_.)^2\}/(m-1).$$

This composite parameter equals zero if, and only if, the location parameters that describe non-empty cells, in other words cells where $n(i) > 0$, are all equal in value.

The upper summation limit $m$ within the expression for $s_m^2$ can be any positive integer larger than 1. Yet it is interesting to evaluate $s_m^2$ for the special $m = 2$ case. A small amount of algebraic manipulation discloses that when $m = 2$, the statistic

$$s_m = (\bar{y}_1 - \bar{y}_2)/\sqrt{\frac{1}{n(1)} + \frac{1}{n(2)}}.$$

Thus, in the $m = 2$ case, a $t$-statistic can be expressed as $s_m/s_p$. Obviously, statistics constructed from three or more means are more complex than their dual mean counterparts. Consequently, even in the null hypothesis case, when the number of cells, $m$, is greater than two, the ratio $s_m/s_p$ can hardly be expected to have simply a Student-$t$ distribution.

To illustrate how the size of m affects this ratio's distribution, consider that the formula for the pooled estimate of variance,

$$s_p^2 = \frac{\sum_{j=1}^{n(1)} (Y_{1j} - \bar{y}_1)^2 + \sum_{j=1}^{n(2)} (Y_{2j} - \bar{y}_2)^2}{n(1) + n(2) - 2},$$

can be generalized easily to the $m > 2$ case to form

$$s_p^2 = \frac{\sum_{j=1}^{n(1)} (Y_{1j} - \bar{y}_1)^2 + \sum_{j=1}^{n(2)} (Y_{2j} - \bar{y}_2)^2 + \ldots + \sum_{j=1}^{n(m)} (Y_{mj} - \bar{y}_m)^2}{n(1) + n(2) + n(3) + \ldots + n(m) - (m-1)}.$$

Under the null hypothesis that $\mu_1 = \mu_2 = \mu_3 = \mu_4 = \ldots = \mu_m$, the square of $s_m/s_p$ has a cdf whose values are tabled under the title, "F distribution with $m - 1$ numerator degrees of freedom and $n - m$ denominator degrees of freedom." (The capital letter F was chosen long ago to honor the statistician R. A. Fisher and, consequently, does not represent some general cdf, but instead, in most contexts, designates a variate whose cdf belongs to a particular family of distributions.)

## 16.2 Invariance and confounding

For many ANOVA models, the conceptual tricks outlined in the previous section can not only be extended to any finite number of levels of a single variable, like the "factor" $\mathcal{A}_2$, but they can also facilitate studies of multiple attribute variables. The reasoning that underlies these extensions is illustrated by the technique called "Model I ANOVA."

Since no matter how many alphabets are thrown into the pot of statistical notation, the soup still calls for some new ingredient, many symbols are called upon to perform dual roles. Hence because the letters $\alpha$ and $\beta$ appear at the beginning of the Greek alphabet, it is small wonder that these symbols represent many distinct quantities. For example, one common notational device that is based on $\alpha$ and $\beta$ is called a *two-factor additive linear ANOVA* model that describes how the measurement $Y_{ij}$ is generated.

The data-generation model $Y_{ij} = \mu_. + \alpha_i + \beta_j + \sigma Z_{ij}$ is, in reality, a recipe that describes the processes thought to have cooked up $Y_{ij}$. Constraints are imposed so that the sum of all $m$ parameters $\alpha_i$ equals zero and the sum of all parameters $\beta_j$ equals zero. These constraints, together with the assumption that all $Z_{ij}$ are generated from a standard model like the Normal, justify the interpretation of $\mu_.$ as a grand mean about which row effects, designated by the letter $\alpha$ and column effects, designated by the letter $\beta$, differ.

We will see in Chapter 20 that it can make a great deal of difference in terms of the applicability of the CLT whether a statistical method is designed with an eye to a parameter or a curve property. Because the data-generation model $Y_{ij} = \mu_. + \alpha_i + \beta_j + \sigma Z_{ij}$ can be rewritten as $Y_{ij} - (\mu_. + \alpha_i + \beta_j) = \sigma Z_{ij}$, the model's validity depends on whether or not it is true that a location parameter changes from cell to cell. In general, when $Y_{ij} - (\mu_. + \alpha_i + \beta_j) = \sigma Z_{ij}$ it follows that $E[Y_{ij} - (\mu_. + \alpha_i + \beta_j)] = \sigma E[Z_{ij}]$. Consequently, for example, were $Z$ a standard or truncated Cauchy variate, the linkup between ANOVA and the CLT cannot be trusted to make up for curve property existence ambiguities.

ANOVA notation quickly exhausts the supply of individual letters from the middle of the alphabet like $i$, $j$, and $k$. For example, $i$ characterizes different rows and $j$ characterizes different columns. By convention, when a *cell* with a specific row-with-column pairing contains multiple replicates, different replicates would be characterized by different indices $k$.

Suppose, for each individual column, variates in all rows are obtained from a data-generation model with a constant location parameter. Then this is tantamount to all $\alpha$-parameter values being zero. In other words, sharing a common location parameter for each individual column implies that even though—because one or more $\beta_j$ might be nonzero—location parameters might differ as columns differ, they do not differ because rows differ. In this sense the $\mathcal{C}_1$ response is unrelated to $\mathcal{A}_2$, even though, when one or more $\beta_j$ are nonzero, it is related to the variable $\mathcal{A}_3$.

When $(\mathcal{C}_1|\mathcal{A}_2,\mathcal{A}_3)$ notation is applied, "row effects" are coded in terms of the values assumed by the variable $\mathcal{A}_2$, while "column effects" are coded by the values assumed by $\mathcal{A}_3$. Specifically, the value of $\mathcal{A}_2$ determines the index or subscript $i$ of $\alpha_i$, while the value of $\mathcal{A}_3$ determines the index or subscript $j$ of parameter $\beta_j$.

Now what is the point of calling upon two Greek letters, like $\alpha$ and $\beta$ within the data-generation model $Y_{ij} = \mu_. + \alpha_i + \beta_j + \sigma Z_{ij}$, or, more generally, two separate variables $\mathcal{A}_2$ and $\mathcal{A}_3$ within the data-type designation, $(\mathcal{C}_1|\mathcal{A}_2,\mathcal{A}_3)$? The density of response variate $\mathcal{C}_1$ is conditioned given values of the two distinct variables $\mathcal{A}_2$ and $\mathcal{A}_3$. The goal of experiments and observational studies is to tease apart effects on $\mathcal{C}_1$ exerted by these two distinct variables. In a crucial sense, one or both of these variables play the role of key variable while the other then takes its turn as a nuisance variable. For example, one variable might denote levels of a nurture factor while the second denotes levels of a nature factor.

Analyses of variance based on complete factorial designs are balanced so that $(\mathcal{C}_1|\mathcal{A}_3,\mathcal{A}_2)$, and $(\mathcal{C}_1|\mathcal{A}_2,\mathcal{A}_3)$, are equally applicable. Specifically, unlike the ANCOVA model for data-type $(\mathcal{C}_1|\mathcal{A}_2,\mathcal{C}_3)$, which is different in many respects from any model for data-type $(\mathcal{C}_1|\mathcal{C}_2,\mathcal{A}_3)$, it makes little difference, at least from a mathematical perspective, which attribute-valued variable, $\mathcal{A}_2$ or $\mathcal{A}_3$, is the nuisance, and which is the key, variable. For this reason, tests targeted to $\mathcal{A}_2$ and $\mathcal{A}_3$ individually are simply referred to as *tests of main effects*.

Tests of main effects are distinguished from separate tests of interaction. The row main effect involving the $\alpha_i$ generation model component treats $\mathcal{A}_2$ as a key variable and $\mathcal{A}_3$ as a nuisance variable. Correspondingly, the test of the column main effect involving the $\beta_j$ generation model component, treats $\mathcal{A}_3$ as a key variable and $\mathcal{A}_2$ as a nuisance variable. On the other hand, rather than partialling out nuisance from key effects, researchers often focus on interaction between effects.

In many experiments, $\mathcal{A}_2$ is the sole key variable and, as a consequence, all other attribute variables are nuisance variables. Thus, when $c$ is greater than 3, the symmetry of factorial main-effects-tests helps assess the relative importance of the contamination due to some particular $\mathcal{A}_2,\ldots,\mathcal{A}_c$. In addition, models more complex than the additive model $Y_{ij} = (\mu_. + \alpha_i + \beta_j) + \sigma Z_{ij}$ commonly contain an interaction term, usually symbolized by double or even more complex subscripted Greek letters such as $\gamma_{ij}$.

Interaction terms quantify how variables act in synchrony, or in a synergistic fashion, to produce a catalytic, or mutual, effect on $\mathcal{C}_1$. For example,

## 16.2. Invariance and confounding

often a significant row main effect implies that some change in the level of a row variate will affect $C_1$. Yet, on the other hand, sometimes row main effects are nonexistent; in other words, for all indices $i$ the parameter $\alpha_i$ is zero-valued. However, a nonzero $\gamma_{ij}$, for at least one $ij$ combination, implies that either the row effect tags along and changes in value as the column effect affects $C_1$ or both effects are required to alter values of $C_1$.

The fascinating concept of interaction is usually given more attention in elementary statistical texts than is given to a more general abstraction, one that permeates a considerable portion of the applied statistical literature. Before introducing the concept, it will be useful to describe the term "invariance."

To render the test-statistic parameter free under the null hypothesis, in simple forms of ANOVA a numerator term is divided by

$$s_p^2 = \frac{\sum_{j=1}^{n(1)} (Y_{1j} - \bar{y}_1)^2 + \sum_{j=1}^{n(2)} (Y_{2j} - \bar{y}_2)^2 + \ldots + \sum_{j=1}^{n(m)} (Y_{mj} - \bar{y}_m)^2}{n(1) + n(2) + n(3) + \ldots + n(m) - (m-1)}.$$

When $m = 2$, the numerator of the Student-$t$ test two-level one-way, in other words, one-factor, special case of ANOVA, can be divided by the square root of $s_p^2$ to accomplish this same goal. In both instances the test-statistic is parameter free under the null hypothesis because this statistic is a variate with values that can be looked up in some set of tables. In the ANOVA case, the set of tables are called *F-tables*. For the Student-$t$ test special case the tables are listed under the heading, "PERCENTILES OF THE $t$ DISTRIBUTIONS," in the Fourth Edition of Dixon and Massey's classic book, *Introduction to Statistical Analysis* (1983). (Much will be said below about Dixon and Massey's choice of the plural, distributions, in place of the singular, distribution.)

If it is not divided by $s_p^2$, or its square root, the numerator of the F or $t$ test could be looked up in a table only when, perhaps by some mysterious secret process, the exact right table were found. Here being the exact right table means that the correct table describes the distribution of the particular variate whose scale is determined by the $\sigma$ component of the $Y_{ij} = (\mu. + \alpha_i + \beta_j) + \sigma Z_{ij}$ model. In other words, such a table can be characterized as being "scale parameter specific."

Table-with-parameter linkage is related to the concept, "invariance." Invariance is defined by Kendall and Buckland (1960) as "a property that is not changed by a transformation." This definition requires considerable elaboration, since the word "transformation" refers to a concept more general than simply the effect produced when one data file is modified in order to construct another new data file in order to improve the match between data and analysis by calling upon an elementary function such as the log, square root, or even a more general function like the Normalized score.

Suppose, for example, two experimenters collaborate on a study and help each other gather the same sample of measurements of a single variate, height. However, for some reason, one experimenter likes to collect data in inches, while the other experimenter likes to collect data in centimeters. Other than this one quirk, both investigators study the same data by means of calculation procedures provided by the same packaged statistical program package. Were one investigator to measure height in inches and the other in miles there might be some numerical accuracy or rounding compatibility problems. But for two measurement scales like inches and centimeters, numerical problems are unlikely.

Yet what about-statistical problems? In the sense that both investigators would distill their data down to exactly the same significance level, the F- and the $t$-test are both invariant with respect to scale. This special case of invariance, in part, is assured by the division of the F or Student-$t$ numerators by $s_p^2$ and its square root, respectively. Confounding suggests the opposite of invariance. For example, consider balanced factorial design analyses of data generated as $Y_{ij} = (\mu. + \alpha_i + \beta_j) + \sigma Z_{ij}$, and the F-test row effect numerator extension of $SS_m$. In the same way that all F-tests are invariant with respect to the $\sigma$ parameter's value, when the new $SS_m$ term is divided by $s_p^2$, the "F-statistic" formed is also functionally independent of all the $\beta_j$ column effect parameter values.

The capacity to tease apart ANOVA main effects is based on methodology called, collectively, the *breakdown of sum of squares*. Conditions that assure the validity are outlined as part of a basic lemma and a related theorem first proved by W. G. Cochran. Cochran's conditions depend on whether or not degrees of freedom, as will be discussed later in this chapter, add up to the sample size. When they do, parameters align themselves with the F-test numerators in a tidy fashion.

When Cochran's conditions are not met, no amount of sophisticated computation can remedy this situation perfectly. However, highly sophisticated designs can, and often do, relegate confounding to high degree interactions, yet retain prized unconfounded main effects and low-degree interactions. In other words, simple but knotty problems are solved, while bad snarls are left alone. The goal is to avoid snarls that involve main effects and low-order interactions—for example, those that involve only $\mathcal{A}_2, \mathcal{A}_3$, among factors $\mathcal{A}_2, \mathcal{A}_3$ and $\mathcal{A}_4$ of data case $(\mathcal{C}_1 | \mathcal{A}_2, \mathcal{A}_3, \mathcal{A}_4)$. There is not much point to assessing how the four variables $\mathcal{A}_2, \mathcal{A}_3, \mathcal{A}_4, \mathcal{A}_5$ of $(\mathcal{C}_1 | \mathcal{A}_2, \mathcal{A}_3, \mathcal{A}_4, \mathcal{A}_5, \mathcal{A}_6, \mathcal{A}_7)$ interact, insofar as their four-way mutual effect on $\mathcal{C}_1$ is concerned.

A careful experimental design, in terms of the allocation of subjects to cells, is an ounce of prevention worth a pound of computational cure. Here the design plays a role similar to that served by preliminaries, such as Normalized-scores data-transformations, which help align partial correlation-based procedures with the assumption of multivariate Normality.

In a balanced design, parameters whose individual interpretation is important are separated from other parameters, so that nuisance variables are filtered from response-with-key variable relationships. Clever experimental designs do this in such a way that the need to consider all of the combinations of all levels with all factors is circumvented.

## 16.3 ANOVA Tables and confounding

A simple ANOVA design is described by the tabular form shown in Table 16.1. The symbol $n(\text{row})$ represents the number of rows and $n(\text{col})$ represents the number of column levels of the $\mathcal{A}_2$ and $\mathcal{A}_3$ variates respectively. In the two-factor case, the meaning of confounding depends on the interpretation of the $\sigma^2_{row}$ and $\sigma^2_{col}$ terms in the "expected mean square" $E(MS)$ column.

Prior to F-table consultation, either the row entry or the column entry in the "sum of squares" SS column is divided by the $SS_{res}$ entry. This division is required because statistics in the F-table estimate the quantities

$$[\sigma^2 + n(\text{col}) \times \sigma^2_{row}]/\sigma^2 \quad \text{or} \quad [\sigma^2 + n(\text{row}) \times \sigma^2_{col}]/\sigma^2.$$

In the first instance, where $\sigma^2_{row} = 0$, or in the second instance, where $\sigma^2_{col} = 0$, these two terms are functionally related to the known quantities $n(\text{col})$ and $n(\text{row})$. They are not related to any of the component parameters that help define data-generation model $Y_{ij} = \mu. + \alpha_i + \beta_j + \sigma Z_{ij}$.

The $\sigma^2_{row}$ and $\sigma^2_{col}$ terms of the E(MS) column serve as the two way, in other words, two factor, ANOVA extensions of the noncentrality parameter

$$\{\sum_{i=1}^{m} n(i)(\mu_i - \mu.)^2\}/(m-1)$$

that was discussed in the previous section. For a balanced design, the $\sigma^2_{row}$ term is dependent solely upon the $\alpha_i$ components of the $Y_{ij} = \mu. + \alpha_i + \beta_j + \sigma Z_{ij}$ data model, while the $\sigma^2_{col}$ term is solely dependent upon the $\beta_j$ components of $Y_{ij} = \mu. + \alpha_i + \beta_j + \sigma Z_{ij}$. Consequently, the model's parameters align themselves neatly with the test-statistic numerators applied to test row and/or column effects.

| Source | SS | df | MS | E(MS) |
|---|---|---|---|---|
| Rows | $SS_{row} = n(col) \times \sum_i (y_{i.} - y_{..})^2$ | $n(row) - 1$ | $SS_{row}/[n(row) - 1]$ | $\sigma^2 + n(col) \times \sigma^2_{row}$ |
| Columns | $SS_{col} = n(row) \times \sum_j (y_{.j} - y_{..})^2$ | $n(col) - 1$ | $SS_{col}/[n(col) - 1]$ | $\sigma^2 + n(row) \times \sigma^2_{col}$ |
| Residual | $SS_{res} = \sum_i \sum_j (y_{ij} - y_{i.} - y_{.j} + y_{..})^2$ | $(n(row) - 1) \times (n(col) - 1) = n(r,c)$ | $SS_{res}/n(r,c)R$ | $\sigma^2$ |
| Total | $SS_{tot} = \sum_i \sum_j (y_{ij} - y_{..})^2$ | $n(row) \times (n(col))$ | | |

**Table 16.1.** Basic two-way factorial ANOVA table with one observation per cell and grand mean $y_{..}$.

Cochran's Theorem assures that when the $Y_{ij} = \mu_. + \alpha_i + \beta_j + \sigma Z_{ij}$ data-generation model is valid, and when no cell measurement of the one-observation-per-cell-design is missing, each of the three quantities, across from the Rows, Columns, and Residuals table rows, is statistically uncorrelated with the other quantities. Unfortunately, this does not necessarily guarantee that these quantities are statistically independent. Suppose the $Z_{ij}$ model components are all iid Normal. Then these three numerators are three independent variates. The two statistics formed by dividing either Row, or Column, sums of squares (SS), by the Residual SS, are F-distributed under the null hypothesis that row or column parameters $\alpha_i$ or $\beta_j$ are all zero. On the other hand, because their denominators are the same statistic, they are not actually independent of each other.

Fortunately, row and column F-statistic dependence or independence is not a major concern. What is important is the alignment of parameters within numerator SS. When the $\alpha_i$ parameters are zero, and yet some or all $\beta_j$ parameters are nonzero, the test of row effects is not altered by the values of the $\beta_j$ parameters. Consequently, the test of row effects is, in this important sense of the word—unconfounded—with the column effects.

There is a close connection between the statistic that was discussed at the beginning of Chapter 8 which estimates the parameter $\rho$ of $\mu_2 + \sigma^2(\rho Z_1 + \sqrt{1 - \rho^2} Z_2)$ and ANOVA confounding. The correlation coefficient r-estimator of $\rho$ is unrelated to scale while, with one notable exception the partial correlation counterpart of r is unrelated to all parts of the data-generation model. The sample partial correlation is only affected by the population partial correlation—a parameter that assesses the effect of a pure key variate $Z_2$ on the response variate.

A balanced design's main-effect-F-statistic resembles a sample partial correlation. Both statistics are designed with the same goal in mind: the control; standardization; adjustment or; partialling of a nuisance variate or variable. What is different about two-factor ANOVA and partial correlation, is the sort of nuisance—for ANOVA an attribute-valued variable, that is dealt with. For example, ANOVA concerns $(\mathcal{C}_1|\mathcal{A}_2, \mathcal{A}_3)$ data, while partial correlation deals with $(\mathcal{C}_1, \mathcal{C}_2, \mathcal{C}_3)$ data. Whether or not a comma, as opposed to a slash, separates the response variate from the two other quantities, determines whether the model-based machinery, multivariate Normal, or the machinery referred to by the name "general linear model," is called for. However, the purpose served by these two machinery choices— the teasing apart of factors or variates—is the same for both $(\mathcal{C}_1|\mathcal{A}_2, \mathcal{A}_3)$ and $(\mathcal{C}_1, \mathcal{C}_2, \mathcal{C}_3)$.

Analysis of covariance, ANCOVA, is a third way to deal with the same confounding issues dealt with by $(\mathcal{C}_1|\mathcal{A}_2, \mathcal{A}_3)$ main effects, and $(\mathcal{C}_1, \mathcal{C}_2, \mathcal{C}_3)$ partial correlation data analyses. However, here one or more nuisance variables are continuous, while as is true in two way ANOVA, key variables are attribute-valued. Specifically, $(\mathcal{C}_1|\mathcal{A}_2, \mathcal{C}_3)$ data is often studied by ANCOVA techniques.

Because there is no simple—in the sense of off-the-shelf software—way to study $(\mathcal{C}_1|\mathcal{C}_2, \mathcal{A}_3)$ data, when $\mathcal{A}_3$ assumes three or more levels, there is a methodological vacuum. Despite much model-based territory remaining uncharted, there are many nonparametric procedures that can be applied to a wide variety of data configurations. For example, the two age-adjustment-death-rate procedures that are discussed in Chapter 11 are both designed with the nuisance control idea in mind. Like two-way ANOVA, and ANCOVA, a nuisance variable is dealt with. While it is difficult to generalize the age adjustment nonparametric techniques to deal with nonbinary valued key variables, at least in theory, these procedures have few other limitations. However, in practice, rigorous tests, such as ANOVA and ANCOVA F-table-based tests, are difficult to formulate in the context of direct and indirect age adjustment.

## 16.4 Contingency tables and the parameter $\nu$

The term "degrees of freedom" recurs in many statistical applications. For example, when it is determined from a sample of Normal variates, the statistic $\bar{x}$ has a Normal sampling distribution. On the other hand, even when it is determined from iid Normal variates, the sampling distribution of the statistic $s^2$ will not itself be Normal; instead, it has a distribution described by a density that is, in turn, parameterized in terms of the degrees of freedom parameter $\nu$.

When $X$ is standard Normal, $X^2$ has a distribution called *one degree of freedom chi-square*. Correspondingly, when $X$ and $Y$ are both independent standard Normal variates, then it follows that $W = X^2 + Y^2$ is chi-square

($\chi^2$) distributed with two degrees of freedom. This process continues, in the sense that the name of the distribution of a sum, $W$, of $\nu$ iid standard Normal variates, all individually squared, is the $\nu$ *degree of freedom chi-square*, where $W$'s density is proportionate to

$$W^{(\nu/2-1)} \exp(-W/2) I_{[0,\infty)}(W).$$

Philosophers had written extensively on the topic of causation for at least two centuries prior to the innovations of the pioneer statistician, Karl Pearson. Even before the eighteenth century, theologians had pondered the intricacies of free will and determinism. Yet Pearson's contributions, through what he called *contingency tables*, and his study of the chi-square $f$-type curve, provided the first quantitative conceptual structure upon which variate-to-variate linkage could be examined rigorously.

Contingency tables describe attribute-valued $\mathcal{A}$-type nominal data. For instance, suppose racial subgrouping was a key variate and AIDS-related-disease type designated a response variate. The counts of individuals with specific values of these two variates could be put in tabular form, with racial groupings as table rows and disease types as table columns.

Pearson determined what would happen when the sample sizes within the table entry with the smallest count increased. He found that his contingency table assessment-statistic approached a variate that had a chi-square distribution with a number of degrees of freedom equal to the product of the number of table columns minus one, $c-1$, and the counterpart quantity, $r-1$, where $r$ denotes the number of table rows.

In many research applications, the sums across rows or columns of a contingency table are not random. These sums are determined by the experimenter who picks a number of individuals, such as individuals belonging to different races or with different disease types. For instance, were three different racial subgroupings studied, then once the counts of two races are known within any column, it follows that the count of the third could be determined by subtracting the sum of the first two from the known researcher-assigned total of all races. The term "degrees of freedom" represents, in a contingency table context, the number of table entries that are random variates.

The reasoning that underlies Pearson's contingency table analysis carries over to analyses based on Student-$t$-statistics. When the denominator of the $t$-statistic is calculated from one size $n$ iid sample of Normal data, then the statistic designated as $(n-1)s^2$, is chi-square-distributed with $\nu = n-1$. The subtraction of the statistic $\bar{x}$, within formulas often applied to compute $s$, is an operation identical to the subtraction of count totals in a chi-square contingency table. Similarly, when $s$ is applied to form a Student-*t-statistic* with density proportionate to $[1+t^2/\nu]^{-(\nu+1)/2}$, it follows that, because $\sigma$ is unknown, and since scale is determined as a deviation from the estimated quantity $\bar{x}$, a small portion of sample size $n$ is diverted for scale-determination purposes.

The subtraction of $\bar{x}$ has the effect of reducing the number of iid Normals from which $s^2$ is calculated. This sort of reduction makes particularly good sense in two degenerate special cases. Specifically, it would be inappropriate to study a contingency table that contained a single count. Similarly, were $n = 1$ it would be futile to estimate the curve property $\text{Var}(X)$, or some scale parameter $\sigma$ model component, by calculating a sample variance $s^2$ or sample standard deviation $s$. The deviation of some unique sample value from the sample mean would, of course, be zero when $n = 1$, no matter what the true value of $\sigma$ happened to be.

## 16.5 Student-t and Cauchy densities

Unlike the statistic $t$, whose density is proportionate to $[1+t^2/\nu]^{-(\nu+1)/2}$, to apply some comparable standard-Normally-distributed $z$-statistic, the value of $\sigma$ must be known. Were the value of the parameter $\sigma$ actually known, there would be no need to assess the reliability of the statistic $s$. On the other hand, when this statistic's reliability is relevant, it is appropriate that the density of the Student $t$-statistic has as one of its parameters, the value $\nu = n - 1$. For each different choice of $n$, a different table must be consulted to evaluate $t$'s cdf. Chi-square tables, like their Student-$t$ counterparts, are multiple subtable assemblages. Like the Student-$t$ tables, the rule—one table per different degree of freedom—applies.

The parameter $\nu$ differs from the parameters discussed previously. Mixing parameter $p$, as well as Greek-letter-designated $\mu, \sigma, \tau$, and $\rho$, all describe a variate like $Y$, of $Y = \mu + \sigma G^{-1}(Z)$, for some inverse cdf $G^{-1}$. Rather than performing this role, $\nu$ serves as part of the model that describes a particular sampling distribution, as discussed in Chapter 18.

When a model like $Y = \mu + \sigma G^{-1}(Z)$ is forwarded as part of what Fisher called the *specification* phase of the statistical process, a paint-by-numbers picture was provided of each individual sample constituent. In its typical form, the Student-$t$ statistic only applies to situations where $n$ is chosen by the researcher, and hence is a known quantity. Consequently, since it is functionally related to $n$, the value of the parameter $\nu$ is not, in actuality, a space to be filled, or painted.

Because $n$ is known, $\nu$ of $[1+t^2/\nu]^{-(\nu+1)/2}$ is also known. It is not a nuisance parameter, in the sense that $\sigma$ of $Z = (\bar{x}^{(n)} - \mu_0)/(\sigma/\sqrt{n})$ is a nuisance; except with regard to the trouble it takes to locate the right table for the Student-$t$, rather than to locate the single $z$-statistic table. The role that $\nu$ plays can be illustrated by comparing the $\nu = \infty$ row or column of the $t$-table to some other row or column, say the one headed by a finite value of $\nu$. Calling upon the $\nu = \infty$ row or column, when a finite value of $\nu$ labels the appropriate column or row, is tantamount to assuming that nature is smoother than she really is, and that $Z = z$, not some Student-$t$ with a finite value of degrees of freedom $\nu$, applies to the problem under study.

Where smoothness comes into play is illustrated by the special, $\nu = 1$, case where a constant times the curve $[1 + t^2/\nu]^{-(\nu+1)/2}$ is identical to the standard form of a Cauchy density. Consequently, in those situations where the null hypothesis is correct, a Student-$t$ statistic that is computed from a size $n = 2$ sample of iid Normally-distributed variates has a Cauchy density. That this $t$ variate's sampling distribution can be described in this way depends on the validity of the null hypothesis. This variate will only have a density summarized by Student-$t$ tables when the $t$-statistic numerator is a constant times a standard Normal variate. Were the $t$-statistic's numerator component a Normal variate with a nonzero location parameter $\mu$, then the composite variate would have a density that is not a Student-$t$ density. It would be the different density called a *non-central-t curve*.

Scientists seldom apply a $t$-statistic with one degree of freedom, but they often encounter variates that are either themselves Cauchy distributed or, more likely, variates that share many of the troublesome features of Cauchy variates. This is because rates and ratios are basic tools of risk analysis vital statistics and demography. Because of their association with Cauchy-like curves, these seemingly simple tools, unless applied with great care, can yield results that are so rough and distorted that they are all but valueless. As was done in Section 5.4, a little simple algebra discloses the connection between rates and ratios, on the one hand, and the $\nu = 1$, Student-$t$-statistic, on the other. It shows that the ratio of the midrange to the range, divided by $\sqrt{2}$, of two iid standard Normal variates both with a distribution described by density $\phi$ is the Cauchy special $\nu = 1$ case of the Student-$t$-statistic.

It happens that when they are computed from the same sample of $n$ iid Normal variates, the sample mean $\bar{x}$, and the statistic $s$, are independent variates. Hence, the fraction that forms any Student-$t$ statistic from an iid sample of Normal variates has a numerator and a denominator that are independent-statistics. With a slight amount of additional argument concerning the fact that the range must be a nonnegative variate, it can be shown that the ratio of two independent standard Normal $\phi$-distributed variates is a standard Cauchy variate. Even in what might ordinarily seem

## 16.5. Student-t and Cauchy densities

to be the best case scenario, were a ratio constructed with a numerator and denominator that are independent $\phi$-distributed variates, this ratio will be Cauchy-distributed with all the parameter-with-curve-property problems described in previous chapters that pertain to Cauchy-distributed variates.

# CHAPTER 17
# Parameter-Based Estimation

## 17.1 Likelihood and BLU estimation

Even today, R. A. Fisher's maximum likelihood (ML) approaches are still mainstays of statistical practice. In everyday English, ML takes the best density estimator to be the estimator that maximizes the chance of seeing what one has already seen. The word "seeing" refers to some joint density pdf taken to be a function within which variate values are taken to be constants and parameters are taken to be variables.

Since the goal of curve maximization is the same if the parameter to be estimated is a location, mixing, or any other sort of parameter, there is little about ML methodology that is parameter-type specific. As could be expected, location parameters are computationally easier to deal with by ML procedures than are most other parameters. Yet, although it is classified as a parametric approach, ML does not take specific parametric characteristics into account.

Best linear estimation (BLU) procedures, unlike their likelihood counterparts, depend on a $\mu$ being a location parameter and $\sigma$ being a scale parameter. BLU methodology is based on the distinction between parameters and curve properties that was discussed in Section 10.7. There Fisz's (1963) comparison of methodology for estimating the parameters $\alpha$ and $\beta$ that minimize the curve property $E\{[Y - (\alpha + \beta x)]^2\}$ was compared to an alternative procedure based on the curve property $E(Y|x)$. However, before considering further $E\{[Y - (\alpha + \beta x)]^2\}$ and the "Gauss-Markoff Theorem" that concerns $E\{[Y - (\alpha + \beta x)]^2\}$, an even more general statistical idea will be reviewed in this section.

When a statistician begins a study, he or she must choose from among mutually exclusive methods of procedure. One possibility is to seek out statistical methods that are known to be robust in the sense of protection against incorrect assumptions. Another possibility is to seek out some procedure that is so sensitive to assumption incorrectness that potential harm will be glaringly obvious. Were disregard for model-with-curve-fit compared to the sweeping of dust under a rug, then potential statistical harm can be compared to tripping over this pile of swept dust.

A surprising capability of the BLU procedures that are discussed in this Chapter is that they can provide both tripping protection and a degree of robustness that is seldom attainable with alternative methodology. They can do this in the sense of providing a researcher with the potential to check for outliers. In addition, unlike the traditional estimates of the standard deviation and variance, being linear estimators, BLU estimators are not overly sensitive to upswept outliers. How BLU procedures manage to harness Gauss's squared error metric in such a way that the squaring of the residual $[Y - (\alpha + \beta x)]$ within $E\{[Y - (\alpha + \beta x)]^2\}$ does not cause the same difficulties that are usually caused by squaring, is the puzzle considered in the remainder of this chapter.

As previewed in Chapter 10, at $x_0$, for some specific set of parameter $\alpha$ and $\beta$ values, the squared residual $[Y - (\alpha + \beta x_0)]^2$ is a new variate formed from a response variate $Y$ that has a density $f_2$ which, in turn, is taken to have expectation $E_2(\alpha, \beta)$, a curve property related to the two quantities, $\alpha$ and $\beta$. In the sense that $\alpha$ and $\beta$, of $\alpha + \beta x_0$, are determined by minimizing all possible curve properties $E_2(\alpha, \beta)$, the $\alpha$ and $\beta$ quantities are themselves curve properties. On the other hand, in the sense that each candidate $f_2$ is defined in terms of $\alpha$ and $\beta$ it follows that $\alpha$ and $\beta$ are parameters. Thus, regression methodology that concerns the minimization of $E\{[Y - (\alpha + \beta x)]^2\}$ is based on the correspondence of parameters $\alpha$ and $\beta$ with the specific curve properties $E_2(\alpha, \beta)$.

## 17.2 Censoring and incompleteness

The term "censoring" that was introduced in Section 15.3 refers to restrictiveness in reference to statistics computed from a sample. It denotes this in the same way that the term "truncation" applies to curves which describe a population. In the recent statistical literature, truncation is sometimes taken to mean the same thing as censoring at a specific predetermined value. However, in this and in the remaining chapters, truncation will retain its earlier strict meaning in reference to a density or cdf curve. So many sections of this book are devoted to truncated curves that switching to the newer usage would cause considerable confusion.

A fundamental idea distinguishes censoring from many other forms of data pathology. Suppose a value is taken to be censored, and other values assumed by a sample of replicate measurements are uncensored. The result is a sample of hybrid variates, some whose values are known exactly and

## 17.2. Censoring and incompleteness

some whose values are merely known to be within some interval of finite length, or some ray that contains points less than or greater than some finite value. On the other hand, a variate whose density is described by a truncated curve will commonly be a continuous variate, and a sample with such variates as members will often be unitary in the sense that this term is defined in Section 13.3.

Insufficient time for the observation of all study outcomes is a common cause of censoring. For example, suppose for a light bulb—survival time— the length of time from initial use to burnout is determined. Also suppose $n = 1000$ light bulbs are lit for the first time at time $t_0$ and then, as bulbs burn out, their individual failure times are recorded, one by one.

The light bulb example illustrates a common cause of censoring. Clearly it is seldom practical to wait until all bulbs burn out and—only then— assess the results of an experiment. Were a researcher to actually wait for all returns to come in, then he or she would be severely affected by delays due to the choice of a large, rather than a small, sample. For example, a sample mean could, in practical applications, be expeditiously computed from a complete small sample of, perhaps all $n = 10$ measurements. Yet it is unlikely that all $n = 10,000$ light bulbs of a large sample of size $n$ will have burned out before the invention of some new and improved variety of light bulb renders the experiment useless.

Laboratory experiments are often designed to measure survival time to death caused by tumor. For such an experiment, death due to the tumor under study is the counterpart of light bulb burnout. Similarly, it is rare for experiments to be continued until all mice or other laboratory animals die. Typically, of the 50 white mice who live through an experiment's first day, from 2 to 20 will be alive two years later. In addition, as will be discussed below, mice and many other laboratory animals may die and yet be found to be tumor-free.

There is still another reason why censoring is commonplace. Extremely long and short measurements, as well as big- and small-valued observations, are often outliers. At extremes, measurement instrument reliability is taxed beyond reasonable limits. Again, as $n$ increases it becomes more—not less—common to encounter outlier-caused censoring problems. The larger a sample, the greater the chances are that at least one measurement assumes an unreliable value. For example, the variate might have a density that is described better by the Cauchy, than by the Normal, elementary function.

Censoring often results from the manner in which survival study subjects are selected. Although a uniform batch of experimental animals, for instance white mice, is often studied beginning at a common point in time, this form of experiment is rarely undertaken with humans. In human survival-after-treatment experiments, subjects have different periods of observation. For example, one patient will be treated early in the course of a study while another enters the study only a short time before the data-gathering phase of the study is completed. Many statistical tools, such as clinical life table and Cox model methodology, are designed to deal with this variety of data incompleteness.

There are two ways of looking at the incompleteness problem. For example:

1. Suppose an experimental subject enters into a study, or is known to have been exposed to a disease agent such as an HIV-contaminated blood transfusion, at time or date $t_0$. Suppose that date of study completion $C$ is measured in terms of the same units, days, applied to express date $t_0$, and suppose that the experimental subject survives $y$ days after either entering the study or being exposed to the disease. Then clearly, because only that part of $y$ equal to $C - t_0$ can actually be known, the problem of incompleteness arises whenever $t_0 + y > C$.

2. Suppose two random numbers are selected where the first, $X$, represents the period over which an individual is studied and the second, $Y$, represents survival time of the person. If $Y > X$, then an "incomplete measurement," as described in (1), is obtained. Consequently, every individual entering this type of study generates two variates: first, the minimum value of $X$ and $Y$ and; second, a variate that discloses whether the minimum value is the complete survival measurement $Y$, or is an incomplete measurement whose known putative as-yet-survival time equals the value assumed by $X$.

A specific tumor or other often-studied possible cause of death is never a laboratory animal's Achilles' heel, i.e., the single possible cause of death. Thus, $X$ can often be taken to be the interval between the study starting date and the date at which an animal dies and yet is found to be tumor free, while only the as-yet-incomplete part of $Y$ can be measured.

## 17.3 Outliers and errors

Throughout the history of science, minor departures from basic tenets have been noticed. For example, there is a slight difference between the observed and the predicted perihelion of the planet Mercury. A planet's perihelion is the point in its orbit where it is closest to the sun. Astronomers noticed that, although the methodology of Newtonian physics had provided accurate measurements of the orbits of planets other than Mercury, this one planet behaved unpredictably.

## 17.3. Outliers and errors

As is well known, the scientific detective work of Einstein and his colleagues led to understanding the limits of Newtonian physics. In true Sherlock Holmes style, what first seemed to be a clue of minor significance, the planet Mercury's aberrant perihelion, later was shown to be of considerable importance.

The law of large numbers is a cornerstone of statistical theory. One version of this theorem states that as the size of any one out of many specific types of samples increases, the mean of sampled variates approaches what is expected of any individual variate within the sample. (Appendix II defines the concept, independently and identically distributed (iid), upon which many variants of the law of large numbers are based.)

As was true of Newtonian physics applied to all of the planets besides Mercury, classical statistical methods based on this law usually behave as expected. However, what one expects of Cauchy variates is an exception to what had seemed to be a rule. The message hidden in Cauchy data aberrance can be deciphered with considerable precision today. This capability was not at the disposal of scientists such as Khintchine, who in 1929 first proved the version of the law of large numbers paraphrased in the previous paragraph. Khintchine's proof required the basic assumption that the curve property $E(Y)$ exists. While it was known in the late 1920s that the Cauchy distribution does not have a well-defined property $E(Y)$, this seemed to be merely a minor aberration.

Today, thanks to improved statistical graphics capacity, we know that when they are viewed closely, problems with the bias of the sample mean calculated from Cauchy data are hardly data analysis exceptions. For example, when a sample mean is computed from a sample comprised of two or more Cauchy or other mixture components, the statistic is usually a biased estimator of *any* one the $c \geq 2$ parameters $\mu_1, \mu_2, \ldots, \mu_c$, when two or more of these location parameters are distinct.

The viewpoint that exceptions to statistical laws are beyond the pale, in the sense of being outside the limit of proper data behavior, is embedded in much of the data-analysis vocabulary. For example, consider how the term "outlier" is defined by Kendall and Buckland: "In a sample of $n$ observations it is possible for a limited number to be so far separated in value from the remainder that they give rise to the question whether they are not from a different population or that the sampling technique is at fault." The outlier as defined in this quotation suggests a departure from a default setting.

The Kendall and Buckland (1960) definition of outlier seems to imply that once the faulty sampling technique is corrected, things can proceed as usual. Sometimes this is correct. Yet unfortunately, what happens often resembles what a carpenter experiences when he or she removes nails or fills knots in lumber. Outlier removal and correction, *imputation*, appears at first sight to merely require that an additional step be added to usual construction procedures. However, through experience one learns that, like a previously inserted nail or a filled knothole, things can pop out when others are inserted. Curves like the truncated Cauchy often do not hold still for outlier removal and correction procedures.

## 17.4 Ordered variates and subscripts

There are many alternative location parameter estimators besides $\bar{x}$. These, of course, include the sample median as well as the sample midrange, where the midrange is the average of the largest and the smallest measurement in the sample. An entire class of statistics that contains, among other statistics, the midrange, median, and mean of a sample is called, collectively, the class of *linear estimators*.

A few examples will help to illustrate what the term "linear estimator" designates. There are no symbolic representations for the median and midrange of a size $n$ sample that, like the symbol $\bar{x}$, have been universally accepted by the statistical community. A reason for this lack of statistical terminology is that notation based on parentheses can represent many statistics in a concise manner (See Appendix I).

Parentheses that enclose a single symbol provide a notational clue that the ordinary interpretation of the symbolic representation is supplanted by some new meaning. Specifically, consider the distinction between the symbols $X_j$ and $X_{(j)}$. The former represents the $j^{\text{th}}$ member of some sequence of variates, where this member's value is not associated in any particular way with the value assumed by the symbol $j$.

For example, since information provided by the values 1 or $n$ in no way distinguishes the particular value assumed by $X_1$ from the value of $X_n$, it follows that $X_1$ could assume a greater, a smaller, or the same value as $X_n$. On the other hand, the symbol $X_{(1)}$ represents the member of a size $n$ sample that assumes the smallest value or, if several observations have identical smallest values, then $X_{(1)}$ represents one of these sample members.

Suppose as in Section 15.2, the values assumed by a random sample are ordered from smallest to largest in such a way that within a set of tied values, members are assigned randomly to appropriate ranks where the subscript $(j)$, of $X_{(j)}$, designates the rank or place occupied by the $j^{\text{th}}$ observation. However, if two and only two variate members of a random sample take on some particular value, such as the smallest, then the equivalent of a coin toss determines which variate is designated as $X_{(1)}$, and which is designated as $X_{(2)}$.

## 17.5. BLU estimators

A *linear estimator* is a weighted sum

$$\sum_{j=1}^{n} a_j X_{(j)}.$$

For any size $n \geq 2$ sample, the statistic called the *sample midrange*, or *midrange*, is defined as $0.5(X_{(1)} + X_{(n)})$ where, within the weighted sum

$$\sum_{j=1}^{n} a_j X_{(j)},$$

the weights $a_1 = a_n = 0.5$ and equal zero otherwise. The range $X_{(n)} - X_{(1)}$ and; for an odd-sized sample, the *sample median* defined as the statistic $X_{([n+1]/2)}$, are also special cases of

$$\sum_{j=1}^{n} a_j X_{(j)}.$$

For an even-sized sample, the sample median is defined as

$$[X_{(n/2)} + X_{(1+n/2)}]/2.$$

For example, when $n = 4$ the midrange is the average of the two extreme values $X_{(1)}$ and $X_{(4)}$, while the median is the average of the two middle values $X_{(2)}$ and $X_{(3)}$, in other words, $a_2 = a_3 = 0.5$ and $a_1 = a_4 = 0$.

The mean $\bar{x}$, median, and midrange, are examples of unbiased linear estimators of the location parameter within the expression for many models. In addition, for a given sample size $n$, the range $w$, multiplied by a constant appropriate for the given sample size and standard model, is an example of an unbiased linear scale parameter estimator. (In the remainder of this chapter, unless otherwise specified, it will be assumed that not only are the properties $E(X)$ and $\sqrt{\text{Var}(X)}$ finite and well-defined, but they have a known and simple relationship with some density expression's location and scale parameters, $\mu$ and $\sigma$.)

## 17.5  BLU estimators

How order statistics are assembled to form statistics, called *best linear unbiased* (BLU) estimators, can be explained in terms of the ordered sequence of variates, $\{Z_{(j)}\}; j = 1, 2, \ldots, n$. Suppose $\{Z_j\}; j = 1, 2, \ldots, n$, designates independent and identically-distributed variates, where the density that describes the distribution of one of these variates is represented by the standard curve $g$. The term "standard curve" usually, but not necessarily, designates an elementary function that, like the Normal model $\phi$, can be represented without reference of any parameter values.

When $n$ unordered sample values are placed in rank order, this process yields the order statistics

$$\{Z_{(j)}\}; j = 1, \ldots, n.$$

Suppose, as will be typical for most choices of standard model $g$, that: curve properties

$$\{E(Z_j) = 0\}; j = 1, 2, \ldots, n, \ \{\operatorname{Var}(Z_j) = 1\}; j = 1, 2, \ldots, n;$$

and that both the parameters $\mu$ and $\sigma$ are constants. Then, when $\sigma$ is nonzero, $Y = \mu + \sigma Z$ is order preserving, in the sense that if the $n$ equations $Y_j = \mu + \sigma Z_j; j = 1, 2, \ldots, n$ are valid, it follows that the $n$ equations $Y_{(j)} = \mu + \sigma Z_{(j)}; j = 1, 2 \ldots, n$ are also valid. This implies for $j = 1, 2, \ldots, n$ that $E(Y_{(j)}) = \mu + \sigma E(Z_{(j)})$, where members of the sequence of means, $E(Z_{(j)}); j = 1, 2, \ldots, n;$ are all known when $g$ is known. It also implies that a sequence of known relationships similar to $E(Y_{(j)}) = \mu + \sigma E(Z_{(j)}), j = 1, 2, \ldots, n;$ also applies to the curve properties defined below that are called *covariances*.

Thanks to these known expectation and covariance relationships, it is possible to use generalized least squares methodology to estimate the parameters $\mu$ and $\sigma$. Specifically, the Gauss-Markoff can be brought into play to take advantage of the linearity of the relationships $Y_{(j)} = \mu + \sigma Z_{(j)}; j = 1, 2, \ldots, n$, where $\mu$ is the $y$-intercept of a line and $\sigma$ is its slope.

The term "order preserving" is illustrated by the line with slope $\sigma$ and $y$-intercept $\mu$ shown in Figure 17.1. The parameters $\sigma$ and $\mu$ must be estimated and, in this sense, they have unknown values. Since $\sigma$ must be positive, it is known that line $y = \mu + \sigma z$ must have a positive—and hence nonzero—slope.

To see what BLU strategy actually is, it is helpful to list the quantities that are known and those that are unknown. Initially, we will assume that all the $n$ values of the order statistics $Y_{(j)}$ are known. For example, in Figure 17.1 one such value is pointed out as an observed $Y_{(j)}$. In addition to these values, $E(Z_{(j)})$ and

$$\{\operatorname{Var}(Z_{(j)})\}, j = 1, 2, \ldots, n$$

## 17.5. BLU estimators

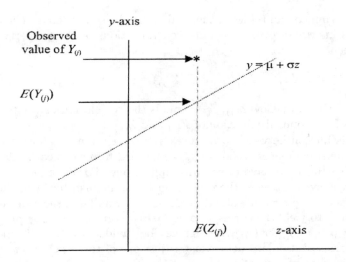

**Figure 17.1.** Sketch of one $Y_{(j)}$'s value and the line $y = \mu = \sigma z$.

are also known, as are quantities called the covariances $\text{Cov}(Z_{(j)}, Z_{(k)}), j = 1, 2, \ldots, n; k = 1, 2, \ldots, n$. (Because the distribution of each of the $n$ iid $Z_j, j = 1, 2, \ldots, n$ is described by a known standard density $g$; if they are well-defined, the values of these curve properties can all be determined.)

Because $\mu$ and $\sigma$ are not known, the $n$ values $E(Y_{(j)}), j = 1, 2, \ldots, n$ are not known exactly. Yet an important fact concerning these curve properties is known, namely that the $n$ points $(E(Z_{(j)}), E(Y_{(j)})); j = 1, 2, \ldots, n$, must all have coordinates placed along a line, such as the line $y = \mu + \sigma z$ shown in Figure 50. Any observed order statistic, such as $Y_{(j)}$, assumes a value equal to the ordinate of a point pair whose abscissa-coordinate is the value assumed by $Z_{(j)}$. Consequently, $\mu + \sigma E(Z_{(j)})$ can be thought of as the part of $Y_{(j)}$ explained by the curve $y = \mu + \sigma z$; hence the Gauss-Markoff Theorem's applicability.

When ordinates vary or deviate from a linear model, the Gauss-Markoff Theorem states a set of conditions under which the least squares approach yields optimal estimates. However, things are never quite this simple. In its original form, the work of Gauss requires some modification before it can be applied to find BLU estimators.

Suppose a known standard model $g$ together with the unknown parameters $\sigma$ and $\mu$ describe the density of each of $n$ unordered iid $Y$-variates. Then, because $g$ is a known function: $E(Z_{(j)}); j = 1, 2, \ldots, n$; $\text{Var}(Z_{(j)}); j = 1, 2, \ldots, n$, and; for $j \neq k$, the covariances

$$\text{Cov}(Z_{(j)}, Z_{(k)}) = E[\{Z_{(j)} - E(Z_{(j)})\}\{Z_{(k)} - E(Z_{(k)})\}]; j = 1, 2, \ldots, n; k = 1, 2, \ldots, n,$$

can be determined and tables of their values can be constructed. The curve properties known as *covariances* are generalizations of the property variance. When

$$\text{Var}(Z_{(j)}) = E[\{Z_{(j)} - E(Z_{(j)})\}^2]; j = 1, 2, \ldots, n,$$

all equal 1, then any $\text{Cov}(Z_{(j)}, Z_{(k)})$ equals the correlation bivariate curve property of the joint distribution of $Z_{(j)}$ and $Z_{(k)}$.

The distinction between covariances and correlations is another of the many bothersome places where slightly modified forms of nearly identical curve properties each have their own applications. For example, it would have been convenient were there no need to distinguish between the Var curve property and a curve's standard deviation. Similarly, the requirement that in order to yield the corresponding $(j, k)^{\text{th}}$ correlation curve property, the curve property $\text{Cov}(Z_{(j)}, Z_{(k)})$ must be divided by the $j^{\text{th}}$ and $k^{\text{th}}$ ordered variates' standard deviations, is just one of those pebbles that can get into the shoe of a practicing statistician. Sometimes it is convenient to work with covariances and sometimes correlations are appropriate. In the case of BLU estimation, computational relationships can be formulated in terms of covariances that would be more cumbersome if expressed in terms of correlations.

A technique developed by Lloyd (1952) yields what are referred to as *normal equations* for unbiased linear estimators,

$$\hat{\mu} = \sum_{j=1}^{n} a_j^{(\mu)} Y_{(j)}, \text{ and } \hat{\sigma} = \sum_{j=1}^{n} a_j^{(\sigma)} Y_{(j)},$$

where $E(\hat{\mu}) = \mu$ and $E(\hat{\sigma}) = \sigma$, that are best in the sense of having minimal variance. These two estimators are both weighted sums formed by first multiplying each consecutive $Y_{(j)}^{\text{th}}$ order statistic by some predetermined tabled weight

$$a_j^{(\mu)} \text{ or } a_j^{(\sigma)}$$

and then adding $n$ products like

$$a_j^{(\mu)} Y_{(j)}.$$

In practice, once the proper table for

$$a_j^{(\mu)} \text{ and } a_j^{(\sigma)}$$

values is located, it is easy to compute BLU estimators.

The Gauss-Markoff Theorem concerns the variance, a particular property of the sampling distribution of the linear estimator. This curve property serves as a metric by which the desirability of the estimator is assessed in terms of its proximity to the parameter it estimates. The term "best linear unbiased" designates the estimators $\hat{\mu}$ and $\hat{\sigma}$, while the term "mean squared error," (MSE) represents the particular metric, in other words criterion, by means of which $\hat{\mu}$ and $\hat{\sigma}$ are assessed.

Here, two distinct MSE are involved. The location parameter estimator—and its scale parameter counterpart—are evaluated by two MSE metrics separately. This is a convenient feature that is unique to BLU procedures. Simultaneous assessment is provided for estimators that shed light on distinctly different parameter types such as location and scale. Statisticians often use multiple comparison methods to provide what they call *simultaneous inference*. However, these procedures make aggregate decisions that concern location parameters exclusively. Viewed from a general perspective, only BLU procedures shed the same wavelength of light on both location and scale parameters.

BLU estimators are not simply alternatives to $\bar{x}$, $s$, the midrange, or the weighted range. For one thing, in the Normal model special case, $g \equiv \phi$, the BLU estimator of location, $\hat{\mu}$, is the estimator $\bar{x}$. Statistics like $\bar{x}$, $s$, the midrange, and the weighted range were computed by many scientists long before theoretical statisticians got around to designing estimators on the basis of clearly spelled out criteria such as $E_f[\hat{\theta} - \theta]^2$; in other words, answers were formulated first, questions later. Therefore it is not surprising that every so often, for occasional choices of a standard curve like the Normal $g \equiv \phi$ some modern procedure is found to be identical to some older, heuristically chosen, speculative procedure.

What BLU estimators are can be visualized by focusing on the impossibility of data analysis from scratch where the mind is a tabula rasa. Any formal or informal statistical procedure blends data with assumptions and criteria. Unbiasedness and $E_f[\hat{\theta} - \theta]^2$ illustrate what is meant by criteria, while the choice of standard curve $g$ is one possible assumption. (In 1690, when John Locke conceptualized the mind as a "tabula rasa," science had not even begun to separate the two concepts, data and criteria.)

## 17.6 BLU estimation and censoring

The equations that determine BLU estimators require merely that two or more order statistics be known. Suppose $n > 4$ and large- or small-order statistic values, such as the values assumed by $X_{(1)}, X_{(n)}$, or both, are deemed to be untrustworthy. Then one or more of these values can be discarded, and yet BLU estimators can still be obtained for both the

Figure 17.2.

density location and scale parameters. In addition, suppose results must be obtained before all measurements can be tallied. Then estimators can be constructed from the early-order statistics, in other words, from those order statistics whose parenthesized subscripts are much smaller than $n$. (Section 19.5 will outline methods that can help determine the tradeoffs between inclusion of a potential outlier and the deletion of this candidate's actual value, in the sense of treating it as a censored observation.)

As the guarantee printed on the package of 100 YEAR NITE • LIGHT Never Buy Bulbs Again illustrates, "...theoretical life of over a 100 years," and many other reliability and risk assessments must often be viewed skeptically. In 1879 when Edison developed the first reliable incandescent light bulb, he did not immediately begin life testing enough bulbs for anyone to know—even one hundred years later in 1979—that his or anyone else's bulbs "won't burn out."

Yet, this example notwithstanding, BLU techniques for adjustment are an appreciable advance beyond the simplistic procedure that merely removes outliers, or takes extant subset values, and then proceeds as usual. The key to this difference is the letter U of BLU.

## 17.6. BLU estimation and censoring

The mere removal of actual or potential outliers almost invariably increases estimator bias. For example, this variety of bias often concerns the difference between the true value of the curve property $E_f(X)$ and the location parameter of the model called upon to represent $f$. If an outlier is removed from a sample of data, and no imputed or adjusted value is inserted in its place, then an $\bar{x}$ location parameter estimator is biased by this process. The U of stands for the word "unbiased." Even BLU estimators based on *fewer* than $(n/2)+1$ order statistics share a feature possessed by any sample median computed from $(n/2)+1$ or more ordered variates. The BLU estimator, like the median computed from $(n/2)+1$ or more ordered variates, is unbiased.

BLU methods are often applied to check for the presence of outliers. A sample is purposely censored, for example, by the removal of $X_{(1)}$ and $X_{(n)}$. Then, based on this censored sample, BLU trial estimates $\hat{\mu}$ and $\hat{\sigma}$ of the location and scale parameters within the formula for the density (of an unordered $X$ variate) are obtained. Finally, the curve that is fitted in this way is processed by statistical machinery that helps assess extreme order statistics $X_{(1)}$ and $X_{(n)}$. How far out these values are in left field is gauged first by clearing them from the field.

The emphasis placed above on order statistic covariances illustrates that even when they are computed from an iid sample, order statistics are dependent variates. For example, suppose it is known that the next-to-largest order statistic $X_{(n-1)}$, within a sample of size $n$, takes on the measured value 10. Then this knowledge implies that the value $X_{(n)}$ cannot be less than 10. For a sample of size $n > 2$, Blom (1962) discusses a simple approximation formula for the $i^{th}$ and $j^{th}$, $i \leq j$, order statistic covariance. Specifically, suppose $p_k$ is defined as $k/(n+1)$, and then the $i^{th}$ and $j^{th}$ order statistics are obtained from an iid sample where each variate has a density $f$. Suppose also that the partial integral of $f$ is an increasing function, so that $F$'s inverse $F^{-1}$ is well-defined. Finally, define $\lambda_k$ equal to $F^{-1}[k/(n+1)]$. Then, as the sample size increases, the $i^{th}$ and $j^{th}$ order statistic covariance approaches

$$p_i(1-p_j)/[(n-2)f(\lambda_i)f(\lambda_j)].$$

Blom shows how the above formula, and a comparable formulas for order statistic expectations and variances, can be applied to determine highly accurate approximate BLU estimator weights. These statistics (called *nearly best linear estimates* by Blom) are particularly appropriate choices for samples whose size is greater than the sample sizes tabled in Sarhan and Greenberg's classic work (1962). Today, MATHCAD, MATLAB, C++, and other programming systems make it easy to calculate BLU estimator weights for most user-specified choices of standard model $g$, and there are many tables like those published by Sarhan and Greenberg (1962) from which BLU estimator weights and the properties of these and other linear estimators can be determined.

BLU methodology requires a choice of a "standard curve $g$." The word "curve" was used in place of the word "model" purposely. Suppose the distribution whose location and scale parameters are estimated is described by a density

$$f(y) = g\left(\frac{y-\mu}{\sigma}\right)/\sigma.$$

Then, as long as the known expectation, variance, and covariance curve properties of all order statistics, specifically:

$$E(Z_{(j)}); j = 1, 2, \ldots, n; \; \operatorname{Var}(Z_{(j)}); j = 1, 2, \ldots, n$$

and; for $j \neq k$, the covariances,

$$\operatorname{Cov}(Z_{(j)}, Z_{(k)}) = E[\{Z(j) - E(Z_{(j)})\}\{Z_{(k)} - E(Z_{(k)})\}]; j = 1, 2, \ldots, n; k = 1, 2, \ldots, n,$$

exist (in the sense that they are finite-valued), it is not necessary that there exist a closed-form model for the common density $g$ of each of the unordered iid variates $Z_j; j = 1, 2, \ldots, n$.

Suppose $G(z)$ designates the cdf whose derivative is density $g(z)$. The density of any single order statistic is proportionate to a product formed by raising either $G(z)$, $(1 - G(z))$, or both $G(z)$ and $[1 - G(z)]$, to integer powers, for example

$$[G(z)]^{i-1} \quad \text{and} \quad [1 - G(z)]^{n-i};$$

and then multiplying a function like

$$\{[G(z)]^{i-1}[1 - G(z)]^{n-i}\} \quad \text{by} \quad g(z).$$

Similarly, for $j > i$,

$$\operatorname{Cov}(Z_{(j)}, Z_{(i)}) = E[\{Z_{(j)} - E(Z_{(j)})\}\{Z_{(i)} - E(Z_{(i)})\}] =$$
$$E(Z_{(j)} Z_{(i)}) - E(Z_{(j)})E(Z_{(i)}), \text{ where } E(Z_{(j)} Z_{(i)}) =$$
$$\frac{n!}{(i-1)!(j-i-1)!(n-j)!} \int_{-\infty}^{\infty} \int_{-\infty}^{u} vu G^{i-1}(v)[G(u) - G(v)]^{j-i-1}$$
$$[1 - G(u)]^{n-j} g(u) g(v) dv du.$$

Because at any value of $z$, $0 \leq G(z) \leq 1$ and $0 \leq g(z)$,

$$\{[G(z)]^{i-1}[1 - G(z)]^{n-i}\}\{g(z)\} \leq g(z)$$

and, for any real values of $u$ and $v$,

$$G^{i-1}(v)[G(u) - G(v)]^{j-i-1}[1 - G(u)]^{n-j} g(u)g(v) \le g(u)g(v).$$

The inequality

$$\{[G(z)]^{i-1}[1 - G(z)]^{n-i}\}\{g(z)\} \le g(z)$$

implies that if $E(Z_j^2)$ is finite-valued ($E(Z_j^2) < \infty$), for all $j = 1, 2, \ldots, n$, then $E(Z_{(j)}^2) < \infty$ for all $j = 1, 2, \ldots, n$.

Taylor (1955, Section 16.3) states that if $h_1(x) \ge h_2(x)$ on the closed interval $[a, b]$ then

$$\int_a^b h_1(x)dx \ge \int_a^b h_2(x)dx.$$

Consequently, the definite integral of $xg(x)$ must be smaller than the definite integral of $g(x)$, when both integrals are calculated over the unit interval $[0, 1]$. This fact, and the inequality

$$\int_1^\infty x^2 g(x)dx + \int_1^\infty x^2 g(x)dx \ge \int_1^\infty xg(x)dx + \int_1^\infty xg(x)dx,$$

imply that if $E(Z_j^2) < \infty$, for all $j = 1, 2, \ldots, n$, then $E(Z_{(j)}) < \infty$, for all $j = 1, 2, \ldots, n$.

Because the unordered observations are iid, the covariance of any two of these observations must be zero. Thus, if $E(Z_{(j)}) < \infty$ and $E(Z_{(i)}) < \infty$, the inequality

$$G^{i-1}(v)[G(u) - G(v)]^{j-i-1}[1 - G(u)]^{n-j} g(u)g(v) \le g(u)g(v)$$

implies that $E(Z_{(j)}, Z_{(i)}) < \infty$. Thus, even if no closed form model for $g$ exists, BLU methodology can be applied to find estimators of the location and scale parameters of $f(y) = g\left(\frac{y-\mu}{\sigma}\right)/\sigma$ from an iid sample whose members have common density $f$ whenever $g$ has a finite variance.

## 17.7 BLU estimators and alternatives

In the sense that even when it is determined from Normal data,

$$s = \sqrt{(n-1)^{-1} \sum_{j=1}^n (X_j - \bar{x})^2}$$

is a biased estimator of the scale parameter $\sigma$, the BLU estimator of $\sigma$ is superior to $s$. Yet, unlike BLU estimators, the statistic $s$ is not a linear estimator. Therefore, for certain densities, when gauged in terms of MSE, the statistic $s$ can also be superior to a BLU-scale estimator. But MSE, like a hypothesis test's power that will be discussed in Chapter 19, must be balanced against robustness. For example, not only is the statistic $s$ a biased estimator of the scale parameter $\sigma$ of the Normal density model, but the squaring operation of any algorithm by which $s$ is calculated causes this statistic to be hypersensitive to any outliers that happen to be contained in the sample. As if this hypersensitivity were not bad enough, there is no known modification that can handle situations where outliers are removed without further increasing bias of the statistic $s$.

Section 19.6 will discuss the construction of curves that can help a researcher gauge BLU estimator tradeoffs. For example, suppose a potential large-valued outlier candidate is detected. One alternative is to treat this candidate as a censored value known solely to be greater than the next largest measurement in the sample. The other alternative is to sweep suspicions under the rug and proceed as usual.

Some loss of power, a criterion discussed in detail in Chapter 19, will occur were the candidate a bona fide value. Consequently, its demotion to censored status could come at a price. But at least by performing the calculations discussed in Section 19.6, it is possible to assess the power-lost-against-robustness-gained tradeoff.

There are a few drawbacks that can occasionally outweigh the advantages of BLU estimators. The ordering operations that underlie BLU estimator determination are often difficult to apply to multivariate procedures or regression analyses that concern multiple variables. Another drawback of these procedures is that the solution of the equations for BLU estimators depends upon prior knowledge of the curve $g$. However, once a standard curve is that has a finite variance is ascertained, the process of determining BLU estimator weights is merely a computational formality sometimes tricky, but always possible.

When location and scale parameters are targeted, BLU methods provide powerful and useful procedures. They alone can deal with outlier issues in a rigorous yet simple fashion. Thus, when robustness with respect to outlier contamination is a concern, BLU methods constitute a statistician's first line of defense.

On the downside however are conceptual and practical dimensionality problems. While the power of BLU methods has so far been found to be high for every choice of $g$ that has been studied carefully, their robustness is a tradeoff, rather than a universal benefit. BLU methods purchase their capacity to effectively deal with outliers at the price of reliance on the

## 17.7. BLU estimators and alternatives

validity of the choice of $g$. In the defense of BLU techniques however, it should be noted that this is also a price paid by Fisher's ML procedures, techniques, that cannot be adapted, as can their BLU competitors, to expeditiously deal with censored samples.

# CHAPTER 18
# Inference and Composite Variates

## 18.1 Curves and composite variates

The type of uncertainty considered in the previous chapter was attributable to bias. Uncertainty that can be traced to measurement error has generated such an enormous literature that this topic requires a specialized and separate monograph. In particular, the (1968) survey paper "Errors of Measurement in Statistics" written by the Cochran of Cochran's Theorem fame, leaves little that requires elaboration.

This and the next two chapters concern an aspect of uncertainty that is sometimes distinguished from measurement error by the term "sampling error." To understand what sampling errors are it is necessary to have a firm grasp of the concept of sampling distribution. Unfortunately, lowercase letters like $s$, $r$, and the $x$ of $\bar{x}$ can give a false impression about these and other useful statistics, as well as about important quantities such as the squared standard error, $\text{Var}(\bar{x})$. Because $s$, $r$, and the $x$ of $\bar{x}$ are lowercase letters, it is easy to forget that entities like $\text{Var}(\bar{x})$ are merely generalizations of curve properties like $\text{Var}([X+Y]/2)$, and that to apply $\bar{x}$, $r$, and $s$ to solve statistical problems, fluency in $E([X+Y]/2)$ and $\text{Var}([X+Y]/2)$ notation is required.

Fluency concerns a property like $E([X+Y]/2)$ that abstracts information about a curve. This curve describes a composite variate's distribution. A property like $E([X+Y]/2)$ can often be expressed as a simple function of curve properties associated with individual variates like $X$ and $Y$. For example, if $X$ and $Y$ are variates that have expectations $E(X)$ and $E(Y)$, it follows from simple calculus procedures that $E(X+Y) = E(X) + E(Y)$, where $E(X+Y)$ represents the expectation of a variate whose value equals the value assumed by $X$ plus the value assumed by $Y$. More generally, for any pair of constants $a_1$ and $a_2$,

$$E(a_1 X + a_2 Y) = a_1 E(X) + a_2 E(Y)$$

while, if $X$ and $Y$ are independent variates,

$$\mathrm{Var}(a_1 X + a_2 Y) = a_1^2 \mathrm{Var}(X) + a_2^2 \mathrm{Var}(Y).$$

What does the formula

$$\mathrm{Var}(a_1 X + a_2 Y) = a_1^2 \mathrm{Var}(X) + a_2^2 \mathrm{Var}(Y)$$

refer to? The three variates: (1), $a_1 X + a_2 Y$; (2) $X$; and (3) $Y$, each have distributions that are described by an individual density curve. Unless stated otherwise, it will be assumed below that each of these density curves has, as a well-defined variance property: (1) $\mathrm{Var}(a_1 X + a_2 Y)$; (2) $\mathrm{Var}(X)$; and (3) $\mathrm{Var}(Y)$, respectively. The term "well-defined " means that the integrals which express each of three curves' Var property all define unique and finite numbers.

Besides linear functions, such as $a_1 X + a_2 Y$, and curve properties, like $E$, there are, of course, many other combinations of functions and properties upon which statistical studies are based. For example, when the population mode, population median, and population mean of the same Normal curve are equal, three properties of a single curve are described. On the other hand, the expression $\mathrm{Var}(a_1 X + a_2 Y) = a_1^2 \mathrm{Var}(X) + a_2^2 \mathrm{Var}(Y)$ concerns a single type of curve property, density curve variance. Yet this formula concerns three curves, where the two curves that describe the densities of the component variates $X$ and $Y$ may or may not assume the same form, while generally the density or any other curve that describes the distribution of $a_1 X + a_2 Y$ differs from counterpart curves that describe the distribution of $X$ or $Y$.

As illustrated by BLU procedures, analyses of densities, taken as a whole, as well as analyses of specific properties of these densities, such as the property $\mathrm{Var}(a_1 X + a_2 Y)$, often help determine optimal ways to combine multiple observations. This is the primary reason why the densities of composite variates like $a_1 X + a_2 Y$ are important. There is a subtle but profound change of emphasis here. Generally speaking, if everything were known concerning the joint distribution or even just the joint density, of all variates of interest, then the statistician's job would be over. Yet perfect knowledge about some density $f$ is an unattainable goal.

Because $f$ is not known perfectly, a researcher must make do with a limited sample of information and seek the best means, no pun intended, available to utilize this sample. Despite lack of detailed knowledge about $f$, he or she is in a position to know certain mathematical relationships, such as those between $f$ and its properties, and between composite variates, like $a_1 X + a_2 Y$, and their counterpart curve properties. Assumptions, such as a curve's finite variance; and procedures, such as random sampling, are

## 18.1. Curves and composite variates

often attended to in order to assure that these mathematical relationships can be trusted.

Linear estimators, such as the weighted average based on weights $a_1$ and $a_2$ of the sum $a_1 X + a_2 Y$, the numerator of the weighted average $[a_1 X + a_2 Y]/[a_1 + a_2]$ of $X$ and $Y$, are common examples of composite variates. When the component variates of some composite variate are commensurable, their values can be added, subtracted, or otherwise combined, numerically. In the sense that a cdf associates composite variate values with probabilities of occurrence, the values thus formed are values of new variates. From this perspective, as long as the $X$-and $Y$-variates have values that can be graphed on some comparable scale, then the value of the variate $X$, in other words $x$, and the value of the variate $Y$, $y$, can be added to yield the value $x + y$ of the variate called $X + Y$. On the other hand, of course, were $x$ the variate, body mass, and $Y$ the variate, height, then it would make little sense to form a new variate $X + Y$.

Consider the density curve that describes the distribution of a composite variate, for example a weighted average. The curve is likely to have well-defined properties like $E$ and Var whenever the component variates' density curves have well-defined properties like $E$ and Var. When this is the case there will be a numerical relationship between certain properties of the component curves and the properties like $E$ or Var of the composite variate's curve. Some relationships, like $E(X+Y) = E(X) + E(Y)$, may seem obvious and be true under almost all circumstances; while some relationships, like $\text{Var}(a_1 X + a_2 Y) = a_1^2 \text{Var}(X) + a_2^2 \text{Var}(Y)$ are not obvious, and are true when certain assumptions, such as in this example—variate independence—are valid.

Why the constants $a_1$ and $a_2$, that are unsquared when they appear within the parentheses of the argument of $\text{Var}(a_1 X + a_2 Y)$, are squared on the right side of $\text{Var}(a_1 X + a_2 Y) = a_1^2 \text{Var}(X) + a_2^2 \text{Var}(Y)$ is an important question. The quantities $a_1^2$ and $a_2^2$ are squared because the integrand on the right side of the definition,

$$\text{Var}(X; u, x) = \int_u^x [z - E(X)]^2 f(z) dz,$$

contains a squared quantity. Since $a_1$ and $a_2$, of $a_1 X + a_2 Y$, are constants, they can both be factored from some generalization of an integral like

$$\int_u^x [z - E(X)]^2 f(z) dz.$$

But, because

$$(a_1 X)^2 = a_1^2 X^2 \quad \text{and} \quad (a_2 Y)^2 = a_2^2 Y^2,$$

when these constants are factored, the same squared weights $a_1^2$ and $a_2^2$ are encountered that are found within the relationships

$$E\{(a_1 X)^2\} = a_1^2 E\{X^2\} \quad \text{and} \quad E\{(a_2 Y)^2\} = a_2^2 E\{Y^2\}.$$

We will now consider a fundamental relationship based on $\mathrm{Var}(a_1 X + a_2 Y) = a_1^2 \mathrm{Var}(X) + a_2^2 \mathrm{Var}(Y)$ that is often called upon when combinations of variates are applied.

## 18.2 Specific sampling distributions

A measurement of a single variate, $Y$, is a single number and provides limited information. Clearly then, a central statistical question is: How can $n > 1$ measurements taken from $n$ multiple randomly chosen, in other words randomly sampled, subjects be called upon? One clue comes from the basic relationship $\mathrm{Var}(a_1 X + a_2 Y) = a_1^2 \mathrm{Var}(X) + a_2^2 \mathrm{Var}(Y)$, for the special case where $a_1 = a_2 = 1/2$, and where $X$ and $Y$ are obtained from two randomly-sampled subjects. Because these subjects are chosen as part of a random sample, it follows that $X$ and $Y$ are independent, and thus $\mathrm{Var}(X/2 + Y/2) = [\mathrm{Var}(X) + \mathrm{Var}(Y)]/4$.

To simplify matters, assume $X$ and $Y$ are identically distributed. It then follows that a particular curve which characterizes the distribution of $X$ is identical to the counterpart $Y$ variate curve. A short step along this line of reasoning then indicates that all properties of counterpart curves must be equal for $X$ and $Y$ which, in turn, implies that $\mathrm{Var}(X) = \mathrm{Var}(Y)$. Therefore the formula $\mathrm{Var}(X/2+Y/2) = [\mathrm{Var}(X)+\mathrm{Var}(Y)]/4$ implies that as long as the variances $\mathrm{Var}(X)$ and $\mathrm{Var}(Y)$ are well-defined and finite, the variance of the average of $X$ and $Y$ equals one half $\mathrm{Var}(X)$ or, equivalently, one half $\mathrm{Var}(Y)$.

By extending the above argument, the following relationship that is familiar to most researchers can be derived. Given that they individually have a well-defined variance, the variance of the mean

$$\bar{x} = n^{-1} \sum_{j=1}^{n} X_j$$

of any $n$ iid variates $\{Xj\}; j = 1, \ldots, n$, with consecutive values $\{x_j\}; j = 1, \ldots, n$, equals the variance of any one variate divided by the number $n$. Also, basic algebra applied to the above expressions indicates that dividing any variate by the square root of its variance will yield a new variate that always has a variance equal to one. For example, suppose the variate $Z_n$ is

## 18.2. Specific sampling distributions

formed by dividing some variate $X_n$ by the square root of $\text{Var}(X_n)$ which, below, will be represented as $\tilde{\sigma}$. The "tilda" symbol $\tilde{\phantom{o}}$ is a reminder that the particular Greek letter $\sigma$ is best thought of as a symbol for a curve property; one that may or may not correspond to a model parameter.

Setting $x = X_n, a_1 = 1/\tilde{\sigma}$ and $a_2 = 0$ within

$$\text{Var}(a_1 X + a_2 Y) = a_1^2 \text{Var}(X) + a_2^2 \text{Var}(Y)$$

shows that the variance of $Z_n$ equals one. It is also true that for any constant $K$, as the upper and lower limits of integration of

$$\int_u^x [z - E(X)]^2 f(z) dz$$

approach plus and minus infinity, respectively, if the integrals that define $\text{Var}(X + K)$ and $\text{Var}(X)$ are well-defined, then these two curve properties assume the same value. Thus the variance of a new variate constructed by adding or subtracting a constant value from any given variate equals the variance of that given variate. Despite its seeming simplicity, an expression like $\text{Var}(X + K) = \text{Var}(X)$ is a compact representation that tends to hide the complexities of not only two variates but, in addition, dual distributions and their associated curves.

Although most readers are undoubtedly already familiar with mean's variance formula (MVF), its proof is considered in detail here. It is included not simply because of the MVF's key role in statistical thinking, but because uncertainties about statistical procedures often call for modified versions of this formula.

An important example of such a modification concerns a *sample quantile*, an evaluation at some value of $p$, where $0 < p < 1$, of a curve which can be thought of as the inverse of the scdf defined in Section 15.2. For any sample of moderate to large size, such as $n > 30$, for all applied purposes the sample quantile at $p$ is the particular order statistic $X_{(\text{ceil}(np))}$.

The Blom (1962) "nearly best linear estimates" that were discussed briefly in Section 17.2 are based on relationships proven by Cramer (1946, Section 28.5). For variates with common density $f$ and inverse cdf curve $F^{-1}$, Cramer explains why as $n \to \infty$ the $p^{\text{th}}$ sample quantile approaches a Normal variate whose location parameter equals $F^{-1}(p)$ and whose scale parameter equals $[f(F^{-1}(p))]^{-1}$ times $\sqrt{p(1-p)/n}$. Cramer also shows in earlier chapters of his book that weighted sums of Normal variates must, themselves, be Normal variates.

The consequence of the two above relationships that are proven by Cramer (and, in other forms by earlier researchers), is that only unusual weighted sums, that for some reason emphasize order statistics with either or both large and small parenthesized subscripts, cannot be taken, for all practical purposes, to approach Normal variates. For example, as $n$ increases, the sample median approaches a Normal variate with associated location parameter equal to the population median, and scale parameter equal to the standard deviation common to each iid variate times $\sqrt{\pi/(2n)}$.

BLU estimator tables, such as those of Sarhan and Greenberg (1962), are accompanied by tabled quantities known as efficiencies. It will be shown in the next chapter that when the sample median or some particular BLU estimator—that may or may not have been adjusted for outliers—is substituted for some sample mean, such as one of the means that appear within the ANOVA and ANCOVA relationships reviewed in Chapter 16, these published efficiencies can be applied as adjustment factors that, if needed, allow conventional $t$-and F-tables to be called upon for many statistical purposes.

There is a modular quality to the above process for dealing with uncertainties which concern statistical curves that is presently unique to these procedures. What this quality often amounts to is the following: In order to control for outlier-induced and other forms of bias, there must be in most instances some known loss of efficiency. The two forms of uncertainty—on the one hand sampling error, as determined by the MVF and Cramer's formula and, on the other, bias—can be traded off against each other to suit the needs of each individual investigation.

To prove the MVF, the formula $\text{Var}(a_1 X + a_2 Y) = a_1^2 \text{Var}(X) + a_2^2 \text{Var}(Y)$ can be applied using mathematical induction, a procedure usually taught in high school algebra classes. The relationship

$$\text{Var}(X/2 + Y/2) = [\text{Var}(X) + \text{Var}(Y)]/4$$

implies that the $n = 2$ case of the mean's variance formula is true. Suppose $\bar{x}^{(n)}$ is the sample mean of a size $n$ iid sample while $\bar{x}^{(n-1)}$ represents the mean of a size $n - 1$ iid sample. Also suppose that the mean's variance formula is true for $\bar{x}^{(n-1)}$. When $x_n$ is the value of an $n^{\text{th}}$ variate $X_n$ that is added to a size $n - 1$ sample, it follows that

$$\bar{x}^{(n)} = [(n-1)/n]\bar{x}^{(n-1)} + x_n/n.$$

Consequently,

$$\text{Var}(\bar{x}^{(n)}) = [(n-1)/n]^2 \text{Var}(X_n)/(n-1) + \text{Var}(X_n)/n^2$$

which, when simplified, proves by mathematical induction that the mean's variance formula applies to any size $n$ sample.

Before considering the MVF further, a notational chore must be attended to. While the actual density curve that has the curve property $\text{Var}(\bar{x}^{(n)})$

## 18.2. Specific sampling distributions

might be designated by placing a subscript beneath the letter $f$, this density should not be denoted by a hatted symbol like $\hat{f}$. A sample mean has a distribution and that distribution might be described in terms of a curve, $f_{\bar{x}}$. That this sample mean might be applied as an estimator of a location parameter implies that $\bar{x}^{(n)}$ could also be designated as $\hat{\mu}$. As a consequence, it would be appropriate, although unusual, to denote $f_{\bar{x}}$ as the subscripted symbol $f_{\hat{\mu}}$.

Uncertainty due to sampling error is often tantamount to a shift in emphasis from the variability of the density of some constituent variate to the variability or some other characteristic of a curve like $f_{\bar{x}}$ in particular, and $f_{\hat{\mu}}$ in general. However, both the symbols $f_{\bar{x}}$ and $f_{\hat{\mu}}$ suggest that either an $E$ curve property or, in the case of $f_{\hat{\mu}}$, a location parameter, is the target of the statistical investigation. One way of looking at this distinction, is that the confidence band procedures introduced in Section 15.5 are designed to provide information that concerns some conditional density generalization of $f_{\hat{\mu}}$, while prediction bands do the same thing for some constituent variate density.

Formulas for the median's and other order statistics' variances as $n \to \infty$, together with relationship $\mathrm{Var}(Z_n) = 1$, as well as the MVF, all may seem to be mere theoretical curiosities because, in practice, it is unlikely that the value of $\tilde{\sigma}$ is known. Nevertheless, these are some of the most important relationships within not only the realm of theoretical statistics but within applied statisticians' tool kits. They are important because they have served as the springboards from which important branches of methodology have been launched.

We have now completed all the preliminaries necessary to define and illustrate how sampling distributions are applied. Even in the simplest of cases, to do this, the distribution of a composite variate, such as $X/2 + Y/2 = (X + Y)/2$, or the mean of $X$ and $Y$, must be distinguished carefully from the individual distributions of the variate parts from which the composite is assembled. The term "sampling distribution" is a reminder that a variate being studied is a composite formed from other variates.

There are subtle differences between the three terms: composite variate; statistic and; estimator. The first term is applied in the context of relationships like

$$\mathrm{Var}(a_1 X + a_2 Y) = a_1^2 \mathrm{Var}(X) + a_2^2 \mathrm{Var}(Y),$$

while the word "statistic," even though a close synonym to the term composite variate, suggests that the quantity referred to is computed from a sample with, or without, any guidance from relationships like $\mathrm{Var}(a_1 X + a_2 Y) = a_1^2 \mathrm{Var}(X) + a_2^2 \mathrm{Var}(Y)$. The term estimator emphasizes the existence of some estimated curve, curve property, or parameter that the estimator estimates, while the word statistic and the term composite variate are not accompanied by a term such as "curve" or "parameter" that suggests what was the purpose of the computation.

Whether a statistic is an estimator of a parameter or, alternatively, of a curve property, depends upon the choice between the selection of a parametric, or of a nonparametric, statistical procedure. Although it is a cornerstone of most parametric procedures, the important formula

$$\text{Var}(a_1 X + a_2 Y) = a_1^2 \text{Var}(X) + a_2^2 \text{Var}(Y),$$

relates three curve properties: $\text{Var}(a_1 X + a_2 Y)$; $\text{Var}(X)$; and $\text{Var}(Y)$. Thus the validity of the $\text{Var}(a_1 X + a_2 Y) = a_1^2 \text{Var}(X) + a_2^2 \text{Var}(Y)$ formula depends, in part, on the existence of the Var curve property of three curves. Its validity does not depend on the assumption that a model, like that of the Normal density $\phi([x - \mu]/\sigma)/\sigma$, fits a variate's $f$-type curve.

In the remainder of this chapter it is convenient to represent the standard deviation of the sampling distribution of the mean of $n$ iid variates as

$$\sigma^{(n)} = \tilde{\sigma}/\sqrt{n},$$

where $\tilde{\sigma}$ represents the standard deviation of a single variate, a curve property that below is assumed to exist and may, in certain special cases, be equal to the scale parameter $\sigma$ of some general density model. The parenthesized letter $n$ appears as a superscript and not a subscript to avoid confusion with order statistic parenthesized subscript notation, for example, where $x_{(n)}$ denotes the value assumed by the largest-valued variate.

In this and later sections, much will be made of the facts that: (1) the standard deviation of the variate

$$Z = \frac{\bar{x} - E(X)}{\sigma^{(n)}}$$

always equals 1; (2) $E(Z)$ always equals 0 and; (3) for the mean $\bar{x}$ of any $n$ iid variates, $\{X_j\}; j = 1, \ldots, n$; the formulas $E(X/n) = E(X)/n$ and $E(X + Y) = E(X) + E(Y)$ imply that $E(\bar{x}) = E(X_j); j = 1, \ldots, n$.

Statistics are often used to estimate properties of a distribution that are not numerically equal to a parameter within some model's formula. Certain life tables and all histograms are examples of this type of estimator. In particular, a clinical life table provides an estimator of a survival curve $S$, where $S$, in turn, equals 1 minus a cdf.

Besides estimating a parameter or curve evaluation, statistics often estimate curve properties like $E(X)$ or $\text{Var}(X)$ that are defined without requiring that the curve in question be modeled. Nevertheless, the most common application of estimators calls upon some elementary function like $\phi$ of the

## 18.3. The mean's variance formula and mixtures

Normal model $\phi([x-\mu]/\sigma)/\sigma$, and parameters of this model like $\mu$ and $\sigma$. Chapter 20 will consider variates whose distributions are described best by the truncated form of a density function, where in its nontruncated form, this density has an infinite variance. For such variates, the usual connections that link individual curve properties to individual model parameters tend to be untrustworthy.

Before turning to the details of this general approach it should be pointed out that the general Cauchy model, $\sigma^{-1}\varphi([x-\mu]/\sigma)$; where

$$\varphi(z) = 1/\{\pi(1+z^2)\},$$

can be defined in terms of a scale parameter $\sigma$, even though $\sqrt{\text{Var}(X)}$ is infinite. Because the MVF applies to curve properties, and not to parameters like $\sigma$, testing and estimation procedures based on this formula cannot, at least without considerable emendation, be applied to complex forms of data such as those that are mixture or Cauchy-distributed.

### 18.3 The mean's variance formula and mixtures

Most of Chapter 20 will be devoted to the limitations placed on sampling error methodology based on the MVF when variate distributions are fit well by truncated forms of long-tailed models like that proposed by Cauchy. As a preliminary, this brief section will illustrate a problem that is somewhat easier to cope with than long-tail truncation.

For $0 < P < 1$, consider the two-component mixture of densities $f_1$ and $f_2$, $Pf_1(x) + (1-P)f_2(x)$. The $E(X)$ curve property of this mixture density is simply

$$PE_{f_1}(X') + (1-P)E_{f_2}(X''),$$

where $E_{f_1}(X')$ and $E_{f_2}(X'')$ represent the expectation curve properties of $f_1$ and $f_2$, respectively. However, the variance curve proprerty $\text{Var}(X)$ is not equal to $P\text{Var}_{f_1}(X') + (1-P)\text{Var}_{f_2}(X'')$ but, instead,

$$\text{Var}(X) = P\text{Var}_{f_1}(X') + (1-P)\text{Var}_{f_2}(X'') + P(1-P)\{E_{f_1}(X') - E_{f_2}(X'')\}.$$

To again emphasize that the $\tilde{\sigma}$ symbol designates the standard deviation curve property, not a scale parameter, the tilda symbol is placed over the sigma of $\sigma^{(n)} = \tilde{\sigma}/\sqrt{n}$. One reason for this choice of symbols is that even in the situation where both $f_1$ and $f_2$ are Normal densities, and hence each individually has a standard deviation curve property that is equal numerically to the corresponding density's scale parameter, whenever $E_{f_1}(X') \neq E_{f_2}(X'')$, $\tilde{\sigma}^2$ will not equal $P$ times the first of these scale parameters squared, plus $(1-P)$ times the second scale parameter squared.

Viewed generally, relationship

$$\text{Var}(X) = P\text{Var}_{f_1}(X') + (1-P)\text{Var}_{f_2}(X'') + P(1-P)\{E_{f_1}(X') - E_{f_2}(X'')\}^2$$

implies that the variance curve property of a mixture variate is related functionally to not only constituent variance curve properties, but also to constituent expectation curve properties and, even more generally, location is confounded with scale. Since ANOVA, ANCOVA, and many varieties of parametric regression analyses are based on the distinction between location and scale, the relationship

$$\text{Var}(X) = P\text{Var}_{f_1}(X') + (1-P)\text{Var}_{f_2}(X'') + P(1-P)\{E_{f_1}(X') - E_{f_2}(X'')\}^2$$

will throw a monkey wrench into ANOVA, ANCOVA, and regression machinery. (As illustrated at the end of the next chapter, BLU procedures can help detect potential machinery foul-ups.)

## 18.4 Inference and a two-valued metric

Useful machinery has been described in the preceding sections but, as yet, how this machinery is applied has not been outlined. Shortcut descriptions of this machinery, while suitable for some simple problems, are not applicable to complex problems. This is because the ramifications of basic statistical distinctions are not anticipated easily. A key concept of a *metric* can clarify some of these ramifications.

Like a function, a metric has arguments that are represented within parentheses or brackets. However, these arguments are not simply numerical values. They are also estimators, parameters, curve properties, and curves like $f$ and $\hat{f}$ in their entirety. These are usually paired so that: ends and means; curves and curve estimators; parameters and statistics and; curve properties and their estimators, are matched.

The simplest metric takes one of these pairs and associates it with one choice from two distinct alternative values. For example, one of these values often designates the numerically positive cost, loss, or price of making a wrong decision, and the other the value zero, which of course signifies when a correct decision is made with no cost or loss incurred. Older formulations of what is known as the *frequentist approach* to two-valued inference, couched the above process in terms such as the acceptance or rejection of a null hypothesis, level of significance, and power, as described later in this and the next chapter. Although helpful in some ways, these terms obscured the two-valued metric's fundamental role. It also made it difficult to compare those procedures that were based on this metric to other procedures.

## 18.4. Inference and a two-valued metric

As was the case for the symbol $\tilde{\sigma}$, symbols with a parenthesized superscript letter, such as the symbol $\sigma^{(n)}$, designate curve properties that may or may not correspond to some particular model component, such as a scale parameter. Based on this notation, two-valued inferential choices are illustrated by an example based on the composite Normal variate

$$Z = \frac{\bar{x} - E(X)}{\sigma^{(n)}}.$$

The key facts are that the standard deviation of $Z$ equals 1, and that $E(Z)$ equals 0. Thus the term $Z\sigma^{(n)}$ on the right hand side of $\bar{x} - E(X) = Z\sigma^{(n)}$ discloses how much the estimator $\bar{x}$ of the curve property $E(X)$ differs from $E(X)$, the estimated curve property.

In statistical applications the value assumed by $Z\sigma^{(n)}$ has an important interpretation. The formula $\bar{x} - E(X) = Z\sigma^{(n)}$ asserts that the composite variate $\bar{x} - E(X)$ can be simulated by multiplying a variate $\Phi^{-1}(\text{rnd}(1))$ by the quantity $\sigma^{(n)}$. The $\sigma^{(n)}\Phi^{-1}(\text{rnd}(1))$ generation model casts light on the important issue: How likely is it to encounter some particular value of $\bar{x} - E(X)$? For example, when only a small proportion of simulated $Z$-times-$\sigma^{(n)}$ values equal or exceed some observed value of $\bar{x} - E(X)$, this casts doubt on the hypothesized null value of $E(X)$ within $\bar{x} - E(X)$. As far as uncertainty about statistical curves is concerned, the above reasoning explains why the tails of sampling distributions are of primary importance.

In particular, suppose $E(X)$ or a model parameter $\mu$ known to equal $E(X)$, has a value that, were it also known, would be central to the decision-making process. Then if a prior, null, or trial, guess $\sigma_0$ about $E(X)$ or $\mu$ were to differ from $\bar{x}$ by more than some predetermined large value of the quantity $Z\sigma^{(n)}$, it would be sensible to infer that this null or trial value is incorrect. Now for the punch line. If $\sigma^{(n)}$ were a known constant, and a table available from which the cdf $F$ of the variate $Z$ could be evaluated, then it would be practical to assess just how reasonable it is for estimator $\bar{x}$ of $E(X)$ to differ from $E(X)$ by the amount observed. In the case of the two-valued metric, when $\bar{x} - \mu_0$ is greater than some particular choice of $Z\sigma^{(n)}$, a decision is then made to reject the null or trial guess that $E(X)$ equals $\mu_0$. We will now turn to the details of processes by which such hypothesis tests are performed.

The single assertion, that $\mu_0$ does not differ from $\bar{x}$ by more than $Z\sigma^{(n)}$, concerns the four components: $\mu_0$ $\bar{x}$; $\sigma^{(n)}$; and $Z$, of a general statistical procedure. Consequently, before considering this statement in terms of its details, it is useful to step back a bit in order to take a wide angle view. When a statistical hypothesis test is performed, the best single word that designates what is actually done to the four components: $\mu_0$ $\bar{x}$; $\sigma^{(n)}$; and $Z$ is "package." The individual pieces $\mu_0$ $\bar{x}$; $\sigma^{(n)}$; and $Z$ are wrapped or bundled together to form a whole package. Like the preliminaries that precede the mailing of a large parcel, convenience plays a large role in hypothesis testing. The padding and string that keeps components in place correspond to the subtraction operation that separates $\bar{x}$ from $\mu_0$ as well as the division operation that separates the quantity $\bar{x} - \mu_0$ from $\sigma^{(n)}$.

It is not coincidental that $(\bar{x} - \mu_0)/\sigma^{(n)}$ resembles the argument inserted within the standard Normal $\phi(z)$ to form the general Normal $\phi([x - \mu]/\sigma, /\sigma$. Like the argument $z = [x - \mu]/\sigma$ within $\phi(z)$, the test statistic $[\bar{x} - \mu_0]/\sigma^{(n)}$ preprocesses or prepares three components so that they can act as a single scalar quantity. Similarly, were $\Phi$ to represent the standard Normal cdf curve, then $\Phi([x - \mu]/\sigma)$ can be thought of as a package where the function $z = [x - \mu]/\sigma$ is wrapped by the function $\Phi$. Testing can be thought of as the unwrapping of the $\Phi[x - \mu]/\sigma$ package.

Consider the order in which two parameters, $\mu$ and $\sigma$, as well as the $\Phi$ function are written in $\Phi([x - \mu]/\sigma)$. This order of operations, specifically the subtraction and then division that are performed in order to evaluate the curve at a given value of $x$, is the opposite of the order of operations needed to compute the generated variate $X$ of $X = \mu + \sigma \Phi^{-1}(Z)$. Suppose within a data-generation model a standard Normal variate $\Phi^{-1}(Z)$ is first simulated. Then in order to construct Normal $X$-variates based on a scale parameter $\sigma$ and a location parameter $\mu$, a multiplication is followed by an addition. Viewed from this perspective, the formula $X = \sigma \Phi^{-1}(Z)$ is a way to unwrap the package $\Phi([x - \mu]/\sigma)$.

The term $\Phi^{-1}(Z)$ of the $X = \sigma \Phi^{-1}(Z) + \mu$ data-generation model has, as a special case, the coefficient 1.96 that is often encountered in hypothesis testing applications. What is known as the "alpha level" or "$p$-value" is associated correspondingly with the argument $Z$ of the function $\Phi^{-1}(Z)$. When a table of the Normal cdf $\Phi$ is utilized, among the steps of hypothesis-test construction, the alpha level or $p$-value is located within the body of the table, and a value like 1.96 is located at the margin of the table. This reverse use of a table, reading from its body to its margin, corresponds to the -1 symbols which separate the $\Phi$ symbol from the $(Z)$ argument among the six symbols that make up $\Phi^{-1}(Z)$.

The test statistic package $(\bar{x} - \mu_0)/\sigma^{(n)}$ preprocesses the components of the testing procedure so that they can be compared to a number like 1.96 or, in the example below, 1.64. The steps required to do this are repeated so often in testing applications that it pays to describe them in detail. Of the three components that are assembled to form $(\bar{x} - \mu_0)/\sigma^{(n)}$, the choice of the $\mu$ parameter, or its $E(X)$ curve property counterpart, serves as the first step of the test construction process. Notice that in many applications it would, of course, be preferable to obtain knowledge about the entire curve $f$, rather than merely about some single $\mu$ within this curve's model, or

## 18.4. Inference and a two-valued metric

some single property like $E_f(X)$ that abstracts the information about $f$.

Choices of parameters or properties to be tested, in general terms—statistical ends—determine the choice of statistics such as the arithmetic mean $\bar{x}$—statistical means—in a dual sense of the term *means*. It is usually the case that a parameter like $\mu$ is associated with a curve property like $E_f(X)$, where this property is, in turn, defined in terms of the cdf $F$. The so-called alpha level or $p$-value $\Phi^{-1}(Z)$ that is looked up in a table, or its counterpart for any other statistical test is, in reality, an evaluation of a particular curve, the inverse cdf $\Phi^{-1}$. The word "alpha" sometimes appears in place of $\alpha$, an entity that should not be confused with the regression model intercept parameter of $\alpha + \beta x$ or an ANOVA Model I row effect parameter.

Although $\mu$ is subtracted from $\bar{x}$ within $(\bar{x} - \mu)/\sigma^{(n)}$, it is not always the case that a statistic and the parameter it estimates are compared by taking their difference. A statistic is often divided by a parameter value. All that is required is some way to combine the means to an end, here $\bar{x}$, with a hypothesized value of that end, here what every practicing statistician knows as the "null hypothesized" value of $\mu$. Obviously significant discrepancies between means and ends correspond to test-statistic values that are inordinately large or small.

The component $\sigma^{(n)}$ of test statistic $(\bar{x} - \mu)/\sigma^{(n)}$ resembles a nuisance variate or variable. The question, whether or not models represent natural processes adequately, applies both to nuisance variates or nuisance variables and to model components that are nuisance parameters. Knowledge of a quantity like $\sigma^{(n)}$ is a necessary preliminary, without which a test statistic cannot be evaluated, in the sense that a tabled value like 1.96 must be supplemented by $\sigma^{(n)}$ itself or this quantity's sampling distribution scale-determining counterpart.

Division of the difference $(\bar{x} - \mu)$ by $\sigma^{(n)}$ accomplishes a great deal. One table of a single Normal cdf, specifically $\Phi$, suffices for any choice from among the many test statistics that, like $(\bar{x} - \mu)$ can be taken to be Normally distributed. Once the package is wrapped, it no longer matters what sample size $n$ of iid Normal variates was employed to compute $\bar{x}$. The package $(\bar{x} - \mu)/\sigma^{(n)}$ is ready to be shipped, via a published $p$-value, to anywhere scientists are located.

One loose end of the packaging process must still be attended to. This is the scale parameter $\sigma$ symbol as it appears on the right side of the mean's variance formula

$$\sigma^{(n)} = \sigma/\sqrt{n}.$$

Suppose temporarily that the symbol $\sigma$ and the curve property $\sqrt{\mathrm{Var}(X)}$ can be interchanged with impunity. Although the sample size, in other words the value of $n$, is usually known, the value assumed by $\sigma$ is usually unknown. Were it the case that the true value of $\sqrt{\text{Var}(X)}$ is known exactly, the parameter $\mu$ or the curve property $E_f(X)$ would, almost certainly, also be known; and were $E_f(X)$ actually a known value why would one want to test a null-hypothesized value of this curve property in the first place?

During the long period that began with Laplace and Gauss's investigations concerning the sample mean, and ended in the early twentieth century, the problem of unknown $\sqrt{\text{Var}(X)}/\sqrt{n}$ remained unsolved. As is the case for the connection of $E_f(X)$ with $\bar{x}$, there is a default statistical estimator of $\sqrt{\text{Var}(X)}$, the statistic $s$. Yet it was observed that the simple substitution of this sample standard deviation within the formula that helps estimate the "standard error" $\sigma^{(n)}$, followed by calculations based on Normal tables for $\Phi$, was only accurate when sample size $n$ was large. Finally, in the early twentieth century, the simple solution to the unknown $\sqrt{\text{Var}(X)}/\sqrt{n}$ problem, based on the construction of specialized tables whose entries change as the sample size changed, revolutionized statistics.

## 18.5 The one tail z-test

This section considers the role played by Normal tables and the relationship these tables have to statistical power and robustness. Suppose the distribution of the variate $Z$ is described by a density model whose parameters are completely determined by knowledge of this model's Var and $E$ curve properties. When such a density is determined completely, it follows that some table or computing algorithm will allow the cdf of $Z$, $F(z)$, to be evaluated. For example, the evaluation $1 - F(z_0)$ discloses how probable it is for the value of the variate $Z$ value to exceed the specified value $z_0$.

Suppose a table discloses which number $z_0$ is exceeded by $1/20^{\text{th}}$ of $Z$'s values. This implies, in turn, that the product $z_0\sigma^{(n)}$ is exceeded by values of the variate $\bar{x} - \mu$ or $\bar{x} - E(Z)$, exactly $1/20^{\text{th}}$ of the time. Then, technically speaking, this one-tailed test should reject the null hypothesis that $\mu = \mu_0$ whenever the "z-statistic," $[\bar{x}-E(X)]/\sigma^{(n)}$, or correspondingly, $(\bar{x} - \mu)/\sigma^{(n)}$, is found to exceed the value $z_0$. In the remainder of this chapter, this z-statistic application is called a *one-tailed z-test*.

Since, for a two-valued metric, one is either right or wrong when one makes a decision, the quantity alpha is equal to the probability of being incorrect when one says what one usually wishes to say, namely that the null-hypothesized value of the property $E(X)$, or the location parameter $\mu_0$, is incorrect. Like a bowling pin, a null-hypothesized value of the parameter

## 18.5. The one tail z-test

or property being studied is set up in the hope that it will later be knocked down. If you think there is something artificial about this, you can join the many Bayesian-oriented statisticians who take a markedly different view of inference. Yet, since it is an approach much in accord with the way real-life decisions are made, the frequentist—reject or accept—method of applying the two-valued metric is admittedly elegant.

When an electrical appliance is purchased, a manufacturer often attaches a small label to the power cord that reads either "Such and Such Laboratory Approved" or "Good Something-or-Other Seal of Approval." Alpha levels and $p$-values play an analogous role when they are included in a technical publication or document. The reader of this publication is interested in finding out whether or not the conclusions reached in the study being documented are safe. Here the word "safe" denotes a standardized degree of certainty.

Today scientific publications usually contain at least one finding based on an alpha equal either to 0.01 or 0.05. These are common choices, since 100% certainty always unrealistic in any inductive setting; 0.5, 50% or 50-50 odds of being wrong provide far too much uncertainty; while 0.0001 is so close to certainty that it is rarely encountered.

Consider the following admittedly implausible situation where many publications that concern a specific research topic are gathered together. Suppose that each individual paper deals with an aspect of a topic that has no bearing on the concerns addressed by the other publications. Furthermore, assume each of these publications is based on only a single-hypothesis test, one that has led to the rejection of a null hypothesis at the alpha = 0.05 level.

Now suppose all assumptions upon which a particular hypothesis-testing process is based, are valid. Then the larger the subset of publications, the less likely it will be for more than 5% of these publications to present incorrect findings. In the idealized extreme case where some completely, and correctly, discredited subject matter is studied—in other words, in the situation where the null hypothesis is true—for every member of the above set of publications there will be nineteen studies that had not obtained significant findings.

Notice that in the above paragraph neither: (1) the exact value of sample size $n$ nor; (2) what the curve properties or expression parameters are that were actually tested, were specified. Were an investigator lucky enough to perform an experiment with so definitive a protocol that very few measurements are needed to reject the null hypothesis at an acceptable alpha level—so much the better. What is of great concern however, is the changing of an experiment's rules in midstream. For example, it is improper for a study conducted with a protocol involving sample of size $n$, that does not result in significant findings, to simply be extended so that more data will carry a test statistic, like $(\bar{x} - \mu_0)/\sigma^{(n)}$, into what is called the *critical region*. Like the number of consecutive downs during a football game, once a team has used up its allotted four downs, it is too late to carry the ball or throw a completed pass into the end zone.

The games of chance that were studied by Pascal and many other mathematicians differ substantially from aggregations of scientific investigations, whose study is now referred to by the term "meta-analysis". On the other hand, meta-analytic statistical science resembles sports journalism, in the sense that both fields take an overview of their respective areas of interest. Games of chance and athletic contests are conducted through repetitive application of the same rules. Yet, even though the designs of most scientific investigations are based on findings made in earlier investigations, rarely is the same study repeated exactly.

Two-stage procedures, such as those of Stein (1945), deal with pilot studies conducted prior to the principal study. Stein's approach allows crucial decisions about the sample size required to conduct the principal study to be based on results obtained in the pilot study. It also makes allowance for the possibility that without proceeding to the second-investigation stage, enough evidence will have been collected in the pilot-study stage to reach a definitive conclusion. On the other hand, while Stein's procedures are applicable to many realistic scientific investigations, because of its simplicity, the simple one-tailed test based on the "$z$-statistic" described above provides a useful starting point for discussion of many important aspects of statistical inference.

Based on the $z$-statistic, two-tailed significance tests can be designed to deal with situations where an alternative true value of a parameter might be either greater or smaller than the null value. Often, when many parameters are tested separately, multiple comparison methods of analysis of variance, ANOVA, provide simultaneous inference. If, rather than a single null value, an entire prior distribution of parameter values is input along with sampled data, Bayesian statistical approaches are called for. Metrics often assess costs with more flexibility than is possible with dichotomous functions. Such metrics often evaluate confidence intervals based on what are called *confidence levels* that quantify both these intervals and the confidence bands introduced briefly in Section 15.5.

From a statistical vantage point, confidence concerns: (1) the degree of certainty that the true value of a curve property is within a specific estimated interval, or; (2) if a valid model is selected, the certainty that the true value of a parameter is within a specific estimated interval. For instance, when the properties under study are curve evaluations, such as a regression line, then this curve can be surrounded by confidence bands

## 18.5. The one tail z-test

so that a pair of 95% confidence bands are designed to contain the true regression line 95% of the time. When slope parameter $\beta$ of the model for a line is zero, it follows that this line is horizontal. Therefore, when confidence bands do not contain any horizontal line ordinate value, this geometric pattern implies that, given the existence of a *linear* regression curve, the null hypothesis, $\beta = 0$, should be rejected.

Both significance levels and degrees of confidence are forms of research certification. Consequently, when so much emphasis is placed on some Laboratory Approved or Seal of Approval certification, it is small wonder that robustness is an important consideration. Why bother with elaborate calculations when the assumptions that guarantee the validity of the calculations are questionable? It requires considerable computation and attention to basic conceptual issues to check assumptions. This effort can exceed the effort needed to perform a one-tailed or other test in the first place—hence the advantage of robust procedures, techniques with the loopholes plugged (loopholes through which statistical reasoning could otherwise be brought into question).

Because of the importance of assumption-validity-check processes, statistical techniques have been divided into broad categories consisting of: (1) those based on a sampling distribution's density model; (2) those based on category (1) approaches that have been redesigned with an eye to robustness enhancement; (3) model-free and nonparametric approaches. When they rely on unquestionably valid statistical foundations, category (3) procedures have been designed to deal with the robustness issue from the ground up.

The role played by the sample size $n$ provides a good reference point from which the strong and weak points of category (1), (2), and (3) procedures can be gauged. For example, irrespective of the size of the available sample, whenever a value of the $z$-statistic is found that exceeds $z_0$, the same decision is made. However, suppose the traditional symbol $\mu_A$, where A denotes "alternative," is known to represent the true value of the property $E(X)$ or, in this case, the parameter $\mu$. Let the symbols $\bar{x}^{(3)}$ and $\bar{x}^{(4)}$ represent the sample *arithmetic averages* or *means* calculated from respective random samples of size $n = 3$ and $n = 4$ from the same population within which the standard deviation of the $X$-variate equals the scale parameter value $\sigma$. Then the corresponding variate values $z^{(3)}$ and $z^{(4)}$ can be defined as

$$z^{(3)} = \frac{\bar{x}^{(3)} - \mu_0 + \mu_A - \mu_A}{\sigma/\sqrt{3}}$$

and

$$z_{(x)} = \frac{\bar{x}^{(4)} - \mu_0 + \mu_A - \mu_A}{\sigma/2}.$$

Suppose
$$z_A^{(3)} = \frac{\bar{x}^{(3)} - \mu_A}{\sigma/\sqrt{3}}.$$

Also let
$$z_A^{(4)} = \frac{\bar{x}^{(4)} - \mu_A}{\sigma/2}, \quad \text{while} \quad \Delta = \frac{\mu_A - \mu_0}{\sigma}.$$

Then the statistic $z^{(3)}$ equals $z_A^{(3)} + \sqrt{3}\Delta$ while, on the other hand,
$$z^{(4)} = z_A^{(4)} + 2\Delta.$$

A comparison of $z_A^{(3)}+\sqrt{3}\Delta$ with its sample-size-4 counterpart, $z_A^{(4)}+2\Delta$ shows that when $\mu_A = \mu_0$, and consequently $\Delta = 0$, a somewhat surprising but important conclusion follows from the $\Delta = 0$ assumption. Despite the computation of $z^{(3)}$ from a sample of size $n = 3$, and the calculation of $z^{(4)}$ from a sample of size $n = 4$, the two curve properties, expectation and variance, of the sampling distributions of the two statistics, $z^{(3)}$ and $z^{(4)}$, are identical. Specifically, both $z_A^{(3)}$ and $z_A^{(4)}$ have densities that, although they might differ in other ways, have identical $E$ and Var properties.

The fact that expectation and variance are equal above does not, of course, imply that all properties of two sampling distributions are identical. It was to stress this point that up to now, underlying iid variates, specifically $\{Xj\}; j = 1, \ldots, n$; that are combined to form the $n = 3$ and $n = 4$ sample means, have not been associated with any particular model. Suppose, however, that each member of a new iid random sample $\{Xj\}; j = 1, \ldots, n$; is Normally-distributed. As shown by using simple mathematical relationships based on the techniques devised by Laplace, Fourier, and other nineteenth century mathematicians, for any choice of sample size $n$, these methods imply that a weighted sum of Normal $\{Xj\}; j = 1, \ldots, n$; variates will, itself, be Normally-distributed (Cramer, 1945). Also, if $V$ represents any Normal variate, then for any choice of constant $K_1$, and any $K_2 \neq 0$, it follows that $(V - K_1)/K_2$ will also be a Normal variate.

The values of $K_1$ and $K_2$ modify the scale and location parameters of the variate $V$, in the sense of determining which particular special case of the
$$f(x) = \sigma^{-1}\phi\left(\frac{x-\mu}{\sigma}\right)/\sigma$$
general Normal model describes the density of the $Z = (V - K_1)/K_2$ variate. In particular, when $K_2$ is chosen so that the $(V - K_1)/K_2$ variate has a variance equal to one, then the $V$-variate is said to be standardized.

## 18.5. The one tail z-test

When $f$ can be expressed validly as standard model $\phi$ this permits confidence or significance levels to be determined through reference to a table for cdf $\Phi$. Moreover, the valid match between $f$ and the Normal density, guarantees that $z_A^{(3)}$ and $z_A^{(4)}$ are not only both Normal variates, but that they are both standard Normal variates, each with location parameter 0 and scale parameter 1. Consequently, even though they are determined from different size samples, the two statistics, $z_A^{(3)}$ and $z_A^{(4)}$ have exactly the same distribution. In fact, if a comparable statistic were to have been computed by substituting the mean $\bar{x}^{(10,000)}$ of measurements obtained from a random sample containing 10,000 iid Normal variates to form

$$\frac{\bar{x}^{(10,000)} - \mu_A}{\sigma/100},$$

then this statistic would still have a distribution that is identical to the distribution of $z_A^{(3)}$ or the distribution of $z_A^{(4)}$.

The assumption that $z_A^{(3)}$ and $z_A^{(4)}$ are variates with density $\phi(x)$, when valid, simplifies statistical processes. Yet, for the Student-$t$ test extension of $z$, the counterparts of $z_A^{(3)}$ and $z_A^{(4)}$ are neither Normally distributed nor even each distributed with the same density. Student-$t$ test extensions are based on estimates of the scale parameter test statistic component $\sigma$. As distinguished from curve location and scale differences, for different $n$, Student-$t$ shape differences do not invalidate inferential processes. Given the truth of the null hypothesis, these densities are determined completely by the known value of the sample size $n$. Consequently, once the $t$-test statistic numerator is divided by a quantity that adjusts for scale, as it is for the standardized $z_A^{(3)}$ and $z_A^{(4)}$, tabled values can be called upon to assess the resulting test statistic, as discussed and generalized at the end of the next chapter for the situation where BLU scale estimators substitute for $s$.

# CHAPTER 19
# Parameters and Test Statistics

## 19.1 The parameter $\Delta$

Before considering problems that can complicate statistical procedures, a few well-known simple techniques and terms will be reviewed in this section. Only the null case, where $\Delta$ actually equals 0, was discussed in the previous chapter. When $\Delta \neq 0$ neither

$$z_A^{(3)} \quad \text{nor} \quad z_A^{(4)}$$

are standard Normal variates. To see what these variates are, the relationship

$$E[z_A^{(4)} - z_A^{(3)}] = (2 - \sqrt{3})\Delta$$

is informative because, in the alternative, as opposed to the null case, what can be expected of the difference between two $z$-statistics calculated from size 4 and size 3 samples, is a positive multiple of the composite parameter

$$\Delta = [\mu_A - \mu_0]/\sigma.$$

Based on the assumption that the expected values of the two $z$-statistics respectively equal some corresponding models' location parameters, it follows that $\Delta$ is the standardized difference between the null and the true location parameters.

To interpret the effect that $\Delta$ has on what will now be defined as frequentist *power*, consider the following goal shared in the many experiments where the null value of $\mu_0$ is some quantity based on a standard or on a control, regimen, treatment, or drug. Alternatively, this value designates that, as far as location parameters are concerned, there is no difference between an experimental group and a control group. Whichever the application, it is typical for an experimenter to hope that some true value $\mu_A$ is found to differ from some $\mu_0$. Consequently, the statistical term "power" refers to the chance that an experimenter will be in a position to say what he or she wants to say when he or she should say it.

The last word of the preceding paragraph, the word "it," denotes that the experimenter says that he or she has rejected the null hypothesis at a respectable significance level or $p$-value. Traditionally, a $p$-value of $1/20$ = 0.05=5% suffices for the publication of relevant findings in most journals, while the value $1/100 = 0.01$=1% is a cause for rejoicing. Being in a position to say it, implies that $\Delta$ is indeed nonzero; in other words, the alternative case prevails. Based on the exact nature of the null and alternative hypotheses, those values that merit rejection by test statistics like $z^{(3)}$ and $z^{(4)}$ are divided into categories. For example, values that point to rejection of the null hypothesis constitute what, as is well-known to most researchers, is called the *critical region of a hypothesis test*. In the particular one-sided or one-tailed case illustrated here, it is assumed that the critical region extends from the critical point $z_0$ to infinity.

Now we come to the punch line. The purpose served by the addition of the fourth sample variate value to flesh out $z^{(4)}$, from what otherwise would have been the test statistic $z^{(3)}$, is to increase the chance of inserting the test statistic into the critical region. This punch line can be delivered without falling flat, of course, only when the null hypothesis should be rejected. Given that it should be rejected, the larger the sample size $n$ of $z^{(n)}$, the deeper the punch will penetrate into the critical region. What might be borderline or even non-significance were a small-sized sample gathered, might be a cause for rejoicing when a test is conducted based on a large-size sample—hence the appropriateness of the term "power."

As stated by John Emerich Edward Dalberg-Acton in his April 5, 1887 letter to Bishop Mandell Creighton (Bartlett, J., 1948, p. 1041), "Power tends to corrupt and absolute power corrupts absolutely." This may be true for political or economic power. However in statistics, absolute power implies a 100% chance of being in a position to say what one wants to say only when one should say it.

It is useful to step back a bit and view significance level and power from a general perspective. Unfortunately, although power has no direct connection with the regression slope parameter, it is often represented in terms of the Greek letter $\beta$ as $1-\beta$. The $p$-value is a quantity that is chosen by the experimenter or observational scientist. On the other hand, $1-\beta$'s value can only be determined indirectly, on the basis of sample size $n$, as well as parameters like $\sigma$ or curve properties like $\sqrt{\mathrm{Var}(X)}$ and—perhaps most importantly—the difference between the true value of a parameter and its null-hypothesized value which, when standardized, is represented by the composite parameter $\Delta = [\mu_A - \mu_0]/\sigma$.

In practice, $\Delta$ is never known exactly. If it were, there would be no point to testing the null-hypothesized value $\mu_0$. For this reason, many applied and theoretically-oriented statistics texts provide power curve graphs. These graphs are designed to help scientists assess the first of two fundamental aspects of the statistical process—power and robustness. (The classic ANOVA text by Scheffe( (1959) reproduces power graphs designed by Pearson and Hartley (1956) as well as Fox (1956).)

Power and robustness are two sides of a balance that pivots about the p-value. For this reason, much statistical research has been devoted to the tradeoffs between these fundamental characteristics of statistical processes. For example, consider a statistical Cassandra, endowed with the gift of prophecy, but never believed. Correspondingly, it would do little good to say what one wants to say and not be believed because, due to robustness problems, research procedures are called into question. On the other hand, one can be overcautious in the sense of relying on a low-powered robust procedure and, as a consequence, one can seldom be in a position to say what one wants to say.

Powerful, but frail, data-with-model pairings often cause the equivalent of drug side effects. No drug is completely safe, as the many people with allergies to aspirin and other simple medicines know. Conversely, no drug is always effective. Attention paid to the robustness and power characteristics of a test is therefore similar to reading the label that accompanies medications.

## 19.2  Power and efficiency

Suppose that the cdf $F$ has a well-defined inverse. It then follows that, for testing purposes, any table consulted in order to associate the value $1-\alpha$ where $0 < \alpha < 1$, with the value of $F^{-1}(1-\alpha)$ is, rounding excepted, the equivalent of a table that has the values of $x$ in its margin and the values of $F(x)$ in its body.

The formula $F(x) = G[(x-\tilde{\mu})/\tilde{\sigma}]$, where $F(x)$ has a well-defined inverse $F^{-1}(y)$, can help explain the purpose for individual test construction steps. Suppose $F^{-1}(y) = \tilde{\mu} + \tilde{\sigma} G^{-1}(y)$. Here, in words, $F^{-1}(y)$ is an inverse cdf whose argument is any $y$ between zero and one. This curve is equal to the standard model inverse cdf $G^{-1}$ evaluated at $y$, times the scale parameter of the cdf $F$, plus $F$'s location parameter. Note that the computational steps beginning with $y$ and ending with $F^{-1}(y)$ are the reverse of the sequence of steps taken when some expression for $G[(x-\tilde{\mu})/\tilde{\sigma}]$ is called upon to evaluate $G[(x_0 - \tilde{\mu})/\tilde{\sigma}]$ at a point $x_0$.

For example, Figure 19.1 illustrates the geometry that underlies the well known $1-\alpha = 0.95$ critical point of a test that is based on a Normally

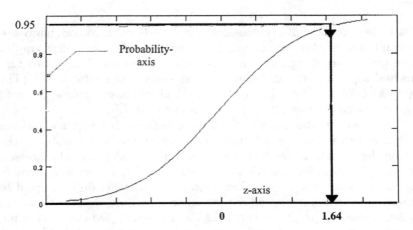

Figure 19.1. Critical $1-\alpha$ point and standard normal $\Phi$. The upper horizontal line leads to the 95% critical point.

distributed test statistic. As a first step, a single critical point is selected. (For a two-sided test, two critical points would have been selected.) Notice that any $p$-value or $\alpha$ level is a probability, and thus one or two values are selected on the ordinate, here the $\Phi(x)$ probability-axis. The determination of such a point is obviously the graphical counterpart of starting with a number within the body of a table in order to determine the corresponding number in the table's margin. If a table of the standard Normal cdf, $\Phi$, is called upon to determine the critical point $F^{-1}(1-\alpha) = \tilde{\mu} + \tilde{\sigma} G^{-1}(1-\alpha)$, then the calculation of the value of $G^{-1}(1-\alpha)$, for $\alpha = 0.05$, is equivalent to tracing the upper horizontal line within Figure 19.1 down to the abscissa.

When $p$-values or, equivalently, levels of significance like $\alpha$, $1-\alpha$, $\alpha/2$ or $1-\alpha/2$ are located within the body of a Normal $z$-table, the function whose values are entered in the margin of this table is the inverse of the cdf of a sampling distribution. For example, consider the case of a critical point, the finite-valued coordinate at which a one-sided critical region begins or ends. For the one-sided test where the critical region begins at $-\infty$, this region *ends* at critical point $F^{-1}(\alpha) = \tilde{\mu} + \tilde{\sigma} G^{-1}(\alpha)$. Figure 19.2 shows a critical region that ends at $\infty$; the critical region of a one-sided test begins at $F^{-1}(1-\alpha) = \tilde{\mu} + \tilde{\sigma} G^{-1}(1-\alpha)$. The critical point is the value, approximately equal to 2, shown on the $x$-axis. In this example, its construction is based on a test statistic that is logCauchy-distributed; in other words, the log of this statistic is Cauchy-distributed, as illustrated in Section 13.2.

Suppose the critical point of a one-sided test, or the dual critical points of a two-sided test, are determined, and we then wish to calculate the

## 19.2. Power and efficiency

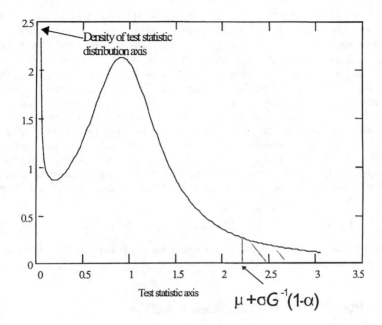

Figure 19.2. Critical point of a non-Normal distribution with standard form cdf $G$ interpreted in terms of the nonstandard $f$-type, here the logCauchy curve.

chance that a test statistic will actually be placed within the critical region. Then it follows that power can be calculated by finding the area under a density curve either:

1. over the region that begins at $-\infty$, and ends at $F^{-1}(\alpha) = \tilde{\mu} + \tilde{\sigma} G^{-1}(\alpha)$;

2. over the region that begins at $F^{-1}(1 - \alpha) = \tilde{\mu} + \tilde{\sigma} G^{-1}(1 - \alpha)$ and ends at $\infty$ or;

3. for the two-tailed test—over a critical region that is the complement of the $[\tilde{\mu}+\tilde{\sigma} G^{-1}(\alpha/2), \tilde{\mu}+\tilde{\sigma} G^{-1}(1-\alpha/2)]$ "acceptance region," where the complement of a region R is defined to include all values not in R. In the case of a one-sided test, for both the critical region that ends at $F^{-1}(\alpha) = \tilde{\mu} + \tilde{\sigma} G^{-1}(\alpha)$, and the region that begins at $F^{-1}(1 - \alpha) = \tilde{\mu} + \tilde{\sigma} G^{-1}(1 - \alpha)$, power can be interpreted as the area under a density curve.

Alternatively, power can be calculated as the evaluation of either a cdf curve itself or, for the $F^{-1}(1-\alpha) = \tilde{\mu} + \tilde{\sigma} G^{-1}(1-\alpha)$ to ( critical region, one minus the evaluation of a cdf curve.

Consider $F^{-1}(1-\alpha) = \tilde{\mu} + \tilde{\sigma} G^{-1}(1-\alpha)$, for $\tilde{\mu} = \mu_0$ that is defined to be the null-hypothesized value of $E(X)$. Here $\tilde{\sigma} = \sigma^{(n)}$ designates a value which, in this context, is called the *standard error of the test statistic*. Suppose, for instance, that $\bar{x}^{(4)}$ estimates $E(X)$. In this situation consider the probability of saying what one wants to say when one should say it, in other words, saying *it* in the circumstance where the alternative value, $\mu_A$, is the true parameter value. This probability equals the area over the critical region under the density curve that describes the test statistic's distribution. One will say that the null hypothesis is rejected when the test statistic is found to occur within the critical region, and one should say this when $\mu_A$, not $\mu_0$, is true. Consequently, for example, in the Normal case, power

$$1-\beta = 1 - \Phi\left([\tilde{\mu} + \tilde{\sigma}\Phi^{-1}(1-\alpha) - \mu_A]/\tilde{\sigma}\right) = 1 - \Phi\left(\Phi^{-1}(1-\alpha) - (\Delta\sqrt{n})\right),$$

which is a function whose arguments can be expressed as the three quantities $\alpha$, $n$, and $\Delta$.

## 19.3 Power and test considerations

The function $\beta(\alpha, \Delta, b) = \Phi(\Phi^{-1}(1-\alpha) - (\Delta\sqrt{n}))$ illustrates the roles that the three quantities $\alpha$ $\Delta$ and $n$ often play. In the early general statistical literature $\beta(\alpha, \Delta, n)$ is referred to as the chance of a type II error or error of the second kind. Similarly, $\alpha$ is referred to as the chance of a type I error or error of the first kind. To see how $\beta(\alpha, \Delta, n)$ changes as its arguments change, it helps to keep in mind that power, being a highly desirable entity, equals one minus type II error $\beta(\alpha, \Delta, n)$. For the one-sided tests associated with $\Phi\left(\Phi^{-1}(1-\alpha) - (\Delta\sqrt{n})\right)$, the quantile $\Phi^{-1}(1-\alpha)$ is the critical region border or the critical point. Because this point is positioned along the test yardstick on the $z$-axis, the symbol $z_0$ can represent $\Phi^{-1}(1-\alpha)$ within

$$\Phi\left(\Phi^{-1}(1-\alpha) - (\Delta\sqrt{n})\right),$$

which yields the curve evaluation at $z_0$ that can be denoted as $\Phi\left(z_0 - (\Delta\sqrt{n})\right)$.

In the sense that $(\Delta\sqrt{n})$ is the location parameter of a cdf

$$F(z) = \Phi\left(z - (\Delta\sqrt{n})\right),$$

the symbol $\Delta$ plays the role of a noncentrality parameter. When $\Delta$ equals zero, it follows that the letter $F$ represents a standard Normal cdf. However—and this is the key idea—as the value of $(\Delta\sqrt{n})$ increases, this change, in

## 19.3. Power and test considerations

turn, translates or relocates the $F$ curve to the right and, as a consequence reduces the type II error value of $\beta(\alpha, \Delta, n)$

For the right-sided single-tailed test, whenever $\mu_A$ is substantially greater than $\mu_0$, a researcher has an appreciable chance of saying what he or she wishes to say when he or she should say it. Sure enough, as the value assumed by $\Delta = [\mu_A - \mu_0]/\sigma$ increases, the quantity $\beta(\alpha, \Delta, n)$ decreases, and consequently power increases. All things being equal, variability associated with planned-experiment-variate-values is less than that of some corresponding observational study's variates (given, of course, that such a correspondence exists). When the $\sigma$ denominator-component of $\Delta = [\mu_A - \mu_0]/\sigma$ decreases, $\Delta$ will increase and, again, as in the increasing noncentrality parameter case, $\beta(\alpha, \Delta, n)$ will decrease. Finally, consider the fundamental trade-off, expend money to decrease $\sigma$ or simply increase sample size $n$. Whenever

$$\beta(\alpha, \Delta, n) = \Phi\left(\Phi^{-1}(1-\alpha) - (\Delta\sqrt{n})\right)$$

is valid, it follows that halving $\sigma$ is equivalent to *quadrupling* $n$.

The relationships

$$\Delta = [\mu_A - \mu_0]/\sigma \quad \text{and} \quad \beta(\alpha, \Delta, n) = \Phi\left(\Phi^{-1}(1-\alpha) - (\Delta\sqrt{n})\right)$$

show why certain variate measurement choices, such as the choice between centigrade and Fahrenheit temperatures, do not affect certain tests' power. Suppose for example, perhaps due to the conversion of Fahrenheit prevailing measurements to the centigrade scale, scale and location units change. A change in location due to variate definition that does not also affect scale, leaves the $\sigma$ term of $[\mu_A - \mu_0]/\sigma$ unchanged. Hence it relocates or translates both of the terms of the difference $\mu_A - \mu_0$ in such a way that $[\mu_A - \mu_0]/\sigma$ is unchanged.

In a similar way, a scale change will multiply both the numerator and the denominator of $[\mu_A - \mu_0]/\sigma$ by the same constant. Hence this constant cancels and leaves $\beta(\alpha, \Delta, n)$ unchanged. Notice, however, that when data are transformed by taking the logarithm or square root of variate values, this transformation changes curve properties in ways that generally will not produce comparable cancellation. A parametric procedure will often have both its power and validity, in the sense that validity denotes assumption trustworthiness, changed by a variate transformation, sometimes for the better and sometimes for the worse.

If, for example, $n = 4$, then the formula

$$\beta = \Phi\left([\tilde{\mu} + \tilde{\sigma}\Phi^{-1}(1-\alpha) - \mu_A]/\tilde{\sigma}\right)$$

and the function

$$\beta(\alpha, \Delta, n) = \Phi\left(\Phi^{-1}(1-\alpha) - (\Delta\sqrt{n})\right)$$

are based on the assumption that the statistic $\bar{x}^{(4)}$ has a Normal density. Even if these observations are not independent and identically distributed, this will be true if all four of the variates from which $\bar{x}^{(4)}$ is computed are individually Normally distributed. However, the validity of the formula

$$\left[\sigma^{(n)}\right]^2 = \sigma^2/n$$

depends on the assumption that the size $n$ sample is comprised of iid variates. Also, although this formula does not depend on the Normality assumption, it does depend on the assumption that standard deviation of each iid variate's distribution is finite. For example, the mean's variance formula does not apply to Cauchy-distributed data.

Even in situations where data are iid with a finite variance $\sigma^2$, an important question remains. Can the cdf of the sample mean's sampling distribution be assumed to be a Normal cdf? According to Cramer (1945), "Whatever be the distributions of the independent variables ...subject to certain very general conditions, ...the sample mean is asymptotically Normally distributed." One of Cramer's examples of Centeral Limit Theorem (CLT) special cases is based on variates that assume values zero and one, in other words are dichotomous or binary-valued. The sums of m such variate values are, of course, binomial variate values.

Besides citing the binomial case of the Central Limit Theorem as an example where—for medium to large-sized samples—the density of the sample mean can be closely approximated by a Normal density, Cramer considers what he calls the case of the "rectangular and allied distributions." The square and the triangle shaped curves shown in Figure 19.3 match what is shown by Cramer as the rectangular and first "allied distribution," respectively. (When $\mu = 0$ and $\sigma = 1$, the uniform density

$$\mathcal{U}(x) = \sigma^{-1} I_{[0,1]}\left([x - \mu]/\sigma\right)$$

is identical to what Cramer calls the rectangular density.)

This standard form of the uniform density is not just any arbitrary rectangle but is also a square. What Cramer refers to as distributions allied to the rectangular distribution have densities that describe variates that are sums of iid uniforms, the $0^{\text{th}}$ and $1^{\text{st}}$ of which are shown in Figure 19.3. Within the interval (0, 2) the sum of two iid variates, that individually have standard form density $\mathcal{U}(x)$, has density $f_2(x)$ equal to $1 - |1 - x|$, as shown by triangular dashed curve of Figure 19.3.

A technique that will be called upon extensively later in this chapter can be applied here to show just how *bell*-shaped independent rectangular triple variate sum densities tend to be. The bell-shaped curve in Figure

## 19.3. Power and test considerations

19.3 corresponds to what Cramer refers to as the density $f_3$ of the sum of three iid standard form $\mathcal{U}(x)$ variates. Unlike the curve shown by Cramer however, the curve shown in Figure 19.3 was obtained by calling upon the density estimation procedures.

First, 1500 independent random samples, each consisting of $n = 3$ density-$\mathcal{U}(x)$-distributed variates, were obtained. Then the corresponding 1500 sums, each of the $n = 3$ independent variate values, were calculated and, finally, the techniques outlined in Section 14.2 and described in detail by Tarter and Lock (Chapters 3 and 4) were applied to estimate the curve that Cramer calls *allied distribution* $f_3$.

Kernel and series density nonparametric methods have difficulty estimating curves with discontinuous derivatives, such as those that Cramer calls allied distributions $f_1$ and $f_2$. Conversely, although the bell-shaped graph shown in Figure 19.3 should resemble a Normal curve, it is slightly more Normal-density-like than it should be in theory. Nevertheless, in the remainder of this chapter it will be assumed that the CLT suffices to assure that the relevant sample mean in question is, for all practical purposes, a Normal variate.

Even when the CLT is applicable, major problems can remain. For example, suppose iid variates have a common density equal to: (1) some constant times; (2) a Cauchy density times; (3) some uniform or rectangular density $\mathcal{U}(x)$. Since it has a finite variance, this new truncated density satisfies the conditions for Central Limit Theorem validity. However, as will be considered in detail in the next chapter, the connections both location

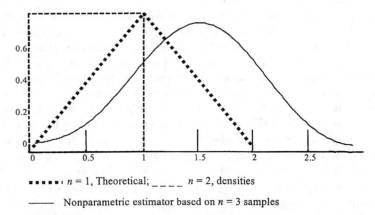

■ ■ ■ ■ ■ ı $n = 1$, Theoretical; _ _ _ _ $n = 2$, densities

——— Nonparametric estimator based on $n = 3$ samples

Figure 19.3. Rectangular, triangular and estimated sum of 3 rectangular variates' densities.

parameters $\mu_0$ and $\mu_A$ have with the density curve $f$'s property $E_f(X)$ are tenuous whenever there is truncation-induced asymmetry.

While power is the chance of saying what one wants to say when one should say it, what one says must be believed if a scientific investigation is to be successful. This brings a basic rule of data analysis into play; specifically, some loss of power must often be traded for some gain in robustness. In the case of tests based on the sample mean, in order to assure that the location parameters being tested actually are associated in a clear way with properties of a density curve such as $E(X)$, modifications of this simple statistic are often required.

## 19.4 The sample mean and the sample median

For any sample of size $n = 2$, the sample mean and the sample median are equal. Consequently, when $n = 2$, the value assumed by $\tilde{\sigma} = \sigma^{(n)}$ is obviously the same for these two statistics. It follows that in this one special, $n = 2$ case, the power of a test based on the sample median, is the same as the power of a test based on the sample mean. (Although the $n = 2$ equivalence of the sample mean and sample median is true for any sample, the remainder of this section will deal exclusively with iid Normal variates.)

For a size $n > 2$ iid sample of Normal variates, each with density $f$, the median's sampling distribution is not Normal. Instead, it is a constant times the product of three quantities: a density $f$; powers of the associated cdf, $F$, and; powers of the survival curve $S = 1 - F$. Yet, as $n$ increases, whenever it is evaluated over any finite value the difference between the density that describes a median's sampling distribution and the best-fitting Normal density, decreases to zero. Convergence suggests that for large-sized samples, once appropriate modifications are made to adjust for scale parameter changes, the sample median can be substituted for the sample mean within the formulas that define test procedures.

When it is computed from a sample comprised of $n = 100$ Normal variates, the density of the sample median has a *shape* that is nearly indistinguishable from the shape of some Normal curve. Thus the standard deviation of this sample median's Normal-density-approximation is the median-based counterpart of the mean's standard error $\tilde{\sigma} = \sigma^{(n)}$. However, this standard-error-of-the-median curve property does not equal the ratio formed by taking the quantity, $\sigma$, here assumed to be a scale parameter $\sigma$ whose value equals the standard deviation of $f$, and dividing this $\sigma$ by $\sqrt{n}$. Instead, this median's standard error equals approximately $\sigma/\sqrt{.637n}$.

## 19.4. The sample mean and the sample median

It follows from the $\sigma/\sqrt{.637n}$ expression that when $n = 100$,

$$\sqrt{.637n} = \sqrt{63.7},$$

which is a quantity smaller than 10, the square root of 100. The fact that the square root of 63.7 is smaller than 10 illustrates that as long as $n$ is greater than 2, the standard deviation of the Normal density that approximates the median's density is always larger than the counterpart standard error of the sample mean.

A number like 0.637 that multiplies the symbol for the sample size $n$ in the formula for the standard error of a statistic like the median, is called this statistic's *asymptotic efficiency relative to the sample mean* (AERSM). For the one-sided test that was discussed previously in this chapter, suppose a median is calculated from a sample larger than size $n = 50$ and then called upon as a substitute for the sample mean. Then after adjustment is made for the distinction between the standard errors of the median and of the mean, the formula

$$1 - \Phi\left([\tilde{\mu} + \tilde{\sigma}\Phi^{-1}(1-\alpha) - \mu_A]/\tilde{\sigma}\right)$$

provides accurate power determinations. For example, in the iid Normal case, a sample mean calculated from a size $n = 637$ sample, yields a test with power nearly identical to that of a test based on the sample median computed from an iid sample of $n = 1000$ comparable variates.

When $n > 1$, the median calculated from an odd-sized sample has an equal number of variate values below and above its value. For $n > 2$ even-sized samples, the median is the average of the two sample members that are closest in value and also have an equal number of variate values below and above their two values. Because, by convention, the sample median is defined differently for odd and for even-sized samples, power determinations for tests based on the sample median vary as sample size $n$ varies from an odd to the closest even value. In a similar way, as $n$ decreases, the median's efficiency approaches that of the mean in a roundabout fashion. Specifically, for a sample of size $n = 2$, the efficiencies of the median and mean are equal while the AERSM is: 0.697 for $n = 5$; 0.723 for $n = 10$; 0.656 for $n = 15$ and; 0.681 for $n = 20$.

Efficiency and power are only two of many research considerations. Some power or efficiency must often be traded for increased robustness and yet, thanks to the Central Limit Theorem, this trade-off is required less for non-Normality reasons, than because there may be a loose connection between a parameter, such as $\mu_0$, and a curve property, such as $E(X)$.

## 19.5  Tables and the details of test construction

Several tricky points concerning the details of test construction still remain. Many of these concern notation. For example, tables of the standard Normal distribution's curves are often difficult to apply because letters like $Y$, $z$, and $A$ are employed without reference to notation for curves like $\phi$, $\Phi$, and $\Phi^{-1}$.

There are two types of Normal tables. The first is set up to deal with two-sided tests conveniently. The second provides a handy way to construct one-sided critical regions. In the first type of table, the letter A represents area. Suppose the capital letter A "heads" the "body" of a table entry located across from a letter $z$ or $X$. Then for a given value of a variate $X$, such as $x$, where $x \geq 0$, the capital letter A often represents

$$\Phi_{\text{two-sided}}(X; -x, x) = \Phi(x) - \Phi(-x).$$

In terms of the standard Normal density curve $\phi$ and the cdf curve $\Phi$, the quantity $A = \Phi_{\text{two-sided}}(X; -x, x)$ is the area beneath the curve $\phi$ that is located over the interval that begins at $-x$ and ends at $x$. Thus, the first type of table is designed to connect the probability, A, of being between two values, $-x$ and $x$; to the interval whose endpoints are these values.

The second type of table provides values of the Normal cdf $\Phi(x)$. In tables of either type, a column headed by A can be preceded by a column headed by the letter $Y$ on its left, and followed by a column headed by 1-A. $Y$ represents the height of the $\phi$ curve at $x$, in symbols, $\phi(x)$ or, since $\phi$ is symmetric about zero, it also represents $\phi(-x)$. For example, because $\phi(0) = 1/\sqrt{2\pi}$, this height is $1/\sqrt{2\pi} = 0.398942\ldots$.

The value zero appears in the area column next to density height column entry 0.398942... . The area column entry is zero because the area referred to is under the Normal, and over the degenerate interval [0,0], an interval that contains the one point $x = 0$. This area of course equals zero. For example, were infinite precision possible, then it would follow that there is zero probability that any standard Normal variate's value equals the mode, mean, or median of the curve $\phi$ exactly.

In many statistical books, tables, and papers the letter $z$ represents a value of the inverse Normal cdf, such as $z = \Phi^{-1}(1 - \alpha/2)$ or $-z = \Phi^{-1}(\alpha/2)$. For example, in certain tables, $\pm z$ designates the interval that extends from $\Phi^{-1}(\alpha/2)$ to $\phi^{-1}(1 - \alpha/2)$. When the complement of this interval serves as a two-sided test's critical region, the area under the curve $\phi$ and over the critical region, equals $\alpha$.

A column that has entries headed by the expression $\mu \pm z\sigma$ is included in a Normal table designed to help calculate the critical points of two-sided

## 19.5. Tables and the details of test construction

tests. It is included to remind the table's user that a two-sided critical region is comprised of two regions; the first of which is a region that ends at $\mu - z\sigma$, and the second that begins at $\mu + z\sigma$.

It is important to keep in mind, however, that the notation used in many Normal tables applies the symbol $\mu$ to represent either the null or the alternative value of the Normal location parameter $\mu$, while $\sigma$ usually represents the theoretical standard error of the statistic that estimates $\mu$. Thus, for example, when a two-sided critical region is constructed, the acceptance region referred to as $\mu \pm z\sigma$ really designates $\mu_0 \pm z\sigma^{(n)}$, while for power calculations, $\mu \pm z\sigma$ is written as $\mu_A \pm z\sigma^{(n)}$, where $\mu_A$ is the location parameter value under the alternative hypothesis.

The function with arguments $u$ and $v$, $F(X; u, v)$, equals $F(v) - F(u)$, where $F$ on the right side of this definition represents a cdf such as $\Phi$. This function can be applied conveniently to designate areas over an interval that is not centered at zero. When Normal tables are designed to evaluate Area $= \Phi(x) = \Phi(X; -\infty, x)$, rather than $A = \Phi(X; -x, x)$, only one value of $z$ is specified in the table margin. Correspondingly, one value, for example $\mu + z\sigma$ is listed under the heading $X$. The value $\mu + z\sigma$ designates either the left or right border of a one-sided critical region, $\mu_0 + z\sigma^{(n)}$ or, for power calculations based on the formula

$$\Phi\left([\tilde{\mu} + \tilde{\sigma}\Phi^{-1}(1-\alpha) - \mu_A]/\tilde{\sigma}\right),$$

this column designates the curve argument $\left([\tilde{\mu} + \tilde{\sigma}\Phi^{-1}(1-\alpha) - \mu_A]/\tilde{\sigma}\right)$.

Because the intervals that form acceptance and critical regions serve as inference-yardsticks, they are marked in order to assess the values attained by the test statistic, and hence calibrated to match the units in which the random variate is measured. Consequently, to construct a test yardstick, a researcher-specified value of $\alpha$ determines a value scaled in random variate units. For this reason, when a number like 1.65 or 1.96 is ascertained with the help of a Normal table, this table is referred to in reverse, in the sense that a value is located within the body of the table and then the corresponding margin value is obtained.

Unlike the construction of the test yardstick based on variate values, power calculations yield numbers that are probabilities. These numbers are probabilities because power, being the chance of saying what one wants to say when one should say it, is obviously a chance or probability. Consequently, to calculate $1 - \beta$ a number must first be located in the margin of a Normal table, and then a probability, in other words a quantity like Area $= \Phi(x) = \Phi(X; -\infty, x)$, must be found in the body of a table. The reason this process does not usually go full circle, and start in the body of a table and return to this same entry in the table's body, is that the power calculation based on $\left([\tilde{\mu} + \tilde{\sigma}\Phi^{-1}(1-\alpha)] - \mu_A\right)/\tilde{\sigma})$ is almost always performed under the assumption that $\mu_A \neq \mu_0$. In the unusual situation where $\mu_A = \mu_0$, it would indeed be the case that table-based calculations would go full circle, in the sense that the level of significance $\alpha$ would equal the test's power, $1 - \beta$, when $\Delta$ of

$$\Phi\left(\Phi^{-1}(1-\alpha) - (\Delta\sqrt{n})\right)$$

is zero.

To see why $\alpha = 1 - \beta$ when $\mu_A = \mu_0$, suppose in a simulation experiment, members of a sequence of alternative-hypothesized values of a parameter differ more and more from the null-hypothesized value. Since, of course, that one should say it follows whenever $\mu_A \neq \mu_0$, as this sequence progresses it is true that there would be an ever-increasing chance of saying what one wants to say. Conversely, were the alternative to differ from the *null* parameter by smaller and smaller amounts, then power would decrease until, ultimately, it reached the level $\alpha$. The quantity $\alpha$ is the chance of being wrong when one says what one wants to say. To be wrong when one says what one wants to say, the null hypothesis must be true. On the other hand, when the null hypothesis is false, it follows that power is the chance of rejecting this null hypothesis in situations where its rejection should be believed.

## 19.6 Power, efficiency and BLU estimators

Power is described by a curve in its entirety. Unlike power, efficiency is almost always taken to be a single number. Consequently, in keeping with the general idea that one or more curve properties serve as abstracts of one or more curves' total information content, there is a major difference between an estimator's efficiency and a test's power. Because statistical efficiency is merely a summary based on a density's variance curve property, there are other properties of a statistic's sampling distribution that, together with the property-variance, can disclose relevant information that concerns the sampling-error-type of statistical uncertainty.

The experiments described in this section were designed to tie together the three concepts: power; efficiency; and BLU estimator robustness, conceptually. Thanks to the modern data-generation procedures discussed in Chapter 7, applied together with the nonparametric curve-estimation techniques introduced in Chapter 13, today a sampling distribution's density can be estimated easily.

For example, what happens when a BLU scale parameter estimator replaces the sample standard deviation $s$ in the simplest of all $t$-tests? Under

## 19.6. Power, efficiency and BLU estimators

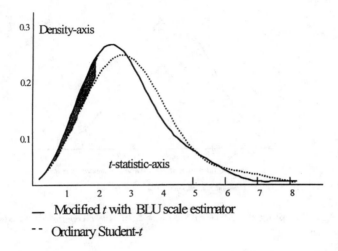

Figure 19.4. Sampling densities given $\mu_A = 1$ of the BLU-modified and conventional $t$-statistic.

the null hypothesis, how Student-$t$-like is the BLU denominator test statistic's density? Under the alternative hypothesis, what does the sampling distribution of this test statistic look like? Does it show an appreciable loss in power? Finally, is it as necessary to apply efficiency correction factors for BLU procedures, as it was for the sample median discussed in Section 19.4?

In the sense that they describe estimated sampling distributions, both the solid and the dashed curves shown in Figure 19.4 correspond to the bell-shaped curve shown in Figure 19.3. Like their Figure 19.3 counterpart, both of these curves were obtained from separate random samples of 1500 Normal variates. However, while the Figure 19.3 bell-shaped curve describes the estimated density of composite variates formed from 3 component variates, the Figure 19.4 curves describe composite variates formed from 8 component variates. This change was made because, unlike larger-sized samples, this sample size shows some appreciable curve difference, and yet is a sample size that is more likely to be found in statistical practice than size $n = 3$.

The iid variates whose datum density is estimated by the dashed curve were generated with location parameter value, $\mu_A = 1$. The critical point for a right tail, $\alpha = 0.05$, critical point of the one-sided test (right-hand side rejection), is given by Dixon and Massey (1983) as 1.895. (To interpret this number, it can be thought of as the 7-degrees-of-freedom counterpart to the $\infty$-degrees-of-freedom table entry 1.645 that is applied commonly.)

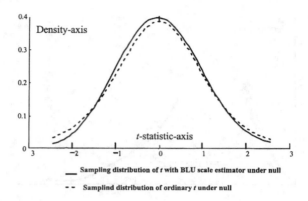

**Figure 19.5.** Sampling densities given $\mu_0 = 0$ of the BLU-modified and conventional $t$-statistic.

The compound variates whose common density is estimated by the solid curve had the same denominator as the noncentral-$t$, but where $s$ appeared, now it is the BLU scale parameter estimator based on table values given by Sarhan and Greenberg (1962) that is divided by the square root of 7. The price in power paid for the robustness afforded by the BLU scale parameter estimator substitution for $s$, is shown as the shaded area between the two curves. The price of BLU protection is this reduced chance of rejecting the null hypothesis.

As far as the null hypothesis and the sampling distributions of the test statistics are concerned, Figure 19.5 shows the estimated density of the BLU-based $t$ as a solid curve and the true density of the ordinary $t$ as a dashed curve. This experiment was repeated five times with 1500 iid samples of size 8. All five experiments showed the same characteristic similarity between curves, and also the slightly tighter concentration of the BLU statistic about the true location parameter 0.

Figure 19.5 suggests that as far as correction for the BLU's scale estimator's reduced efficiency relative to $s$ is concerned, there is little to be lost by calling upon BLU procedures in this simple application of $n = 8$ Normal variates. When it comes to the full potential of BLU procedures based on censored data, comparable experiments can provide validation. However, several computational and statistical matters must be attended to.

On the plus side of the ledger, it helps that power is related closely to a sampling distribution cdf. As first shown by Figures 2.1 and 2.2 of Kronmal and Tarter (1968), and then shown for more complex curve estimators by Tarter, Freeman, and Polissar (1990), there are theoretical

## 19.6. Power, efficiency and BLU estimators

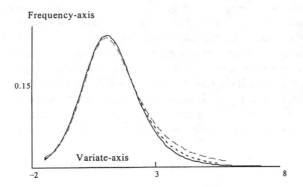

**Figure 19.6.** Noncentral $t$ densities, noncentrality parameter 1, solid $n = 8$, short dased $n = 6$ and long dased $n = 4$.

and computational reasons why in most instances cdf's can be estimated with more precision than is possible for corresponding pdf's. However, on the minus side of the ledger, the simple noncentral-$t$ for low degrees of freedom, like both the $F$-statistic and the noncentral $F$-variates that are to ANOVA hypothesis alternatives what the noncentral $t$ is to single- and dual-sample $t$-tests, have a pronounced asymmetry.

Figure 19.6 shows noncentrality parameter ($\Delta = 1$) densities for samples size $n = 4, 6$, and 8. To conduct experiments comparable to those whose results are shown in Figures 19.4 and 19.5 for BLU table values for $n = 8$ or less, there are two closely related computational strategies, both of which can increase power-determination accuracy. The first of these methods is illustrated by Tarter and Hill (1997) and is based on what are called *pilot* estimators. A pilot estimator is the model-free curve estimation counterpart of the preliminary transformation by a log or square root function that was discussed in Sections 15.1 and 15.2. The second method can be based on the substitution of a weighting function into the metric that underlies the assessment of curve estimation accuracy. This weighting function can be chosen to favor accuracy over the right-tail regions at the expense of some loss of accuracy elsewhere.

Of the two approaches, pilot transformations and metric weighting, the former is simpler computationally than the latter. Therefore, if many trials are required, for example with different-tabled efficiency correction factors, the pilot-transformation approach is much more feasible than the metric-weighting approach. On the other hand, the metric-weighting approach can be applied to emphasize the far right-tail curve region with a degree of control that is not possible for the pilot transformation method.

When nonpiloted approaches based on the conventional mean integrated square error metric, and piloted approaches based on a weighted metric, are applied to simulated data as illustrated in Figures 19.4 and 19.5, there is one modification of basic curve estimation strategy that has been found to be particularly useful. The margins that separate data-extreme values from the interval over which estimates are obtained should be kept large. In most situations, large margins will help compensate for curve asymmetry by, in essence, providing room for curve tails to approach zero at different rates of decline. We will now turn to general issues that concern curve edges.

# CHAPTER 20
# Curve Truncation and the Curve e(x)

## 20.1 Expectation as a limit and the effects of truncation

The Central Limit Theorem (CLT) discussed in the previous chapter can give a false sense of security that, as far as test procedures are concerned, all is well. In reality, for all practical purposes, the mean of a random sample of $X$ variates may be trusted to be distributed Normally. Yet there may be a loose connection between the curve property $E(X)$ and the location parameter within the actual model that describes $X$'s distribution.

Parameter-with-property-connection looseness can be particularly pernicious when density curves are represented accurately by truncated models. In practice, it would be an unusual truncated model of a density $f$, within which location parameter $\mu$ is known to be equal numerically to $E_f$. Yet sometimes this is a problem and sometimes location-expectation differences can be disregarded.

There are two fundamental considerations that pertain to truncated models. The consideration that will be discussed first below is the geometric shape of curve tails. After disposing of this conceptually simple issue, the remainder of this chapter will be concerned with a more troublesome problem. When should a data analyst be concerned about a location parameter and when should he or she be interested in the expectation curve property? (This question will be considered first in the context of a curve designated by the lowercase letter $e$ and then looked at from a general perspective.)

The expectation $E_f(X)$ of a variate $X$ is the limiting,

$$E_f(X : -\infty, \infty); x \to \infty, u \to -\infty,$$

case of the expression

$$E_f(X : u, x) = \int_u^x z f(z) dz.$$

(We will see later in this chapter that the variant of $E_f(X)$, the life table target function's evaluation $e(u_0)$, is based on $E_f(X : u, x)$ where $u_0$ is a specific finite-valued point. As usual, the subscript zero can be thought of as freezing the value of the variable $x$, as in absolute zero (-273.15C or -459.67F).

Loose location-expectation connections can be hidden by phraseology encountered commonly in the early statistical literature. In many seminal publications, a property of a distribution's curve was referred to by terms that include the name of the particular estimator obtained. For example, $E(X)$ was called a distribution's *mean* or "*the* distribution mean." This choice of terms suggests that there is a more direct connection between $E(X)$ and the sample mean $\bar{x}$, than there is between $E(X)$ and other statistics like the sample median, some BLU estimate of location obtained from non-Normal data, or a BLU estimate obtained from censored Normal data.

As outlined in Section 3.5, the frequentist approach to probability can be traced back to seventeenth century gambling strategy analyses. Games involving roulette, cards, or dice obviously are based on the reoccurrence of combinations from among a *finite* number of possible outcomes, where it is fitting for a population to be viewed as having a structure similar to that of a random sample. For example, the Kendall and Buckland dictionary (1960) defines the term "population" as: "... any finite or infinite collection of individuals." In the context of a finite collection of individuals it does make sense to define $E(X)$ in a way that corresponds to the definition of the sample mean $\bar{x}$.

When distributions are described by curves constructed over each point of the entire real-axis, $\bar{x}$ and $E(X)$ can be radically different concepts. It is no more difficult to compute the sample mean $\bar{y}$ of variates that are iid with a Cauchy density,

$$c(x) \equiv \sigma^{-1}\varphi\left([x - \mu]/\sigma\right),$$

where

$$\varphi(z) = 1/[\pi(1 + z^2)],$$

than it is to compute the statistic $\bar{x}$ from any sample of Normally-distributed data. Yet the so-called population mean, calculated as a limit of the expression

$$E(X : u, x) = \int_u^x z\, c(z)\, dz,$$

can be made to take on any particular value by allowing $u$ to decrease, and $x$ to increase, at different rates. (Like the Greek letter $\phi$ in the remainder

## 20.2. Truncation symmetry

of this section the Greek letter $\varphi$ designates a particular standard density curve, not a parameter.)

The stricture that, realistically, data points must attain values within a finite interval, makes the above problem no less a serious one. The reason why this is the case is complicated. Consider first the

$$2I_{[0,\infty)}(x)//\{\pi(1+x^2)\}$$

folded version of the Cauchy density, a curve that is equal to $2/\{\pi(1+x^2)\}$ over zero and the positive part of the $x$-axis, while equal to zero elsewhere. When care is taken to designate the integral by which it is defined, the expected-value property of the unfolded Cauchy density $\varphi(x) = 1/\{\pi(1+x^2)\}$ can be defined to equal zero, and yet the folded Cauchy's expectation curve property is always infinite. Thus, the restriction placed on the support of this density to the ray that extends from zero and proceeds over the positive portion of the abscissa, raises even more difficulties than those that apply to the unfolded and untruncated Cauchy density.

When a heavy tailed curve like the Cauchy density $\varphi\left(([x-\mu]/\sigma)\right)/\sigma$ is truncated at one or both ends, truncation alters the property $E(X)$'s relationship with the location parameter $\mu$ within this curve's formula. In addition, with the exception of computational considerations, the problems of Cauchy $E(X)$'s are passed down to the sample mean. Because of these problems, the interpretation of $E(X)$ has been generalized from its being a mere extension of the concept, sample mean, to its being a curve property that can be considered, when appropriate, without the need to refer to some denumerable population. This perspective provides a framework within which lopsided curves can be studied effectively. In particular, because the curves encountered by risk and reliability engineers, demographers, and many applied statisticians are skewed, arithmetic averages have limited appeal for these scientists.

## 20.2 Truncation symmetry

No instrument has an unbounded measurement range. For this reason all natural variates have distributions that, in reality, should be described by truncated curves. The problem is, however, that not only are truncated curves universal but, in addition, there is no particular reason why truncation has to be symmetric. How, for example, when the value of $E(X)$ is unknown, is a researcher to know that a curve is symmetric about the point $E(X)$? If the value of $E(X)$ was actually known, then why would a researcher go to the trouble of estimating this curve property by calling upon some statistic like $\bar{x}$? Often when the exact nature of the truncation process is not known, the value of $E(X)$—even when known—is of limited value. In general, its value is limited when the location parameter is the target of a statistical investigation and, in addition, $\mu$ does not have a simple relationship with the curve property $E(X)$.

For the above reason it is critical that a scientist know if the distribution of his or her data are better described by the Normal rather than by the Cauchy or, as another alternative, some truncated curve or extended modification of one of these models, such as several of the elementary functions surveyed in Appendix I.4. Unfortunately, proving data are Normally distributed rather than, for example, Cauchy distributed, can be more daunting than an application of a robust procedure like BLU estimation.

For simulation applications, some choice of density model must be made. Consequently it is necessary to focus in on the edges of some simple models like the Normal and the Cauchy with precision. To do this the functions

$$E_f(X:u,x) = \int_u^x xf(x)dx;\ F(X:u,x) = \int_u^x f(z)dz$$

and,

$$\Phi(X:u,x) = \int_u^x \phi(z)dz$$

help disclose the roles these edges play.

It is convenient to designate $\phi([x-\mu]/\sigma)/\sigma$ as $\phi(x;\mu,\sigma)$. The curve property $E_f(X:u,x)$, that helps to describe a truncated modification of the Normal density $f(x) = \phi(x;\mu,\sigma)$, equals

$$\mu\{\Phi([x-\mu]/\sigma) - \Phi([u-\mu]/\sigma)\} + \sigma[\phi(u;\mu,\sigma) - \phi(x;\mu,\sigma)].$$

In general, suppose $F(x) = G([x-\mu]/\sigma)$ designates a cdf whose derivative is $f(x)$. When a new density is formed by the truncation of $F$ and $f$ outside of the interval $[\tau_l, \tau_u]$, then within the $[\tau_l, \tau_u]$ interval the new density with $[\tau_l, \tau_u]$ as its support region equals

$$f(x)/\{G([\tau_u-\mu]/\sigma) - G([\tau_l-\mu]/\sigma)\}.$$

This adjustment is made because the expression for the untruncated curve, $f(x)$, must be divided by the area under the old density over the interval $[\tau_l, \tau_u]$. (The subscripts $l$ and $u$ were chosen to denote the words, lower and upper, respectively.)

The Normal density $f(x)/\{G([\tau_u-\mu]/\sigma) - G([\tau_l-\mu]/\sigma)\}$ that is truncated outside of the interval $[\tau_l, \tau_u]$ has an $E$ curve property equal to

$$\mu + \sigma[\phi(\tau_l;\mu,\sigma) - \phi(\tau_u;\mu,\sigma)]/\{\Phi([\tau_u-\mu]/\sigma) - \Phi([\tau_l-\mu]/\sigma)\}.$$

## 20.2. Truncation symmetry

As $\tau_l$ decreases and $\tau_u$ increases, first consider the numerator $\phi(\tau_l; \mu, \sigma) - \phi(\tau_u; \mu, \sigma)$ component of the fraction with numerator $\phi(\tau_l; \mu, \sigma) - \phi(\tau_u; \mu, \sigma)$ and denominator $\Phi([\tau_u - \mu]/\sigma) - \Phi([\tau_l - \mu]/\sigma)$. This numerator component decreases rapidly.

On the other hand unlike the Normal, which tapers off to zero at the edges of its support region quickly, tapering is much less rapid in the Cauchy density case. For this reason the determination—which of these two curves best describes some true density—is often germane to the problem at hand because of the importance of the question: How rapidly does density height $f(x)$ taper to zero near the support region edges $\tau_l$ and $\tau_u$?

In health science applications, the answer to this question is often simple. Demographic curves, such as the population pyramid and survival density, do not often taper off leftward to zero smoothly near the values, age and survival time, equal to zero. If such a density was compared to a hill, rather then the traditional bell, this hill would hardly be prefixed accurately by the word "rolling." Instead, to the immediate right of zero there would be a precipice, even though the right tail exhibits a long steady decline.

As both $\tau_l$ decreases and $\tau_u$ increases, since the Normal density decreases to zero rapidly as its argument increases, the term $[\phi(\tau_u; \mu, \sigma) - \phi(\tau_l\, \mu, \sigma)]$ decreases to zero rapidly and, as a consequence, in the Normal case the effects produced by asymmetric truncation diminish to imperceptible amounts as $\tau_l$ decreases and/or $\tau_u$ increases. Suppose, for instance, for some $\tau_l < 0$ the upper support endpoint $\tau_u$ is set equal to $-\tau_l + K$ so that $K$'s size determines the extent of "truncation asymmetry." Then even when $K$ is a large positive number, as $\tau_l$ decreases, $E(X : u, x)$ still converges rapidly to the nontruncated Normal density's expected value.

The following conceptual experiment illustrates the insensitivity of the Normal expectation to $K$'s value. Suppose the area subtended by a density $f(x)$ is cut from thin plywood or thick cardboard, in such a way that the left endpoint of this cutout is assigned the negative abscissa coordinate $\tau_l$. Correspondingly, suppose that for some positive value $K$, the right-hand side upper endpoint is defined as $\tau_u = -\tau_l + K$. Obviously, $K$'s value determines lopsidedness or tendency-to-tip-over while, when $K = 0$, it follows that a curve which is symmetric about the origin with curve support region endpoints $\pm\tau_l$ or, equivalently $\pm(-\tau_u)$, will not tip over when pivoted about the origin.

When pivoted against a stiff flat sheet of plywood or comparable material, a physical model balances at the curve property called a *centroid or center of gravity*. The truncated Normal's centroid equals $E(X : \tau_l, \tau_u = -\tau_l + K)/\Phi(X : \tau_l, \tau_u = -\tau_l + K)$. The further cutting proceeds in both directions from the location parameter $\mu$ the less care need be taken that

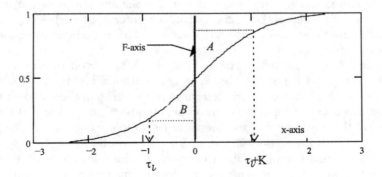

Figure 20.1. Truncation lopsidedness geometry.

the two curve ends are truncated symmetrically. For the truncated Normal, the importance of keeping $K$ moderately-sized diminishes as the value of $\tau_l$ decreases. Thus, when $\tau_u = -\tau_l + K$ is sufficiently large, lopsidedness is not a problem. (The animal kingdom provides a living example of lopsidedness insensitivity since, as is well-known, members of certain lizard species lose their tails to predators and yet live subsequently to grow a new one. Obviously, only a masochistic lizard relishes lopsidedness. Yet, asymmetric truncation at one end need not have dire consequences.)

Suppose that in place of the Normal, an experiment in lopsidedness tolerance is conducted with a cutout model of the Cauchy density. The geometric interpretation of the property $E(X : \tau_l, \tau_u = -\tau_l + K)$ shown in Figure 20.1 illustrates why truncation sensitivity is different for the Cauchy than it is for the Normal. The $E(X : \tau_l, \tau_u = -\tau_l + K)$ property is equal to the difference between two areas. The first area is that of the region marked with a letter $A$ in Figure 20.1. This region lies above the value $F(0)$ and has the cdf $F$, the ordinate $z$-axis, and a horizontal line with height $F(-\tau_l + K)$ as its borders. The second region, $B$, lies below the value $F(0)$ and is bordered by the cdf $F$, the ordinate-axis, and the horizontal line whose height is $F(\tau_l)$.

For the Cauchy distribution with cumulative

$$F(x) = \frac{1}{2} + \pi^{-1} \tan^{-1}\left[\frac{x - \mu}{\sigma}\right],$$

area A equals

$$\frac{\mu}{\pi} \tan^{-1}\left(\frac{\tau_u - \mu}{\sigma}\right) - \frac{\sigma}{\pi} \ln \cos \tan^{-1}\left[\frac{\tau_u - \mu}{\sigma}\right],$$

while area B equals

$$\frac{\mu}{\pi} \tan^{-1}\left(\frac{\tau_l - \mu}{\sigma}\right) - \frac{\sigma}{\pi} \ln \cos \tan^{-1}\left[\frac{\tau_l - \mu}{\sigma}\right].$$

As $\tau_u$ or, equivalently, $-\tau_l + K$, increases, the function $\tan^{-1}[[\tau_u - \mu]/\sigma]$ approaches $\pi/2$. But the cosine of $\pi/2$ equals zero. Therefore the $\cos\tan^{-1}$ term approaches the cosine of $\pi/2$ which equals zero. Consequently, the log of the whole expression approaches $(-\infty)$ which means, since this whole expression is subtracted from the bounded term $\frac{\mu}{\pi}\tan^{-1}\left(\frac{\tau_u - \mu}{\sigma}\right)$, that as $\tau_u = -\tau_l + K$ increases, area $A \to \infty$. Correspondingly, as $\tau_l$ decreases, area $B \to \infty$.

Unlike the case of the Normal lopsidedness, in the Cauchy case if the curve balancing point $A - B$ is assigned a value equal to any x-axis coordinate, there will always exist a value of $K$ such that upper truncation point $\tau_u = -\tau_l + K$ assures that the curve will balance at this coordinate. For some value of $K$, if cuts are made unequally so that $\tau_u$ is kept this particular $K$ units beyond $-\tau_l$, then the value of $E(X : \tau_l, \tau_u = -\tau_l + K)$ can be set equal to any predetermined positive value. Consequently, unlike the truncated Normal case, no matter how far one cuts in both directions, care must still be taken to truncate symmetrically in order to assure that $E(X : \tau_l, \tau_u = -\tau_l + K)$ equals its nontruncated expectation counterpart. When data are Cauchy distributed even small amounts of truncation asymmetry can cause big trouble.

## 20.3 Truncation and bias

An important variety of the curve property discussed in many previous chapters under the heading "bias," equals the difference between the expected value $E(X)$ and the target parameter $\mu$ of some specific model. Consideration of this variety of bias interpretation and importance will be postponed to this chapter's last section. Given that it is relevant to the problem at hand, in many applications that concern hypotheses about, and estimators of, location parameter values, the term *bias* denotes the difference between a curve property

$$E(X : \tau_l, \tau_u = -\tau_l + K)/F(X : \tau_l, \tau_u = -\tau_l + K)$$

on the one hand, and some parameter designated as $\mu$ within a model-based expression on the other.

One form of the CLT asserts that a particular function of the sample size $n$, for example $H(n)$, can be found so that when, as $n \to \infty$, the statistic $\bar{x}$ is multiplied by $H(n)$, the resulting product will have a "limiting distribution" described by the general Normal density model. The term "limiting distribution" is shorthand for the assertion that, as $n$ increases, any distinction between the density of the product, $H(n)$ times $\bar{x}$, and a particular Normal density, becomes imperceptible.

Suppose within a size $n$ iid sample, each variate has a distribution described by an $f$-type curve that is modeled in terms of some location parameter, for example $\mu_f$. According to the CLT, given some large value of $n$, $\bar{x}$ calculated from these variates can be taken to have distribution described by a Normal density curve whose property $E_N(X)$ equals the Normal model's location parameter $\mu_N$. (The value assumed by $\mu_f$ can be thought of as a partial description that is common to each component part, while $\mu_N$ helps describe the composite whole assembled from such component parts.)

The CLT notwithstanding, bias often occurs when due to asymmetric truncation, and any one of many other possible causes, the so-called population mean curve property

$$E(X : \tau_l, \tau_u = -\tau_l + K)/F(X : \tau_l, \tau_u = -\tau_l + K)$$

and the location parameter $\mu_f$, differ. For example, for any nonzero value of $K$, even were $f$ a truncated Normal curve with location parameter $\mu_f$, the parameter $\mu_f$ does not equal $E(X : \tau_l, \tau_u = -\tau_l + K)/F(X : \tau_l, \tau_u = -\tau_l + K)$ since $E(X : \tau_l, \tau_u = -\tau_l + K)/F(X : \tau_l, \tau_u = -\tau_l + K)$—not $\mu_f$—equals $\mu_N$.

Given finite $E$, bias that does not approach zero—even as $n$ increases without limit—implies that a statistic is an *inconsistent* estimator. (Kendall and Stuart (1979, Chapter 17) provide clear discussions of: (1) the distinctions and connections between consistent and unbiased estimators and; (2) the existence of an unbiased estimator that is not consistent.) Until it is discussed in the last section, suffice to say that inconsistency is the statistical counterpart to the vernacular quality of hopelessness.

For $K \neq 0$, suppose $\bar{x}$ designates the mean of an iid sample of variates, each of whose distribution is described by a density model

$$\lambda_\tau(x) = C[I_{[\tau_l, \tau_u = -\tau_l + K]}(x)][\lambda([x - \mu_f]/\sigma)]/\sigma$$

where $\lambda(x)$ is some elementary or other function known to be symmetric about 0, while $C$ is the multiplier defined so that the area subtended by $\lambda_\tau(x)$ equals 1. Because the expectation curve property of the limiting distribution of $\bar{x}$ is not $\mu_f$, when it is computed from any sample of variates that individually have density $\lambda_\tau(x)$, the statistic $\bar{x}$ is an inconsistent estimator of $\mu_f$.

## 20.3. Truncation and bias

When $K > 0$, the lopsided curve $\lambda_\tau(x)$ does not balance at the location parameter $\mu_f$ that appears among the symbols that comprise this curve's model. The term "central," represented by the first letter of the abbreviation CLT, refers to a balancing point, not a location parameter. This example shows that the CLT may apply validly to a statistic like $\bar{x}$, and yet this statistic may still be an inconsistent location parameter estimator. The argument that all curves are truncated, and therefore the finite Var curve property condition for CLT applicability holds, offers a researcher merely a temporary escape; an escape from the infinite Var frying pan to the $\mu_f - \mu_N$ inconsistency fire.

While $\mu_f - \mu_N$ inconsistency is a common and difficult problem to deal with, many useful statistics are actually biased, yet consistent, estimators. For example, when a valid Normal model for a sequence of iid $X$ variates' prototypical datum-density contains a scale parameter $\sigma$ this parameter is numerically equal to the population standard deviation property of $X$'s distribution. Yet although $E(s^2) = \sigma^2$, $E(s) \neq \sigma$. Thus, even when a distribution can be modeled validly in terms of the Normal density, the usual estimator of the sample standard deviation is biased.

Corrections that yield approximately unbiased estimates of scale parameter $\sigma$ are presented by Dixon and Massey (1983). For example, for sample size $n > 5$, the statistic $[1+1/\{4(n-1)\}]s$ will approximate an unbiased estimate of the scale parameter $\sigma$ within the Normal model $\phi([x-\mu]/\sigma)/\sigma$. (As $n$ increases there will be less and less need for any correction, since if an untruncated $\phi([x-\mu]/\sigma)/\sigma$ is a valid model, the statistic $s$ is a consistent estimator of $\sigma$.)

As discussed in Chapter 16, many useful procedures are based both on weighted averages of iid $X$ variates and weighted means of squared residuals from such averages. Truncation-related bias can seep into these $X$ variate sample means as well as into both ANOVA and ANCOVA sums of squares. Due to the nonexistence of the Var curve property of the Cauchy density, the nontruncated Cauchy does not meet the conditions for Central Limit Theorem validity. As far as ANOVA and ANCOVA are concerned, this is a statistical danger sign. On the other hand, natural data must be described by curves that are "doubly-truncated," in the sense that $\tau_l$ and $\tau_u$ are finite-valued. This does, indeed, imply that a sample mean of such truncated Cauchy data will satisfy the conditions for Central Limit Theorem validity, and that such a sample mean will approach a variate that is Normally distributed with a $\phi([x-\mu_N]/\sigma)/\sigma$ density. Yet thinking that $\mu_N$ is also the location parameter within this truncated Cauchy model is little like thinking that all creatures recuperate from tail amputations.

Many data-analysis difficulties can result from asymmetry of densities over their support regions. Among these, Kendall and Buckland (1960) state that: "In a sample of $n$ observations it is possible for a limited number to be so far separated in value from the remainder that they give rise to the question whether they are not from a different population or that the sampling technique is at fault. Such values are called outliers."

Outliers are extreme values whose membership alongside other sample values is brought into question. Since most conventional statistical methods are hypersensitive to the values of data extremes, more harm than good can result from the removal of outliers without attention to the prosthetics of BLU outlier adjustment. These outlier-related issues are analogous to issues that concern density asymmetry. If, for example, two outliers were present in a sample of data, the first located at point $\mu + 10\sigma$ and the second at $\mu - 10\sigma$, then it could easily be the case that $\bar{x}$ computed from this sample, outliers and all, would be a more powerful or efficient estimator of $\mu$, than a modification of $\bar{x}$ that equals the average of those values left after outliers are removed.

Like the rarity of symmetric truncation, outliers are seldom placed symmetrically. For every fortunate arrangement of two outliers it may, alternatively, be the case that when one outlier's value is $\mu + 10\sigma$ its twin's value equals $\mu + 20\sigma$. The question of curve symmetry is as important when outliers are concerned as it is for outlier-free samples. Unlike Normally-distributed data, Cauchy data are likely to not only contain extremely large or small elements, but the placement of these elements is likely to tip the balance in an uncontrolled and arbitrary fashion. There is no guarantee that this instability will be attributed correctly to one of the distinct causes listed by Kendall and Buckland (1960): the existence of a different population; and faulty technique or; it will be traced to the variate density's close resemblance to the shape of a truncated Cauchy.

Figure 20.2 illustrate the substantial difference between the truncation-caused-bias induced in the Normal and the Cauchy cases, respectively. In order to prepare a level playing field for this comparison, the standard Cauchy density was modified. Not only its median, but also its interquartile range, was set equal to that of the standard density Normal $\phi$ with cdf $\Phi$ so that when this particular Cauchy cdf is designated as $C(x)$, then $C(0.25) = \Phi(0.25)$, $C(0.5) = \Phi(0.5)$, and $C(0.75) = \Phi(0.75)$.

Rather than being based on the Var curve property, the interquartile form of standarization was applied because the Cauchy Var curve property is infinite. The Cauchy density that has interquartile range $\Phi(0.5) - \Phi(0.25)$ and the same median as $\Phi$ is $f(x) = \varphi(x/0.6745)/0.6745$ when $\varphi(z) = 1/\{\pi(1+z^2)\}$.

To obtain the bias curves shown in Figure 20.2, the left truncation point $\tau_l$ was set equal to -2 and thus, when the bias ordinate of both figures equals zero at $K = 0$, this value corresponds to the right truncation

## 20.3. Truncation and bias

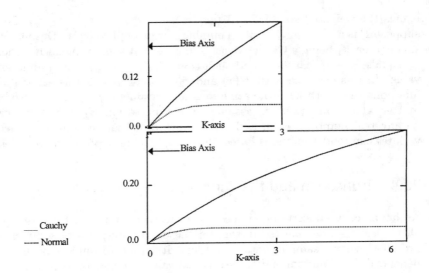

Figure 20.2. Top: Cauchy and Normal bias over the interval [0, 3]. Bottom: Cauchy and Normal bias over the interval [0, 6].

point $\tau_u = 2$. The truncation-lopsidedness-setting $K$ extends from 0 to 3 in Figure 20.2 (Top) and from 0 to 6 in Figure 20.2 (Bottom). The bias ordinate axis extends from 0 to 0.262 in Figure 20.2 (Top) and from 0 to 0.408 in Figure 20.2 (Bottom). For example, when lopsidedness $K = 6$, the doubly-truncated Cauchy has roughly six times more $E_f(X) - \mu$ bias than does the counterpart doubly-truncated Normal.

Problems with location-parameter-expectation differences can, of course, be due to causes other than and including asymmetric truncation. Suppose for mixing parameter $P$, where $0 < P < 1$, it happens that 100 times $P$ percent of the time, the variate $X$ has a density that can be described by a Normal model $\phi([x - \mu]/\sigma)\sigma$, while the rest of the time $X$'s density should be represented by the Cauchy model $\sigma^{-1}\varphi([x - \mu]/\sigma)$. Then if $f(x)$ represents the density of $X$, it follows that

$$f(x) = P\phi([x - \mu]/\sigma)\sigma + (1 - P)\sigma^{-1}\varphi([x - \mu]/\sigma).$$

Of course, it is far-fetched to suggest that the location and scale parameters of the two mixture components of $f$ have the same values. However, for the sake of simplicity, suppose that this is indeed the case.

What is $E_f(X)$ for the above mixture model? For any mixture density defined as $f(x) = Pf_1(x) + (1 - P)f_2(x)$, the curve property $E_f(X : \tau_l, \tau_u)$

equals the weighted sum $PE_{f_1}(X : \tau_l, \tau_u) + (1-P)E_{f_2}(X : \tau_l, \tau_u)$. Now suppose $P$ is a very small positive number, for example 0.0001. Despite $P$'s small value, $f_1$ being a Cauchy density implies that some value of $K$, where $\tau_u$ equals $K - \tau_l$, can be found such that $E_f(X : \tau_l, \tau_u)$ can assume any value. In this example both $f_1(x)$ and $f_2(x)$ were modeled as elementary functions that both have the same location parameter $\mu$. Nevertheless, no matter what finite value $\tau_u$ assumes, the expectation $E_f(X : \tau_l, \tau_u)$ can be any real number. Even a slight amount of contamination by Cauchy variates can render methods based on the $E_f(X)$ curve property worthless.

## 20.4 Truncation and the curve e(x)

As far as curve uncertainty issues are concerned, it is not the case that all truncated-curve effects are the population counterparts of some effect produced when a sample contains outliers. It is hard to think of outliers as beneficial. Yet one tool of great statistical value depends on truncation for its effectiveness. This tool focuses on a curve designated by the lowercase letter $e$, as in the first letter of the phrase "expected length of life remaining." Both in order to later illustrate basic uncertainty issues and, because this curve has important applications in many types of scientific investigation, this section will describe some simple examples of the curve $e(x)$.

The curve $e(x)$ is related both to the survival curve $S(x) = 1 - F(x)$ and to the adjustment procedures described in Chapter 12. However, as far as truncation is concerned, the $e(x)$ curve has several distinct characteristics. For example, survival studies sometimes are based on samples of survival times that contain a few measurements whose recorded—but of course not true—values are negative. Such variate values can occur when an individual dies of an illness before he or she has been diagnosed precisely. Hence, it may take several days for post-autopsy information to be studied and subsequently lead to a decision concerning cause of death.

Since $S(x) = 1 - F(x)$, the estimation of $S(x)$ is one of the many situations where a cdf curve that is taken to satisfy the identity $F(x) = F(X : -\infty, x)$, where $F(X : u, x) = \int_u^x f(z)dz$, must be defined carefully in terms of the exclusion or inclusion of values. However, assuming that negative survival values can be ignored, it follows that once a population and its support region, for example $[u, v]$, are defined precisely, it makes little practical difference whether the curve $F$ is defined as $F(X : -\infty, x)$ or as $F(X : u, x)$. The area "beneath" a density, and to the left of the corresponding distribution's support region, equals zero and, consequently, for such a distribution and for $x > u$, both $F(X : -\infty, x)$ and $F(X : u, x)$ can be taken to denote the same curve.

## 20.4. Truncation and the curve e(x)

In demographic, applied statistics, and economic publications, at the specific value $x$ a specialized trace of curve properties is sometimes represented by the symbol $e_x$. However, in this chapter the notation $e(x)$ will be used so that subscripted notation, such as $e(x_0)$, can be seen clearly. Before turning to the definition and interpretations of this curve, it should be mentioned that the lowercase letter $e$ of $e(x)$ has no necessary connection to the real number $e = 2.71828...$. If $x$ is a running variable, then $e_x$ and, alternatively, $e(x)$, are symbolic representations of a specialized regression curve based on a modification of the $E$ property of a particular conditional density curve $f(y|x)$.

The conditional density whose properties are assembled to form $e(x)$ is a density that has many applications in actuarial, econometric, and demographic studies. Policy premium prices are often based on estimates of the curve evaluation $e(x_0)$ that help assess the expected length of time a policy's purchaser will pay premiums after this purchaser reaches the exact age $x_0$.

When applied in this context, $e(x)$ can be thought of as a curve that prepares the information content of some density curve for actuarial applications. Suppose, for example, that the curve evaluation $e(x_0)$ designates $E(X|x_0 : x_0)$, the expected years of life remaining once some individual survives to $x_0$. The first $x$ of $x$: $x$ notation is shorthand for the phrase *given survival to $x$*, and the second $x$ represents the point where the support region $(x, \infty)$ begins. (It is this second symbol $x$ that underlies the association of the curve $e(x)$ with the general statistical idea, truncation.)

Life insurance policies are sold to living people. Consequently, when an individual who is $x$ years old is sold a policy, expected survival to some age less than $x$ makes no sense. But since the area beneath any pdf must equal one over this curve's support region, some adjustment is required so that the area under an appropriately modified curve over its support region equals 1.

When a chunk of a density curve is removed or truncated, the remaining portion that is not blotted out can, of course, determine a new density. For instance, suppose $f_{\text{new}}$ is a density formed by dividing some untruncated $f$-type curve by the area under $f$ and over the support region of $f_{\text{new}}$. As before, this is done to assure that the area beneath $f_{\text{new}}$ to be found over its modified support region, equals 1. However, when expectation is construed as the place where a curve balances, it makes no difference whether or not the representation of that curve is multiplied by some positive constant, since such a constant will not affect the curve's overall shape. For this reason, the curve $e(x)$ can be thought of as the trace or locus of values, each of which has a height above the point where a curve is truncated. This height equals the distance beyond the truncation point $x$ over which one,

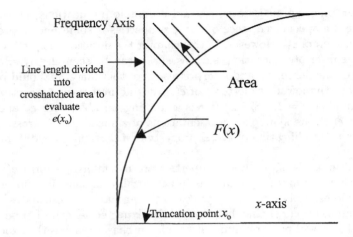

Figure 20.3. Geometric evaluation of $e(x_0)$.

out of a succession of truncated-density-curve pieces that are all bounded from below, balance.

The curve evaluation $e(x_o)$ can be interpreted geometrically as the crosshatched area (bordered by $F(x)$) shown in Figure 20.3, divided by the length of the vertical line segment that extends from ordinate $F(x_o)$ to ordinate 1. In life table applications the crosshatched area corresponds to the total premiums paid. The line length that divides this total corresponds to the number of people left alive at $x_o$ to pay premiums.

Three simple examples illustrate the geometry of the curve $e(x)$:

1. Suppose the insured passes away the exact instant, $x_0$, when the insurance company adds him or her to their list of insured clients. If, by some miracle such as being swallowed up by the Red Sea, this happens to an entire population of newly-insured individuals, then the curve $F$ will jump to the value 1 from its value immediately before $x_o$. The crosshatched area shown in Figure 20.3 will then equal zero and, consequently, in this biblical example, $e(x_o) = 0$.

2. Suppose at date $t$ the opposite occurs, and a new company insures some identically $x_0$-aged group who, together, form what is usually called a *cohort*. Suppose also that all cohort members live so long that the company happens to go out of business at some date $t + c$, before any member dies. Since no deaths are accumulated over the interval $[x_o, x_o + c)$, it follows that the $F$ curve is horizontal until it

## 20.4. Truncation and the curve e(x)

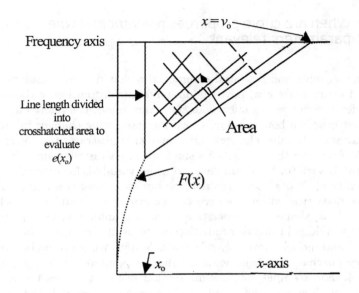

Figure 20.4. Geometric evaluation of $e(x_0)$ with uniformly distributed deaths.

steps up to the value 1 beyond $x_o + c$. In this example, $e(x_o)$ equals the fraction formed by taking the quantity, area $A = c(1 - F(x_o))$, within a rectangle of length $c$ and height $1 - F(x_o)$, and dividing this quantity by $1 - F(x_o)$. Specifically $e(x_o) = A/(1 - F(x_o)) = c$.

3. Suppose, as in Figure 20.4, the curve $F$ contains a diagonal line segment that connects the point $(x_o, F(x_o))$ to the point $(v_o, 1)$, where $x_o < v_o$. In this example, the area of a triangle is equal to the product of its base, $1 - F(x_o)$, times its height, $v_o - x_o$, divided by 2. Consequently, in the uniform mortality case shown by Figure 20.4, $e(x_o)$ equals $(v_o - x_o)/2$. As far as the density $f$ is concerned, for the uniformly-distributed deaths of case (3), between $x_o$ and $v_o$ the slanted *straight* line $F$ curve has a constant slope $f$.

Therefore, instead of the rectangular crosshatched region formed from part of the *cdf* $F$ curve of example (2), in example (3) the *pdf* is rectangular. Specifically, this density is a uniform rectangular block that balances midway between its ends, where, for this density, $e(x_o)$ equals the midpoint $(v_o - x_o)/2$.

## 20.5 When are curve properties relevant and when are model parameters relevant

The demographic and actuarial communities that devote considerable attention to curves like $e(x)$ have seldom paid much attention to the conceptual difference between a curve like $e(x)$, and any of the many curves like $\mu(x)$ that describe how a model's location parameter changes in value as some variable $x$'s value changes. Of course $e(x)$ abstracts the information content of a curve that describes a single variate's distribution, while $\mu(x)$ helps link a variate, for example $y$, to some variable, for example $x$; and thus, unlike $e(x)$, $\mu(x)$ concerns not one but two quantities $y$ and $x$.

Why does truncation, a curve characteristic so troublesome when it comes to bias, shape the cornerstone of demographic and vital statistics? A message hidden in this contradiction concerns uncertainties about curve-property and model-parameter relevancy. Many variates resemble the age-at-policy-purchase actuarial variate value $x_0$ pointed out in Figures 20.3 and 20.4. For example, because an amount of profit or loss forms a concrete bottom line, economics and business problems resemble the problems tackled by demographers and vital statisticians that concern premium payments. When $E(X)$, times some known total number, equals total profits or losses, who cares whether $E(X)$ happens to equal some $\mu$ parameter? Since the CLT concerns a sampling distribution's relationship to a model whose location parameter $\mu$ happens to equal some $E(X)$, it may make little or no difference whether or not this curve property $E(X)$ corresponds to the location parameter $\mu$ of $f$'s model.

On the other hand, suppose that unlike a curve piece that lies over points $< x_0$ in the situation in which insurance is sold to a person who has already reached age $x_0$, a missing curve piece is just as relevant as the part of the curve that is not truncated. Being out of sight, in the sense that a curve piece is truncated, need not imply being out of mind, in the sense of relevance to an issue of scientific importance. Given that $\lambda(x)$ is a function known to be symmetric about 0, the model

$$\lambda_\tau(x) = C[I_{[\tau_l,\tau_u]}(x)][\lambda([x-\mu]/\sigma)]\sigma$$

where $C$ is the multiplier defined so that the area subtended by $\lambda_\tau(x)$ equals 1, helps distinguish two types of uncertainty. The two tau parameters, $\tau_l$ and $\tau_u$, limit what can be seen, yet the two parameters, $\mu$ and $\sigma$ within $\lambda([x-\mu]/\sigma)/\sigma$ are taken to be constant valued no matter what values are assumed by $\tau_l$ and $\tau_u$.

The innocent-looking letter $C$ that helps express the structure of the truncated model $\lambda_\tau(x)$, in reality is a function of all four parameters;

$\mu, \sigma, \tau_l,$ and $\tau_u$. Yet it is not a function of the variate value $x$, the value that serves as the argument of both $[I_{[\tau_l,\tau_u]}(x)]$ and $\lambda([x-\mu]/\sigma)/\sigma$. From this perspective, uncertainties that concern the relevancy of $E(X)$ can be resolved by determining whether or not the two $\tau_l$ and $\tau_u$ parameters are nuisance parameters or, alternatively, whether they have some importance in their own right, apart from the effects their values produce on estimates of $\mu$ and $\sigma$.

For example, in educational studies there is often a truncation point, such as $\tau_l$, below which an applicant's grade point average (GPA) does not suffice for admission consideration. For this reason, even given the questionable assumption that for the population of all high school graduates, GPA is a variate that can be modeled as $\lambda([x-\mu]/\sigma)/\sigma$ where $\lambda = \phi$; a college that keeps applicant GPA records will know little about the part of this variate's density over values less than $\tau_l$. But does it care? If it doesn't, then the CLT will apply. If it does, then modifications based $C[I_{[\tau_l,\tau_u]}(x)][\lambda([x-\mu]/\sigma)]\sigma$ are relevant.

# APPENDIX I
# Models and Notation

## I.1 Notation historical background

In 1914 the psychologist Thorndike selected the lowercase letter $r$ as the symbol for what he called the *median* of certain specialized ratios, a quantity that happened to equal what is now called *Pearsonian sample correlation*. In keeping with today's convention that codes variates as uppercase Latin letters which occur towards the end of the alphabet, it would have been a stroke of good luck if either Thorndike or Pearson had called upon the capital letter $R$, rather than $r$, to denote the most popular estimator of the parameter that is now denoted by the Greek letter $\rho$. Yet, at least when compared to some other tortuous notational pathways, the history of $r$ is straightforward.

For example, three years prior to the publication of Thorndike's book, an influential entry entitled "Probability" was published in the famous 1911 Edition of the *Encyclopedia Britannica*. In Section 137 of this entry, the symbol $y'$ designates a sample mean while twenty lines later in Section 138, this same $y'$ symbol designates a population median. Yet even this inconsistent notation seems the essence of clarity when compared to the statement "...if $\sigma$ be the average number of trials..." on a page that follows the passage, "...if $s$ be the average number of trials..." that, in turn, is followed, seven lines later with the passage, "Therefore s = the sum of...." (Whitworth, 1959).

The application of a bar sign to characterize an average can be traced at least as far back as the beginning of the twentieth century. For example, in a book that was first published in 1901 by Bowley, the symbol $M_w$ designates a weighted average, while the symbol $\bar{m}$ denotes an *unweighted average*. This usage continued what had long been the traditional practice of representing a mean by the lowercase Latin letter $m$. For example, in his lectures that were later edited by E. S. Pearson and republished in 1978, Karl Pearson discussed a mean he represented as $m_a$. When viewed by today's conventions, another notational chowder is served up by a 1920 Fisher paper in which $\bar{x}$ and $\sigma$ are used to denote the sample mean and Normal distribution scale parameter, respectively, while the Normal distribution location parameter is represented by the Latin letter $m$ instead of the Greek letter $\mu$.

As far as mathematical notation is concerned, the French mathematicians Viète, Fermat, and Descartes promoted the shorthand notation where central symbols like $f$ or $x$ serve as a mnemonic device that discloses the role played by the symbol. These central symbols are accompanied by modifying symbols such as $\prime$, the "prime" symbol, or subscript sub-one, of $f'$ and $x_1$, respectively.

As is almost universally known today, letters like $f$ or $x$ that appear within a model's formula help disclose which entity is represented, in much the same way that a football uniform's color and pattern discloses team membership. Similar to numbers that are printed on these colored and patterned uniforms, because the number of distinct letters of the alphabet is limited, notational systems are forced to surround a central symbol with other numerical and/or alphabetic symbols.

Exponent notation, such as the two of $x^2$, was applied by Descartes in the middle of the seventeeth century to replace what had previously been written as repeated symbols like $xx$. As the need for higher valued exponents increased, the simple repetition of a symbol became unwieldy. There is much more to Viète's, Fermat's, and Descartes' innovation than the mere counting of symbols. They cleverly devised what seems today to be the *obvious* placement of the symbol two for the term squared, not only to the right of the $x$, but *above* the $x$. Until recently, almost all computer notation, such as the four symbols $X**2$ that represent the square of the quantity $X$, placed symbols on a single line. Yet as far back as the seventeenth century, mathematical shorthand took advantage of a two-dimensional configuration or graphical design to distinguish $x^2$ from $x$ times 2 expressed as $x2$ and thus, in this way, this shorthand system called upon both pictographic and alphabetic-numerical conventions.

Once the position to the right and above a central symbol was put into play, an obvious next consideration concerned the position to the right and below, this symbol, as in $x_1$. Here the one, to the right and slightly below the $x$, provided mathematicians with a solution to a basic problem. Since Descartes and his contemporaries had selected letters at the beginning of the alphabet to represent constants, together with letters at the end of the alphabet to represent variables, and since curves were to be represented by $f, g$, and $h$, while integer values or indices were to be represented by $i, j$, and $k$; without the assistance of some supplementary notational device the supply of available letters would have been exhausted.

## I.1. Notation historical background

The one of $x_1$ distinguishes the quantity represented by $x_1$ as the first such $x$ quantity. Here, one is an ordinal, rather than a cardinal number, since it discloses the order or succession of the symbol; in much the same sense that King Louis XIV of France was crowned after King Louis XIII of France. Just as there is no reason to believe that King Louis XIV was taller, heavier, or wiser than was Louis XIII, there is no reason to believe, without information supplementary to $x_{14}$, that this quantity is in any respect greater than $x_{13}$.

The numerical value assumed by a subscript, such as the 13 of $x_{13}$, does not have any natural connection with the numerical value or any other characteristic of the central symbol $x$. This is a universal convention. However, as elaborated upon in Section 17.4, what if a researcher actually does want to signify symbolically that some indexed quantity is greater than some second quantity, and override what is called in computer jargon a *default setting*. Such a setting or representation serves as a convention which assures that, in absence of any notation to the contrary, an interpretation is correct. An obvious example of default notation called upon in most mathematical publications (but not by a program language or system like MATHCAD), places two symbols, such as $x$ and $y$, next to each other to designate that the quantities represented by these symbols are multiplied together.

But how are exceptions or departures from default settings represented? *Ordered indexing* and most other forms of default-overrides call upon parentheses or brackets for this purpose. For instance, statistical notation such as $x_{(13)}$ designates the thirteenth *smallest value*; demographers use $P^{(1920)}$ to represent the population of individuals alive in midyear, 1920, and mathematicians distinguish the fifth derivative of the curve $f$ from $f$, raised to the fifth power, by placing a parenthesized superscript 5 after the letter $f$ of $f^{(5)}$.

Unfortunately, a troublesome exception to exception-denoting parenthesis-based override default notation occurs when statistical data-generation models are expressed in terms of the inverse of the cumulative distribution function $F$, such as the curve $F^{-1}(x)$. The placement of the superscript minus one immediately after the $F$ symbol is the only clue which indicates that minus one refers to the curve $F$ and not the reciprocal of an expression. (To avoid confusing the inverse $F^{-1}$ with a reciprocal, it would have been handy if, in the position where the negative one of $F^{-1}(x)$ now always appears, the default override negative one enclosed in parentheses of $F^{(-1)}(x)$ had been inserted.)

The inverse of a curve like $F$ can be visualized in a simple way. Assume: (1) $F$ actually has a well-defined inverse; (2) $F$ takes on the value zero on the far left side of a graph and; (3) $F$ takes on the value one on the far right side. When $F$ is traced on thin paper that is turned over then, by transmitted light, the image of $F$ will be reproduced on the reverse side of the paper exactly where the transmitted light is blocked by $F$'s curve on the non-reversed side. Now if the reversed paper is placed on its edge in the sense of being turned 90°, so that what had previously been the ordinate from zero to one is now a positive portion of the abscissa, then the traced image will be the curve represented symbolically by $F^{-1}$.

The distinction between negative one within the representation $F^{-1}(x)$, and negative one as part of notation such as $x^{-1}$ or $(1+x^2)^{-1}$, underscores a basic idea. Even though $F^{-1}(x)$ and $(1+x^2)^{-1}$ are both shorthand forms of mathematical notation, curves themselves, such as those denoted by the symbols $F$ and $F^{-1}$, differ in several key ways from expressions for curves or functions. One of the most valuable applications of symbols like $F$ and $f$ is the use of these symbols combined with expression notation such as $(x - \mu)/\sigma$. This permits the construction of representational composites that facilitate the substitution of one symbol for a curve, such as $F_1$, for a second symbol that designates a particular curve, such as $F_2$, while, at the same time, the interpretation of symbols like $\mu$ and $\sigma$ can be retained. For example, when a statistician sees an expression like $F(x) = G[(x-\mu)/\sigma]$, he or she immediately knows that two possible distinct cdf curves are equated in such a way that based on the standard curve $G$, the curve $F$'s location parameter is represented by the symbol $\mu$ and its scale parameter by $\sigma$.

There is much that can be gained from the notational richness made possible by curve and expression composites like the 6 letters that form $F(x) = G[(x - \mu)/\sigma]$. Yet, unfortunately, a few mathematical shorthand conventions were devised so early in the history of science that precedents were set before ambiguities could be resolved. As a consequence, today's emphasis on generality is sometimes at odds with yesterday's notation. Fortunately, there is a subtle clue to spots where notational confusion is possible.

A parenthesis that does appear—where this parenthesis at first glance seemingly is not necessary—can provide a subliminal warning of notational variance. Sometimes such a warning is encountered so commonly that it is taken for granted. For example, $F^{-1}(x)$ is not expressed as $F^{-1}x$ because when parentheses enclose $x$, but do not enclose $F$ or $F^{-1}$ within the shorthand representations $F(x)$ or $F^{-1}(x)$, these parentheses serve as a clue that $F$ or $F^{-1}$ denote entities substantially different from $x$'s referent. Similarly the three symbols, $^{-1}($ , indicate that -1 does not simply mean that a quantity is raised to the power negative one. On the other hand, the brackets of $[\sin(x)]^{-1}$ help the reader to keep in mind that $[\sin(x)]^{-1}$ denotes $1/\sin(x)$, not the inverse of $\sin(x)$.

The negative one to the right and slightly above the three letter term "sin" helps to specify a curve while, technically speaking, the negative one above the right bracket of $[\sin(x)]^{-1}$ designates an operation similarly to the way that the + sign of $(1 + x^2)^{-1}$ alerts us that this is a sequence of operations. But when we see $[\sin(x)]^{-1}$, only the brackets help disclose that the operation—take the reciprocal—is designated.

## I.2  Specific models, the Normal

To focus attention on general principles, rather than on computational details, a density or some other curve is often expressed symbolically as $f_1$, $f_2$, or $g$ without reference to a specific model or formula. In many situations what is required of $f_1$, $f_2$ or $g$ is that curve properties like $E$, the MEDIAN, or the MODE exist and are well-defined. However, as illustrated in Chapters 4 and 5, for many curve properties such as $E$ and Var, being well-defined is not a minor concern. Quite the contrary, for the Cauchy density model and other models that are discussed in this book, the curve property $E$ is *not* well-defined, and the population standard deviation together with the AD curve property defined in Chapter 3 are both infinite.

A *well-defined* curve is a curve that is represented unambiguously. Because the four symbols from which $z^2 = 1$ is assembled could represent the horizontal line that has a $z$-intercept at positive 1 or—alternatively—the horizontal line that has a $z$-intercept, negative 1, the line $z = 1$ is not defined clearly by the equation $z^2 = 1$; while, because the curve $y = -e^x$ never assumes a positive value, and yet the log curve only yields a finite real value for positive arguments, over any real value of $x$, the $z$ of $z = \log(-e^x)$ has no value. It is not just that $z$ happens to equal zero. In the sense that mathematical program systems will not return a real value of $z$ when a real value of $x$ is entered as input, $z = \log(-e^x)$ is not well-defined.

Unlike $z = \log(-e^x)$, the standard Normal density $\phi(x) = e^{-x^2/2}/\sqrt{2\pi}$ is well-defined over any real value of $x$. However, despite this and other convenient properties, thorny issues that concern the Normal curve have formed the subject matter of a considerable number of mathematical and statistical publications. Nevertheless, some ideas that motivate the selection of many classical parametric models can be best illustrated by using the standard Normal density $\phi(x)$ as an example.

There are no parameters specified within the $\phi(x)$, or indeed as components of any other standard model's formula. This implies for this Normal curve that any curve property which exists and is well-defined takes on a single numerical value. For instance, the standard Normal $\phi$ has a standard deviation equal to 1 and an AD equal to 0.7988. The function $\phi$ balances at the point 0, which implies that $E_\phi(X) = 0$ and since $\phi(-x) = \phi(x)$, it follows that the standard Normal curve is symmetric about zero. Thus the area to the left of zero under $\phi(x)$ must equal $1/2$, from which it follows that, like the property $E_\phi(X) = 0$, the population median curve property of $\phi$ equals zero. It is easy to see from its equation that the standard Normal density assumes its highest value at mode zero. Consequently, if the numerical value zero for some strange reason is taken to designate a particular suit of cards, the hand consisting of the three properties: $E$; median and; mode, would correspond to a flush.

In real-world economic, social, or biological science, as compared to artificial or simulation applications, there is little chance of drawing an $f$ curve that has equal mean, median, and mode in a given hand. This implies that there is little chance that some realistic variate is perfectly Normally distributed, and there is even less of a chance that all variates that are studied have densities that are standard Normal; hence the question, how can this model be generalized? Here the term "generalized" refers to an expression that has the Normal as a special case and which can also represent other densities that have some practical value.

For the sake of simplicity, suppose a new $f$ curve is formed from the standard Normal by altering one curve property, but retaining the properties that the absolute deviation (AD) defined in Chapter 3 equals 0.7988 and the new curve's population standard deviation equals 1. Suppose to form the new curve, the old curve is translated along the abscissa so that its $E$ property equals the value $\mu$ where $\mu$ represents some real number not necessarily equal to 0. Then the explicit model for this density curve will be:

$$f(x) = e^{-(x-\mu)^2/2}/\sqrt{2\pi}.$$

Besides the explicit model $f(x) = e^{-(x-\mu)^2/2}/\sqrt{2\pi}$, the implicit representation $\phi(x; \mu)$, could designate the translated Normal with location parameter $\mu$. To avoid confusion, the | symbol indicates which arguments of $f$ condition this curve; in the sense, for example, that $x$ of $f(y|x)$ is an argument called a "variable," as distinct from the "variate-value" argument $y$. Symbols for individual variates are separated by commas, as are both individual variables and individual parameters. Finally, although the parameters of a formula's, model's, or expression's representation are separated from each other by commas, taken together they are separated from variate and variable values by a semicolon, as in the $4^{th}$ of symbols that form $\phi(x; \mu)$.

Together with $e^{-(x-\mu)^2/2}/\sqrt{2\pi}$ and $\phi(x; \mu)$, a third notational device shares characteristics with both explicit and implicit forms of representation. The notation $\phi(x - \mu)$ represents explicitly the computational process

## I.2. Specific models, the Normal

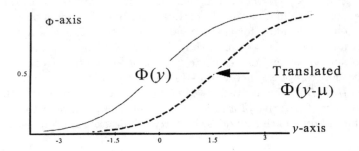

Figure I.1. Standard and translated Normal cdf.

by which the argument of the standard Normal is determined. Yet like the implicit form of representation $\phi(x;\mu)$, $\phi(x-\mu)$ calls upon the single symbol $\phi$ to designate that some new curve is a generalized form of the standard Normal density. It is this third form of representation that we will see in the next section can be continued further in several distinct directions.

Besides being applied in many applications to represent a density curve $f$, the intermediate form of representation is encountered when alternative curves, such as the cdf $F$ are represented. For example, suppose the standard Normal cumulative distribution function $\Phi(y)$ equals

$$\int_{-\infty}^{y} \phi(z)dz.$$

(The value assumed by the function $\Phi$ at $y$ discloses the area under the standard Normal over the interval including, and to the left of, $y$.) Then provided that the curve's standard deviation equals 1, the nonstandard and translated Normal cdf is $\Phi(y-\mu)$.

While the standard Normal density's highest value is assumed over the point called the *mode* whose own value is zero, a variate with cdf $\Phi(y-\mu)$ has its mode at the "location parameter" value $\mu$. In the sense that when $y$ equals $\mu$ the argument of $\Phi$ equals 0 which, for the Normal density $\phi$ is the point about which $\phi$ is symmetric, $\mu$ determines the location of the Normal distribution's pdf curve. In Figure I.1 the standard Normal cdf is shown as a solid curve and a translated Normal cdf is shown to the right of the standard curve for a $\mu > 0$.

To generalize $\Phi$ still further, the standard Normal cdf is often modified to form the expression $\Phi([y-\mu]/\sigma)$. In a similar fashion, the standard Normal pdf symbol $\phi$ often appears in the general Normal curve representation

$f(y) = \{\phi([y-\mu]/\sigma)\}/\sigma$. Notice that the scale parameter is listed twice in the general formula for the Normal density, but it appears only once in the counterpart formula for the general Normal cdf. The solo appearance of $\sigma$ within $\Phi([y-\mu]/\sigma)$ implies that the lower and upper bounds on the values assumed by this function are not changed by changes in the value of $\sigma$. Its encore appearance within $\{\phi([y-\mu]/\sigma)\}/\sigma$ is required because $\{\phi([y-\mu]/\sigma)\}/\sigma$ is the derivative $\Phi([y-\mu]/\sigma)$. (A quick review of the Calculus concept, derivative of a function of a function, will help clear up the reason for the $\sigma$ term that divides the curly left bracket of $\{\phi([y-\mu]/\sigma)\}/\sigma$.

## I.3 Specific models, lognormals and related curves

The symbols $\Phi$, $y$ and $\mu$ within $\Phi(y-\mu)$, can be thought of as together constituting the crossroads through which three statistical pathways converge. The destination that applied statisticians wish to reach along these pathways is the same, and yet the directions taken are very different. The common destination is a simple generalization of the Normal pdf and cdf curves that has more representational scope than either of the curves $\{\phi([y-\mu]/\sigma)\}/\sigma$ or $\Phi([y-\mu]/\sigma)$. To explain these approaches from a general vantage point, it is useful to assume initially that $\sigma = 1$, and examine the three symbolic signposts: $\phi$; $y$; and $\mu$, of the symbolic representation $\phi(y-\mu)$.

The most obvious, but not necessarily the most appropriate, way to seek alternatives to $\{\phi([y-\mu]/\sigma)\}/\sigma$ and $\Phi([y-\mu]/\sigma)$ with more representational scope, is by route of an alternative standard function substitute for $\phi$ and $\Phi$. Since a standard cdf is some particular partial integral counterpart of

$$\Phi(y) = \int_{-\infty}^{y} \phi(z)dz,$$

once the choice of a substitute for $\phi$ is made, the new choice of standard cdf is determined completely.

The Laplace model $f(y) = e^{-|[y-\mu]/\sigma|}/(2\sigma)$ was forwarded by the pioneer French astronomer and mathematician Pierre Simon, Marquis des Laplace. It was proposed in the context of problems for which the Normal curve was later utilized. Strangely, the Normal at one time was called the "Second Law of Laplace." It was first specified by the mathematician Demoivre in 1753, and thus is not called the Gaussian curve because Gauss was the first to investigate its properties. Its identification instead is due

## I.3. Specific models, lognormals and related curves

to Laplace's preference for the model that does, indeed, bear his name, in place of the curve that his German contemporary, Gauss, emphasized. Unlike its Gaussian counterpart, the Laplace model $e^{-|[y-\mu]/\sigma|}/(2\sigma)$ does not involve the square of the term $[y - \mu]/\sigma$. Consequently, as its argument, say $y$, increases in absolute value, the Laplace curve approaches zero at a slower pace than does its Gaussian counterpart.

When a curve shows a reluctance to become imperceptibly valued as $|y|$ increases, it is called *heavy-tailed*. Because so many curves encountered commonly are heavy-tailed, applied statisticians soon become connoisseurs of function appendages. In this regard, the Laplace would be judged to be merely a little husky when it is stacked up against a brother or sister curve. This problem-child of a curve bears the name of the prolific researcher Baron Augustin Louis Cauchy. Since this nineteenth century French mathematician provided the first firm mathematical foundation for the concept—continuity—and laid the cornerstones of both the theory of complex numbers and the mathematical theory of elasticity, perhaps he can be forgiven for conjuring up a curve that is every bit as troublesome within the portals of statistics as the Norse fire god Loki was supposed to be within the halls of Asgard.

The two symbols, $\mu$ and $\sigma$, designate the Laplace, $e^{-|[y - \mu]/\sigma|}/(2\sigma)$, model's scale and location parameters. In many texts the parameters within the general forms of specific models are given their own individual Greek letters. There are many alternatives to the Normal or Gaussian curves, such as alternatives proposed by Laplace and Cauchy. Thus the number of different Greek letters used in this way would approximate the number of different models times the number of model parameter types. For this reason, it is convenient to first define a standard model, such as the standard Laplace density $\lambda(z) = e^{-|z|}/2$, and then define the more general model with location-type parameter $\mu$ and scale-type parameter $\sigma$ as, $\lambda(y; \mu, \sigma) = \{\lambda([y - \mu]/\sigma)\}/\sigma)$. (There is no single Greek letter or other symbol reserved in common usage for the Laplace density or Cauchy density that corresponds to the symbol $\phi$ that represents the Gaussian, in other words, Normal or bell-shaped, density. Hence, although the letter $\lambda$ will be used here to designate the Laplace density, it does not necessarily represent this curve in other books and monographs.)

A second route through the Normal model crossroads—$\Phi(y - \mu)$—extends from the single letter $y$ within the argument of $\Phi$. Often there is no particular reason why the variate $Y$ itself should be the quantity that is Normally distributed. Perhaps it is the log, square root, or reciprocal of this variate's value that can be best assumed to be Gaussian distributed.

The two expressions whose graphs are shown in Figure I.2 are based on the substitution of the log of a variate value within a standard model. The

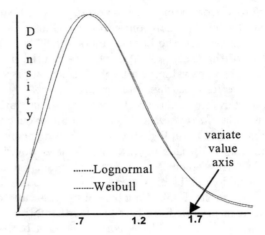

Figure I.2. Weibull and lognormal $f$.

dotted curve is called a *lognormal density* and the dashed curve is called a *Weibull density*. The latter curve is obtained by substituting the log of $y$ for positive values $y$ that appears in the formula for a cdf model known as the "extreme-value" cumulative, $1 - \exp\{-\exp([y - \mu]/\sigma)\}$, and then graphing the Weibull density that serves as the derivative of this new curve.

Mann (1968) asserts that "...the extreme-value distribution is to the two-parameter Weibull distribution as the normal distribution is to the two-parameter logarithmic normal and one could conceivably, therefore, refer to the Weibull distribution as the log-extreme-value distribution." Both lognormal and Weibull, densities and cdf's have been applied to represent distributions of commonly studied variates such as yearly income, product lifespan, or post-diagnosis survival. Yet, as is illustrated in Figure I.2, despite the major differences between the actual formulas for these two curves, these substantially different expressions can yield nearly indistinguishable curves.

Different methodologies estimate the unknown parameters of these two models. That two very different pathways can lead to nearly identical curves is unfortunate because much can depend on the difficult decision: Which of the two alternatives, the lognormal or the Weibull, is preferable? Strangely, the natural and social sciences have shown a marked preference for the lognormal pathway, while the physical and engineering sciences have tended to move along Weibull routes.

The $\Phi(y)$ component of the univariate Normal cdf model, $\Phi([y - \mu]/\sigma)$, is often altered in such a way that the argument $[y - \mu]/\sigma$ is inserted into

## I.3. Specific models, lognormals and related curves

Laplace's, Cauchy's, the extreme value, the logisitic $1/[1 + \exp(-y)]$, or some other standard cdf curve. There is no reason why $\Phi$ or another choice of standard functions, is any more or less complex than Laplace's, Cauchy's, or any other statistician's or mathematician's alternative nominee.

Besides taking $[y - \mu]/\sigma$ to be an argument of some cdf other than $\Phi$, or replacing the term $y$ within this argument with the term $\log(y)$ or even $\sqrt{y}$, parametric models have been extended one step further. Specifically, when the curve $v(y)$ is defined as $\log(y)$, then a "lognormal" cdf equals $\Phi(v(y) - \mu)$. Yet it often has been found that the curve $v(y)$ only provides a suitable generalization of a curve like $\Phi(y - \mu)$ when $v(y)$ is expressed in terms of its own parameter, as in $v(y) = \log(y - \tau)$ or $\sqrt{y - \tau}$. Here $\tau$ is a *threshold parameter*.

When models are generalized, the inclusion of a threshold parameter is of great value. In terms of either the function $v(y) = \log(y - \tau)$ or the function $\sqrt{y - \tau}$, this pathway extends beyond some standard model, in the sense that it broadens to form an additional lane of great importance. As its name implies, a threshold parameter helps model the real world limitations below which values of variates are known to be impossible.

When the transformation function $v(y)$ has a well-defined derivative $v'(y)$ over some region, it then follows that within this region the derivative of curve $\Phi(v(y) - \mu)$ equals $v'(y)\phi(v(y) - \mu)$. The complexity of density $v'(y)\phi(v(y) - \mu)$ indicates is that even when a seemingly simple variate transformation, like the log of the variate $Y$, is taken to be Normally distributed, the density of $Y$ is the product of the curve like $v'(y) = 1/y$ times the composite function $\phi(v(y) - \mu)$.

When a cdf-type curve is represented in terms of the two parameters, $\mu$ and $\tau$ as $\Phi(v(y-\tau)-\mu)$, its derivative equals $v'(y-\mu)\phi(v(y-\tau)-\mu)$. Notice that like the scale parameter $\sigma$ of the density model $\{\phi([y-\mu]/\sigma)\}/\sigma$, the threshold parameter $\tau$ appears both within the argument of the function $v'(y-\tau)$ that is multiplied by $\phi(v(y-\tau)-\mu)$, and also within the argument of $\phi$. In a similar fashion, the derivative of the three-parameter cdf $\Phi[(v(y-\tau)-\mu)/\sigma]$ equals $[v'(y-\tau)/\sigma]$ times $\phi[(v(y-\tau)-\mu)/\sigma]$. For example, the pdf of the three-parameter lognormal model equals $[\sigma(y-\tau)]^{-1}\phi[(\log(y-\tau)-\mu)/\sigma]$, while this same relationship applies to the Weibull and loglogistic models when the standard Normal density $\phi$ is replaced by the derivative of $1 - \exp\{-\exp(y)\}$ and $1/[1 + \exp(-y)]$, respectively.

In addition to lognormal, Weibull, and loglogistic densities, another interesting model can be based on the replacement of the standard Normal model $\phi$ by the standard Cauchy model. In keeping with the Cauchy model's quirks, some examples of logCauchy densities are the sort of curves referred to as bathtub functions that resemble the curves discussed in Section 10.5. In survival analyses as well as product reliability and demographic studies, infant mortality, or product break-in density curves have a mode curve property at, or near, the threshold parameter value of $\tau$. For example, a very high value immediately following birth is followed by the curve's decline after adolescence while, after this decline, curve structure is bell- or lognormally-shaped.

If the threshold $\tau$ decreases without limit, it is possible to continually rescale the lognormal model so that its cumulative distribution function is well-defined. In this circumstance, the three-parameter lognormal approaches some two-parameter Normal curve. This shows that certain lognormal curves resemble some Normal curve, in much the same way that certain other lognormal curves, such as that shown in Figure I.2, resemble some Weibull curve. Unlike the class of Weibull models, which contains members visually similar, although mathematically distinct from some lognormal models, any Normal is the limit approached by some sequence of lognormal models as their parameter values change. In this sense, any Normal model can be thought of as a lognormal *extended model* special case.

## I.4   Model families

Two extended model classifications have the Normal as a special case. The first, known as the Pearson family (PF) of distributions, is expressed as a simple model for the derivative of a pdf's log . For example, $\phi(x)$ is the solution to the PF differential equation obtained by setting this derivative set equal to $-x$. A second classification called the Gram-Charlier expansion is based on the product of $\{\phi([x-\mu]/\sigma)\}/\sigma$ times a polynomial linear combination of terms such as

$$z^k, z^{k-1}, z^{k-2}, z^{k-3}, ..., z^1, z^0;$$

where $k$ is an integer $> 1$, and $z = [x-\mu]/\sigma$. The coefficients of the linear combination serve as additional parameters that supplement the representational flexibility provided by the two parameters $\mu$ and $\sigma$ within the model $\{\phi([x-\mu]/\sigma)\}/\sigma$.

A simple algebraic relationship illustrates an unfortunate limitation of the Gram-Charlier representation. The sole polynomial that does not approach plus or minus infinity, $+/-\infty$, when $x \to +/-\infty$, is the horizontal line $y = p(x) = C$, where C is any real number. (Below, this function will be called a horizontal polynomial, and the symbol $+/-\infty$ will designate plus or minus infinity while $\pm\infty$ designates plus infinity and minus infinity, as in the interval with borders $\pm\infty$.) Except for $p(x) = C$, any product

## I.4. Model families

$p(x)\{\phi([x-\mu]/\sigma)\}/\sigma$ is a product of two terms, the first of which approaches $+/-\infty$ and the second of which decreases to zero as $x \to \pm\infty$.

This formulation does not cause great difficulty for the Normal, which has tails that descend towards zero height rapidly. However, suppose a polynomial were to be multiplied by the Cauchy model density $\sigma^{-1}\varphi([x-\mu]/\sigma)$, where $\varphi(z) = 1/\{\pi(1+z^2)\}$. Then, as $x \to +/-\infty$, when the polynomial contains a third- or higher-degree term with a nonzero coefficient, this product approaches $+/-\infty$.

If $f(x)$ were a curve that increases without limit as $x$ increases, the corresponding cdf, $F(x)$, $\to \infty$, and yet the chance that any realistic variate's value is finite equals 1. Among other reasons, this limit of 1 implies that it is impossible for the value of $f(x)$ to approach $\infty$ when $x \to +/-\infty$. Consequently, while it is feasible for the product of a non-horizontal polynomial times a Normal or Laplace density model to be a density, this is not true for a non-horizontal polynomial times a Cauchy model. That the Gram-Charlier approach cannot generalize the Cauchy model, illustrates that, so far, no single representational scheme includes all interesting models as special cases. (The closest any scheme comes to this form of universality is the scope of the Fourier/wavelet approach discussed in Chapter 13.)

The standard Cauchy density is the ratio of two polynomials, specifically $p_1(z) = 1$ divided by $p_2(z) = \pi(1+z^2)$. Ironically, the entire PF of curves is also based on ratios of polynomials. However, unlike the Cauchy density, which is itself the ratio of two polynomials, the PF is based on the derivative of the log of a density. It is this $d(\log f(x))/dx$ curve that is modeled as a polynomial ratio.

Today, the PF family of models for statistical curves is seldom applied for data analysis purposes. Yet it has found a niche. Because, besides the Normal, several other useful statistical models are special PF cases, PF curve representation often provides alternatives when methods based on the Normality assumption are checked for robustness.

Even though during the first half of the twentieth century many papers had been written about PF and Gram-Charlier approaches, with the possible exception of robustness-study-model alternatives, neither of these extended model families is of much interest today. One reason for this change of emphasis can be seen from the lognormal, $v(x) = \log(x)$, and other special cases of the expression $v'(y-\tau)\phi(v(y-\tau)-\mu)$. Neither the lognormal nor any other generalization of the Normal based on any variate transformation is known to be a Gram-Charlier special case.

Neither the lognormal nor any other transformation-based extended model is a PF member. The two pathways to a more general form of representation are: (1) PF and Gram-Charlier approaches; and (2) log and other transformations based on what are sometimes called, "Johnson-system" curves. There are few connecting roads that link these pathways, although Tarter and Hill (1997) discuss a route that connects the log transformation with Fourier and wavelet methodology.

## I.5 Mixtures and Bayesian statistics

In the quest for representational breadth, the parameter $\mu$ component of the basic univariate model $\phi(y-\mu)$ is often generalized by calling upon the assumption that $\mu$ has its own distribution. As is discussed in many sections of this monograph, one of the simplest of these generalizations for some $\mu > 0$ and for two densities defined as $f(y) = \phi(y-\mu)$ and $g(y) = \phi(y+\mu)$, assumes that a density equals $[f(y)+g(y)]/2$. In other words, $f$ describes a mixture of Normal distributions with mixing parameter $P = 1/2$. In this example, the location parameter(s) are thought of as binary-distributed with support over the two values, $\pm\mu$.

The term "mixture" modifies either the word "curve" or the word "distribution." Representation by a two-component mixture, like that described by the density curve $[f(y) + g(y)]/2$, implies the population under consideration contains members that can belong to one of two subpopulations. For example, in the Normal $\sigma = 1$ case, those from the first subpopulation have a distribution described by the density $f(y) = \phi(y-\mu)$, and those from the second subpopulation have a distribution described by the density $g(y) = \phi(y+\mu)$. At each measurement step, there are equal chances of selecting from $f$ or from $g$ when the variate that has the mixture distribution has the density $[f(y) + g(y)]/2$.

Sampled variate values are distributed in such a way that, were additional information available, these values could be allocated to subpopulations individually described by $f$ or $g$. It is this sort of missing additional information that is usually thought of as the value assumed by a lurking variable. When mixture representation is applied, subpopulation membership information is taken to be unattainable; in the sense that which density, $f$ or $g$, describes the distribution of any single sampled subject is not known. For example, the mixture density $0.4f(y) + 0.6g(y)$ shown in Figure I.3 has two Normal components, $f$ and $g$, where each component has scale parameter that equals 1. Because $f$'s location parameter equals 0, $g$'s location parameter equals 3, and the mixing parameter's value equals 0.4, the density's outline happens to resemble a rug placed over piles that correspond to the two superimposed curves $f$ and $g$. However, were mixing parameter $P$ small enough, only one pile might be seen.

In the sense that this rug obscures the individual $f$ and $g$ curves, there is a major difference between statistical mixtures and chemical mixtures.

## I.5. Mixtures and Bayesian statistics

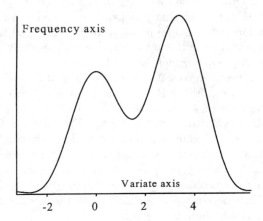

Figure I.3. Mixture of Normal pdfs.

If a molecule of a substance was taken from a mixture of two distinct compounds, then it is feasible, at least in theory, to determine the chemical make-up of the molecule. However, when statistical mixture data are sampled, the resulting measurements are valued in such a way that it is impossible to tell which particular subpopulation yielded any single sampled number. In the engineering literature this circumstance is sometimes referred to as "unsupervised learning," as distinguished from supervised learning in which "training sets," in other words sampled data where subpopulation attribution is known, are available.

Suppose that for $1 < c < \infty$, the $k^{th}$ of $c$ subpopulations is accurately described by a curve model that contains a parameter, such as $\mu_k, k = 1,...,c$; where $c > 1$. The pdf curve that describes $Y$'s distribution is then called a *finite mixture density*. When one or more parameters like $\mu$ are assumed to have a distribution described by a cdf that is either continuous or a step function, a variate with this distribution is said to have a *contagious distribution*. Here the word "contagious" refers to the tendency to spread from one to another, as in the spreading of the $f$ and $g$ components of Figure I.3. In addition to finite mixtures and the general class of contagious distributions, the fertile branch of statistics known as "Bayesian inference" is based on what are called *prior distributions* of parameters.

The contagious- or mixture-model approach is a generalization of a simple parametric model like $\Phi(y - \mu)$. Bayesian methodology quantifies the knowledge accretion process in terms of a blend of prior knowledge concerning a parameter's density with the information conveyed by sampled data. At the culmination of this process, a modified prior, known as a *posterior*, is obtained. For example, according to Kendall and Buckland (1960), "...a probability $p_1$ at the outset of an experiment might be modified to $p_2$ in the light of [an experiment's] result."

The distinction between contagious modeling and Bayesian approaches is subtle, but important. In the Bayesian context where variates are modeled, all data elements of a sample are taken to have been generated by the choice of a single value of each parameter. Parameter values are taken typically to vary between samples, but not within any single given sample. Viète's, Fermat's, and Descartes' parametric notation took parameters to be constant valued for a given problem. However, their values might or might not change from problem to problem. Similarly, in the Bayesian context, statistical parameters are thought of in terms of subjective opinions that change from one sample to the next.

The exact opposite is true of contagious and finite-mixture modeling. For example, in the finite-mixture case, the chances of each element being attributable to a specific subpopulation depends upon a finite number of mixing parameters. Since the element must be attributable to some subpopulation, the sum of these mixing parameters must be 1. Mixing parameters describe how parameter indices vary in the sense of being distributed in the context of a single sample.

Suppose the number of subpopulations is 2 and the mixing parameter odds of sampling from the first subpopulation are 0.9 to 1. Even when the sample size $n$ is large, it is possible for all sample elements to be attributable to the second subpopulation, even though this would be an unusual sample. In the long run, the proportion of data within the sample attributable to the second subpopulation will approach 0.1, the value of the mixing parameter $P$. The trick here is that $P$, as well as the parameters of the two mixture components, for example: $\mu_1; \sigma_1; \mu_2; \sigma_2$, are constant for a given problem in the same sense that Viète's, Fermat's, and Descartes' parameters were constants. What does vary however, is which pair; $\mu_1, \sigma_1$ or $\mu_2, \sigma_2$; goes with which observation? Thus, it is in this sense that the subscript, one or two, of the location parameter and scale parameter are distributed according to the rules discussed in detail in Section 7.4.

Table 2 of Finkel (1990), summarizes a modeling procedure by means of which risk is taken to be a product of seven variates divided by three other variates. As is often the case for this type of model, each of the ten variates is assumed to have a distribution described by density function that is not a mixture but, instead, is a curve like $\phi([y - \mu]/\sigma)/\sigma$. Since neither $\mu$ nor $\sigma$ is taken to have its own prior distribution, this approach cannot be said to be Bayesian. However, there would seem to be much to be said for applying the mixture generation techniques discussed in Chapter 7 to handle the possibility that the densities involved may have characteristics that differ between different disease states and demographic characteristics such as race.

## I.6  Notational conventions about moments and variates

In statistical applications, the curve property $\text{Var}(X)$ is so important that it is worthwhile to review several notational conventions. For example, in some early texts the three lowercase letters, var, without any following parentheses, denote what is now almost always denoted as Var(variate), where the word "variate" denotes an uppercase letter or combination of letters. In much the same way that the letters which form $E()$, $f()$, and $F()$ are assembled, it is advisable to place the parentheses after the letters Var without any space.

In some mathematical statistics texts, the terms $E[h(X)]$ and $\text{Var}[h(X)]$ designate the expected value and variance operators for the function $h(x)$. In practice, what this means is that a new or transformed variate is defined in terms of $X$. Specifically, some function of the variate $X$; $Y = h(X)$, where $h$ might be a log function if $X$ assumes positive values; has a distribution described by a density curve with curve properties $E[h(X)]$ and $\text{Var}[h(X)]$. Because $X$ and $Y$ are uppercase letters, $Y = h(X)$ indicates that the values assumed by the variate $Y$ are determined by evaluating $h$ at corresponding values of the variate $X$.

Care must be taken when $h$ is a function such as $h(x) \equiv x^2$, a function that is defined for both positive and negative arguments even though it yields only nonnegative values. Since, for nonzero $x_0$, the square function transforms negative values like $-|x_0|$ and positive values like $|x_0|$ to the same value, a variate $X$ whose support region—before squaring—stretches from zero in both a positive and a negative direction, will be transformed into a new variate $Y$ that can be said to be folded, in the sense that it is identical to some function multiplied by $I_{[0,\infty)}(y)$, as defined in Section 7.4.

The first three letters of the word "expectation" match the first three letters of the word "exponent." Hence, by convention, the symbolic representation prefixed by three letters as exp $z$, is always taken to designate the number $e$ raised to the $z$ power. The exp $z$ notation would be particularly confusing were $\exp(X)$ or $\text{Exp}(X)$ to appear in place of the notation $E(X)$. For this reason the notation Var calls upon the first three letters of the word variance, while the expectation curve property is denoted by the one letter, $E$.

It is commonplace today for parentheses, as in $\text{Exp}(z)$ or $\exp(z)$, to enclose what was called the argument of "exp $z$" in older texts. The partial integral of the curve $(2/\sqrt{\pi})\mathrm{Exp}(-t^2)$, taken from zero to $z$, is referred to currently as the *error function* erf $z$ or erf($z$), while the partial integral of erf($z$), taken from $z$ to infinity, is called the *complementary error function*, erfc ($z$).

Users of mathematical programming languages like Mathematica, MATHCAD, or MATLAB, must be careful to distinguish the curve erf($z$) from the curve denoted in MATHCAD as cnorm($z$). It is cnorm($z$) that is identical to the standard Normal cdf curve partial integral of $\phi$ from minus infinity to $z$, that is denoted symbolically as $\Phi(z)$. For positive values of $z$, the standard Normal cdf $\Phi$ satisfies the identity

$$\Phi(z) \equiv (1 + \mathrm{erf}[z/\sqrt{2}])/2.$$

This formula can be thought of a relationship between the standard Normal cdf that is called upon in statistical applications, and a cdf standardized so that it facilitates descriptions of non-statistical processes. Not only is the (1/2) term missing within the exponent of the integrand that defines the partial integral erf($z$) but, in terms of a concept introduced earlier in this section, the erf function can be said to denote a folded curve.

# APPENDIX II
# Variate Independence and Curve Identity

## II.1 Independence and identical distribution

Variate independence is defined by a relationship that will be described in this section between the different curves; $f(x), f(y)$ and $f(x,y)$. In contrast to the mathematical definition of *independence*, the term *randomness* is difficult to define without also including a detailed description of how samples are selected. Unfortunately, more labor is often required to describe the practical details of a process which is a means to an end than is required to describe the end or goal itself. According to Kendall and Buckland's *A Dictionary of Statistical Terms:*

> random selection is a method of selecting sample units (subjects) such that each possible sample has a fixed and determinate probability of selection. Ordinary haphazard or seemingly purposeless choice is generally insufficient to guarantee randomness when carried out by human beings and devises such as tables of random sampling numbers, or analog machines, are used to remove subjective biases inherent in personal choice.
>
> The random selection of members of a sample involves the choice of subjects so that the chance of a given subject being selected is equal to the chance of any other subject being selected. It also implies that the choice of a given subject has no bearing on the probability of any other subject's inclusion.

Those values of variates associated with one subject are often assembled as a list or vector. If two subjects are selected randomly, then corresponding variates measured from two distinct subjects selected this way are independent while, within any single vector of variates, those associated with a single subject may or may not be independent of each other. Consequently, although usage in reference to a single subject is preferred,

## II. Variate Independence and Curve Identity

when two different symbols, like $X$ and $Y$, represent two variates, these symbols can either represent two different variates associated with a single subject or, alternatively, the same variety of measurement taken from two different subjects.

The entities that are referred to as independent are not curves or even distributions themselves. Yet the concept *independent variates* is based on three or more *curves*. In this section, roulette spins together with throws of a pair of dice will be used to illustrate the crucial connection between curves on the one hand, and variate dependence and independence on the other.

Even though a roulette wheel is circular, the numbers along its edge that characterize specific bets define a one-dimensional scale (Lake 1901). They do this in the sense that the edge of the roulette wheel could be removed and numbers placed along a line positioned along this edge to form $m$ "bins" which record the occurrence of individual random outcomes that correspond to the same variate value. If $rnd(1)$ designates a uniformly distributed variate on the half-open interval $[0, 1)$ then bin selection can be determined by the formula

$$\sum_{j=1}^{m} j I_{[(\{j-1\}/m), (j/m))}(rnd(1))$$

discussed in Section 7.4. The m bins mark the possible occurrences of *single-spin* roulette turns. They do this in much the same way that the 6 sets of dot patterns of a single die determine what events can occur when a single die is rolled.

Suppose that like the throw of a *pair* of dice, a roulette wheel is spun *twice*. Then, when the word "event" refers to one single spin, dual roulette

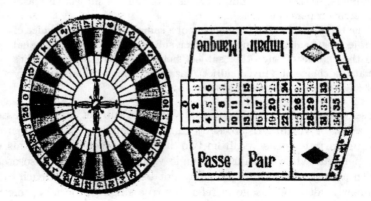

Figure II.1.

## II.1. Independence and identical distribution

wheel spins obviously determine the outcome of joint events so that two event outcomes correspond to two measurements that, in turn, provide the values of two variates. For dice, obviously the roll of the first die provides the value of the first variate and the second roll determines the second variate value.

That an $f$-type curve describes a joint distribution of *two* variates is indicated by separating the two symbols for the values assumed by these variates by a single comma, as in $f(y_1, y_2)$ while, similarly, were three variates measured on a single subject then $f(y_1, y_2, y_3)$ might designate the *joint* density curve that describes these variates' distribution. In practice, when $f(y_1, y_2)$ and $f(y_1, y_2, y_3)$ are graphed, dimensional economy is a major concern. For example, the curve $z = f(y_1, y_2)$ is a three-dimensional surface, while graphs of $z = f(y_1, y_2, y_3)$ require four dimensions, one for $z$ and three for the function arguments $y_1, y_2$, and $y_3$.

Because of the need for dimensional abridgment, topographical devices, such as the isobars and isotherms shown on a weather map, are applied commonly. The subdivisions of a weather map resemble a conceptual method of abridgment known as a sample space. A sample space can also be thought of as a theoretical generalization of a gambling table. Consider, for instance, a green felt-covered table that is divided into regions upon which bets are placed by piling chips or other markers. These subdivisions, when viewed abstractly, help provide two-dimensional abstracts of complex events.

As another example, suppose that in some far-off galaxy there is a strange city that only resembles the cities of our planet in one relevant way. As is true for us, in this city the phrase—*other side of the tracks*—applies to the economic distribution of urban population. Moreover, in this imaginary community houses are laid out on a perfectly rectangular grid where, except at the community's edges, all houses are equally distant from four adjoining dwellings. In addition, border homes are spaced the same distance from each other and from neighboring, non-border, houses, and there are two sets of railway tracks which pass through the town that are each laid out within the town parallel to the city's horizontal and vertical borders.

In this strange city each house has been built for occupancy by one married couple, for which it is the wealth of the husband and wife that determines on which side of the tracks their house is located. Rich husbands married to rich wives live in the Northwest quarter, while the dwellings of poor husbands married to poor wives are in the Southeast quarter. Rich husbands married to poor wives live in the Northeast quarter and poor husbands with rich wives have Southwest residences.

An aerial view of the city layout is like a sample space that describes husband and wife wealth variate distribution. Suppose also that marriages

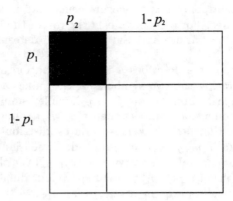

Figure II.2. A simple sample space

occur independently of, in other words without regard to, wealth, and let $p_1$ designate the probability of a given husband being wealthy by the standards of this strange community, and $p_2$ designate the probability of a given wife being wealthy.

Finally, suppose a household is selected at random by calling upon a pair of random numbers, such as those generated using rnd(1). If all houses are occupied by a single couple, then simple geometry implies that the probability that a randomly selected household will be in the Northwest quarter of this city equals $p_1$ *times* $p_2$.

In general terms, suppose $f(y_1, y_2)$ represents the joint distribution of the two variates $Y_1$ and $Y_2$, while $f(y_1)$ and $f(y_2)$ represent the "marginal" density curves of these *individual* variates. Then the identity

$$f(y_1, y_2) \equiv f(y_1) f(y_2),$$

a relationship that concerns three curves, $f(y_1, y_2), f(y_1)$, and $f(y_2)$, implies that the variates $Y_1$ and $Y_2$, whose distributions are described by these curves, are statistically independent.

## II.2 Regression notation

The regression model selected to represent $\mu_x$ within $Y|x = \mu_x + \sigma F^{-1}(Z)$ is often expressed by some generalization of the simple model $\mu_x = \alpha + \beta x$.

## II.2. Regression notation

These generalizations can be confusing because the terms "linear regression" and "linear model" apply the word *linear* in different ways. For example, page 211 of Dixon and Massey's (1983) classic text presents the formula for "linear regression,"

$$\mu_{y.x} = A + B(x - \bar{X})$$

Here $\mu_{y.x}$ denotes a $\mu(x)$ or, equivalently a $\mu_x$ curve; the letters A and B represent the $v$-intercept and slope of the line $v = A + Bu$, and the symbol $\bar{X}$ designates the arithmetic mean of all $x$ variable values.

The formula $\mu_{y.x} = A + B(x - \bar{X})$ is a strange blend of notation. A Greek letter like $\mu$ traditionally symbolizes a parameter, while the modified Latin letter $\bar{X}$ traditionally represents an *estimator* of a parameter or curve property, where the bar above X of $\bar{X}$ is employed in much in the same way that one hat symbol of $\hat{f}$ is chosen to designate that the curve $\hat{f}$ is based on sampled data. Nevertheless, due to the difference between variates, as entities that correspond to random choices, and variables, as entities whose values, even when assumed as the result of some chance occurrence, are taken as given; although, $\mu_{y.x} = A + B(x - \bar{X})$ may appear inconsistent, it actually is proper notation.

The average $\bar{X}$ symbol within $A + B(x - \bar{X})$ represents the arithmetic mean of *variables*. For example, these variables often are "user control characteristics" of the *population* from which Y variate values are sampled. Thus the lowercase $x$ in the subscripts of the symbols $\mu_x, \mu_{y|x}$ or $\mu_{y.x}$, as well as both the uppercase $X$ and the symbol $\bar{X}$ within $A + B(x - \bar{X})$, all designate values taken to have been determined by the researcher and not by nature. The leeway to choose these values stems from the right that an investigator has to delimit the exact nature of the population that he or she wishes to study.

This leeway to delimit a population does not amount to a guarantee that a model like $\mu_{y.x} = A + B(x - \bar{X})$ either applies, or provides an appropriate "fit" of the data being studied. Perhaps, for example, a parabola in the $X$-variable's value $x$, and not a linear or "first degree" polynomial, describes the relationship between the $Y$-variate's model's location parameter $\mu_{y.x}$ and the value $x$. Alternatively, perhaps for some value of the parameter $\tau$ it is a $Y_2 = log(Y - \tau)$ variate, and not $Y$ itself, that is related linearly to $X$. Finally, it could be the case that $Y$'s distribution is described accurately by a mixture of two distinct Normals, each with its own distinct location parameter. When one of these circumstances occurs, a parabola or *any* other single-valued function, one that *could* replace $\mu_{y.x} = A + B(x - \bar{X})$ validly, does not exist.

There is also considerable confusion concerning the word "linear" applied as part of the phrase, *linear regression function or curve,* and the word *linear* in the designation "linear model." For historical and mathematical reasons it is linearity with respect to the parameters, here $A$ and $B$, and not with respect to some $Y$-variate and $X$-variable, that characterizes the statistical approach based on linear *models*. For example, even when $C \neq 0$, the expression $\mu_{y.x} = A + B(x - \bar{X}) + C(x - \bar{X})^2$ is called a linear model with parameters $A, B,$ and $C$. It is a linear model even though, were $\mu_{y.x}$ graphed over the $x$-axis, the resulting curve is a second-degree polynomial, not a line.

When $C = 0$, the expression $\mu_{y.x} = A + B(x - \bar{X}) + C(x - \bar{X})^2$ represents a linear regression curve that is expressed as a linear model. The models $\mu_{y.x} = A + B\Phi(x - \bar{X})$ and $\mu_{y.x} = A + B(x - \bar{X}) + C \exp[(x - \bar{X})^2]$ are other examples of linear models. On the other hand, both $\mu_{y.x} = A + \Phi[B(x - \bar{X})]$ and $\mu_{y.x} = A + B(x - \bar{X}) + \exp[C(x - \bar{X})^2]$ are not linear models.

The strange discordance between linear regression and linear models is due to the computational procedures that are often applied to estimate parameters like $A, B,$ and $C$. The most popular of these is based on the solution of a set of "linear equations." As a preliminary to the construction of the equations applied to estimate the parametric components of a model such as $\mu_{y.x} = A + B(x - \bar{X}) + C\Phi(x - \bar{X})$, terms like $(x - \bar{X})$ and $\Phi(x - \bar{X})$ are dealt with as wholes, in the sense that they provide a set of equation coefficients. As far as the solution of these equations is concerned, it is model parameters like $A$, $B$, and $C$ that are "unknown variables." Consequently, when it designates some given quantity like X within a model like $\mu_{y.x} = A + B(x - \bar{X}) + C\Phi(x - \bar{X})$, the term "variable" is applied in a sense that differs from the meaning of the term "unknown variable" applied in reference to one or more equations.

A subscript like the $x$ subscript of $\mu_{y.x}, \mu_{y|x}$, or $\mu_x$, or curve notation such as $\mu(x)$, helps denote parametric variation symbolically. That a parameter can vary from one sampled subject to another sampled subject contradicts a basic feature of the concept of distribution, whose history can be traced back to Viète, Fermat, and Descartes. To retain the venerable convention that parameters are problem-specific constants, the variates whose distribution is described are collectively designated as *not* being necessarily *identically* distributed. Alternatively, these variates are sometimes said to be identically distributed only in the "null case."

Until recently, parameters whose constancy is an open issue were expressed in terms of other parameters, like intercept $\alpha$ and slope $\beta$ that were themselves taken to be problem-specific constants. However, in recent years methods have been devised to estimate parameters as semiparametric curves that need not be modeled in terms of other parameters.

The dilution level setting at which a measurement is taken, or the age of a white mouse, are examples of two variables' values. The slope $\beta$ of a

## II.2. Regression notation

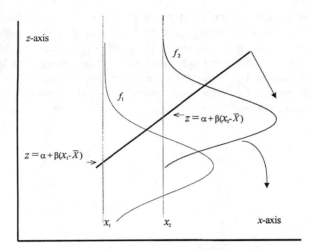

**Figure II.3.** As a regression line approaches a horizontal position and $f(y|x)$ curve changes accordingly

curve, in particular a line, is a typical parameter. In order to estimate or evaluate the value assumed by a parameter like $\beta$, nonconstant conditions are necessary. It follows that in regression-based investigations, at least one variable must, of course, vary. This idea is so central to statistical thinking that the term *identically distributed* was coined to characterize the slicing of a distribution when position and shape are *not* changed when a variable varies. As indicated above, when a variable varies, the "associated variates" are only identically distributed in a particular example of what statisticians collectively call the null case. In much the same way that variation in a costume or a disguise can help a person change his or her identity, in the non-null or "general case", a variable's variation can be said to change the distinguishing characteristics of a variate's distribution.

In the context of linear regression and linear models, the geometry that underlies the term *null* is illustrated by Figure II.2. As is well-known, the parameter $\beta$ of the population regression line shown in this figure determines the line's slope. Consequently, as indicated by the straight arrow-headed line shown at the top of the $z = \alpha + \beta(x - \bar{X})$ curve, were this curve rotated clockwise until it was horizontal, then this rotation would lower the $f_2$ density curve. Were $f_2$ and $f_1$ "congruent," in other words, were the shapes of $f_1$ and $f_2$ such that one curve could be superimposed perfectly on the other then, when $\beta = 0$, the curve $f_2$ would merely be offset to the right of $f_1$.

Suppose $f_2$ lines up with congruent $f_1$ in the $\beta = 0$ null case. Then this configuration implies geometrically that knowledge about the $x$-variable's value being $x_2$ and not, for example, a different value $x_1$, is knowledge that does not disclose anything about the distinguishing characteristics of the distribution of the variate under study, for example $Y$. Then, $Y$ is "unrelated" to $x$, where the term *unrelated* can be thought of as "variate-variable *independence*."

# APPENDIX III
# General Statistical and Mathematical Notation

| | | |
|---|---|---|
| $A$ | regression intercept (old notation) | 351 |
| $A$ | alternative hypothesis subscript | 287 |
| $A = \Phi(X; -x, x)$ | area under standard normal density between $-x$ and $x$ | 303 |
| AD | absolute deviation | 26 |
| $AD_3$ | absolute value counterpart of kurtosis | 50 |
| AERSM | asymptotic efficiency realtive to sample mean | 301 |
| ANOVA | analysis of variance | 237 |
| $a_j^{(\mu)}$ | $j^{th}$ weight of linear location parameter estimator | 264 |
| $a_j^{(\sigma)}$ | $j^{th}$ weight of linear scale parameter estimator | 264 |
| $[a, b]$ | closed interval | 28 |
| $[a, b]$ | indicator function subscript | 46 |
| $(a, b)$ | open interval | 46 |
| $[a, b)$ | interval closed on left and open on right | 46 |
| $0.5(a + b)$ | midpoint of the interval that begins at $a$ and ends at $b$ | 28 |
| $a_i$ | often the $i^{th}$ weight of a linear sum | 274 |
| $\mathcal{A}$ | attribute-valued variate | 199 |
| $\alpha$ | regression intercept parameter | 129 |
| $\alpha$ | alpha level (Type I error) | 285 |
| $\alpha_i$ | $i^{th}$ row effect | 248 |
| $B$ | regression slope (old notation) | 351 |
| $B_k$ | $k^{th}$ Fourier coefficient | 209 |
| BLU | best linear unbiased | 255 |
| $\beta = \rho\sigma_2/\sigma_1,$ | regression slope parameter | 179 |

| | | |
|---|---|---|
| $\beta = 1 - Power$ | Type II error | 292 |
| $\beta(\alpha, \Delta, n) = \Phi(\Phi^{-1}(1-\alpha)(\Delta\sqrt{n}))$ | 1-power one-sided (right tail) $z$ test | 296 |
| $\beta_1 = r_{12 \bullet 3}\sqrt{1-\rho_{13}^2}/\sqrt{1-\rho_{23}^2}$ | regression coefficient in terms of partial correlation | 185 |
| $\beta = 0$ | null hypothesized regression slope parameter equals zero | 352 |
| $\beta_j$ | $j^{th}$ regression parameter (regression coefficient) | 129 |
| $\beta_j$ | $j^{th}$ column effect | 248 |
| $\beta_1$ | moment ratio $(E\{[(X-E(X)]^3\})^2/(E\{[(X-E(X)]^2\})^3$ | 233 |
| $\beta_2$ | moment ratio $(E\{[(X-E(X)]^4\}/(E\{[(X-E(X)]^2\})^2$ | 233 |
| $C$ | often represents a constant | 43 |
| CLT | central limit theorem | 69 |
| $Cov(X_j, Y_j)$ | covariance of $X_j$ and $Y_j$ variates | 114 |
| $Cov(Z_{(j)}, Z_{(k)})$ | covariance of $j^{th}$ and $k^{th}$ order statistics | 263 |
| $C(x)$ | standard Cauchy cdf | 318 |
| $CPA^{(x,y)}/(s^{(x)}s^{(y)})$ | sample correlation in terms of cross product average | 115 |
| $\mathcal{C}$ | continuous variate | 199 |
| $c$ | often represents a constant but sometimes represents the value assumed by a continous variate | 130 |
| $c$ | number of mixture components | 343 |
| $c$ | number of columns in ANOVA factorial | 250 |
| cdf | cumulative distribution function | 21 |
| ceil | ceil function as index generator | 98 |
| $\chi^2$ | Chi-square test, distribution, table, statistic or variate | 250 |
| $D_j$ | difference between ranks | 115 |
| $D_s$ or $d_s$ | number of deaths within $s^{th}$ group | 162 |
| $_nD_x$ | number of deaths of individuals $x$ to $x+n$ years old | 162 |
| $\mathcal{D}$ | discrete-valued variate | 199 |
| $d(x)$ | age at death density | 161 |
| dist $X$ | the distribution of the variate $X$ | 112 |
| dist $X \equiv$ dist $Y$ | variates $X$ and $Y$ are distributed identically (for example, if $f$ represents the density of $X$ and $g$ the density of $Y$ then $f \equiv g$) | 112 |
| $\Delta$ | noncentrality parameter (general) | 288 |
| $\delta$ | noncentrality parameter (ANOVA) | 239 |

# Appendix III. General Statistical and Mathematical Notation

| | | |
|---|---|---|
| $E()$ | expected value | 42 |
| $E_f(X)$ | expected value of variate with density $f(x)$ | 79 |
| $E_{(f+g)/2}(Y)$ | expected value of a mixture of densities f and g | 83 |
| $E_f(Y\|x)$ | the expectation of univariate conditional density $f(y\|x)$ | 82 |
| $E_f(Y\|x)$ | Fisz's Type I regression model | 155 |
| $E_f(Y\|u,x)$ | the expectation of univariate conditional density $f(y\|u,x)$ | 84 |
| $E_f(X:x,t,x)$ | expected life remaining after living to age $x$ at date $t$ | 309 |
| $E[h(X)]$ | expectation of transformed variate $h(X)$ | 20 |
| $E(MS)$ | expected mean square | 248 |
| $E(s) \neq \sigma$ | $s$ is is a biased estimator of $\sigma$ | 317 |
| $E(s^2) = \sigma^2$ | $s^2$ is an unbiased estimator of $\sigma^2$ | 317 |
| $E(X+Y)$ | the expectation of the composite variate X+Y | 273 |
| $E([X+Y]/2)$ | the expectation of the composite variate [X+Y]/2 | 273 |
| $E\{[Y-(\alpha+\beta x)]^2\}$ | Fisz's type II regression criterion | 157 |
| $E(Y\|x)$ | expectation of variate $Y$ given variable $x$ | 82 |
| $E(Y\|x_0:x_0)$ | the expected years of life left after surviving to x₀ | 320 |
| $E[\{Z_{(j)} - E(Z_{(j)})\}^2]$ | variance of $j^{th}$ ordered $Z$ variate | 264 |
| $E(Z_j)$ | expected value $j^{th}$ unranked variate | 262 |
| $E(Z_{(j)})$ | expected value $j^{th}$ ranked variate | 262 |
| $E(\alpha, \beta)$ | expectation as a function of the parameters $\alpha$ and $\beta$ | 255 |
| $E_f(\hat{\theta}-\theta)^2$ | mean square error of estimator $\hat{\theta}$ (of parameter $\theta$) | 265 |
| $E(\hat{\mu}) = \mu$ | $\hat{\mu}$ is an unbiased estimator of $\mu$ | 317 |
| $E_2(\alpha, \beta)$ | expected squared residual function of intercept and slope | 256 |
| $e$ | 2.718281828...... | 321 |
| $e$ | error symbol | 143 |
| $e_i$ | the $i^{th}$ expected number of observations | 175 |
| erf $z$ | archaic notation for the error function | 346 |
| erf $(z)$ | preferred notation for the error function | 346 |
| erfc $z$ | complementary error function | 346 |
| exp $z$ | archaic notation for the exponential function | 346 |

… Appendix III. General Statistical and Mathematical Notation

| Symbol | Description | Page |
|---|---|---|
| $\exp(z)$ | preferred notation for the exponential function | 346 |
| $e_x$ | expected length of life regression curve | 320 |
| $e(x)$ | expected length of life regression curve | 321 |
| $e^z$ | e raised to the $z$ power | 75 |
| $\in$ | $Z \in f$, $Z$ is distributed with density $f$ | 143 |
| $\varepsilon$ | archaic symbol for a model's error term | 142 |
| $\varepsilon$ | usually the variate component $\sigma\Phi^{-1}(Z)$ | 91 |
| $F$ (Fisher) statistic | Common variate (particularly in ANOVA) | 248 |
| $F(X)$ | cdf-transformed variate | 75 |
| $F(x)$ | general cdf function | 21 |
| $F(x) \equiv F(X : -\infty, x)$ | the usual definition of the cdf | 303 |
| $F(x) = G\left[(x-\mu)/\sigma\right]$ | general cdf $F$ in terms of standard $G$ | 322 |
| $F^*(x)$ | sample cumulative distribution function | 220 |
| $F^*(x_{j'}) = j/n$ | $j'$ is unranked index of the $j^{th}$ ordered variate | 220 |
| $\hat{F}^{-1}(1/2)$ | usually the sample median | 222 |
| $F^{-1}(0.25)$ | first quartile | 222 |
| $F^{-1}(0.25\vert x)$ | first quartile regression curve's target | 229 |
| $F^{-1}(1-\alpha)$ | critical point of right sided one tailed test | 294 |
| $F^{-1}(Z)$ | random variate generating top hat function | 47 |
| $f(x)$ | density (pdf) function | 10 |
| $f_2(x) = 1 - \vert 1-x\vert$ | Cramer's second density allied with $\mathcal{U}(x)$ | 298 |
| $f_c(1\vert 85)$ | age 85 specific rate of community $c$ | 169 |
| $f_F$ | female side of population pyramid | 153 |
| $\text{floor}(x)$ | floor function | 98 |
| $f_M$ | male side of population pyramid | 153 |
| $f_m(1)$ | crude rate target curve evaluation | 163 |
| $f_s(1\vert 85)$ | age 85 specific rate of standard $s$ | 169 |
| $f(x, y)$ | joint density of $X$ and $Y$ | 349 |
| $f(y; \sigma\vert x; \omega)$ | density with two types of parameters | 125 |
| $f(y\vert x)$ | univariate conditional of variate $Y$ given variable $x$ | 129 |
| $f(y_1, y_2) \equiv f(y_1)f(y_2)$ | necessary condition for bivariate independence | 350 |
| $f(y_1, y_2\vert x_1, x_2, x_3)$ | bivariate conditional density (given 3 variables) | 126 |
| $f_{\bar{x}}$ | density of the sample mean | 273 |
| $f_{\hat{\mu}}$ | density of the location parameter estimator $\hat{\mu}$ | 273 |

# Appendix III. General Statistical and Mathematical Notation

| | | |
|---|---|---|
| $\hat{f}(x)$ | estimated density curve | 17 |
| $\hat{f}(1\|x)$ | estimated binary conditional density | 161 |
| $[f(F^{-1}(p))]\sqrt{p(1-p)/n}$ | asymptotic scale parameter of $p^{th}$ quantile's model | 267 |
| $\Phi(y) = \int_{-\infty}^{y} \phi(z)dz$ | normal cumulative (cdf) function | 19 |
| $\Phi(z) \equiv (1 + erf[z/\sqrt{2}])/2$ | standard normal cdf in terms of error function | 346 |
| $\Phi(y)$ | Heaviside step function | 95 |
| $\Phi(\text{rnd}(1) - p)$ | switching function | 97 |
| $\Phi^{-1}(z) + 5$ | probit function | 93 |
| $\Phi^{-1}F^*$ | normalized scores transformation to normality | 221 |
| $[\phi([x-\mu]/\sigma)]/\int_u^\infty \varphi([x-\mu]/\sigma)dx$ | left truncated and translated standard normal | 312 |
| $\phi(x) = e^{-(1/2)x^2}/\sqrt{2\pi}$ | standard normal density | 19 |
| $\phi((x-\mu)/\sigma)/\sigma$ | general normal density | 19 |
| $(y;\mu,\sigma)$ | shortened symbol for general normal density $\phi([y-\mu/\sigma)/\sigma$ | 99 |
| | | |
| GCE | Gram-Charlier-Edgeworth | 193 |
| $g_c(x)$ | population pyramid edge density of community c | 169 |
| $g(x)$ and $h(x)$ | two functions | 84 |
| $g(X)$ and $h(X)$ | two transformed variates | 20 |
| $\gamma_1 = \mu_3/\mu_2^{3/2}$ | skewness, $E\{[(X-E(X)]^3\}/(E\{[(X-E(X)]^2\})^{3/2}$ | 233 |
| $\gamma_{ij}$ | interaction term in ANOVA model | 244 |
| | | |
| HT | hypothesis test | 55 |
| $h(x)$ | often a third function or pdf | 169 |
| $h(x)$ | Laplace transform | 47 |
| $h(X)$ | variate after transformation (for example, after squaring) | 20 |
| $h_{c,s}(1)$ | theoretical age adjusted rate | 170 |
| $\hat{\ }$ | hat symbol placed above other symbol to denote "estimator of" | 17 |
| | | |
| $I_{[a,b)}(b) = 0$ | the half open interval indicator function at $b$ | 46 |
| $I_{[a,b]}(x)$ | the indicator function equal to one over the interval $[a,b]$ | 46 |
| $I_{[a,b]}(x) = \Phi(x-a)\Phi(b-x)$ | indicator function in terms of Heaviside function | 96 |
| $I_{[\tau,\infty)}(y)$ | the indicator function equal to one at and beyond $\tau$ | 98 |

| | | |
|---|---|---|
| $I_{[0,1)}(z)$ | uniform density | 51 |
| If$[x,y,z]$ | **Mathematica** conditional | 98 |
| i | often an index | |
| iid | identically and independently distributed | 126 |
| $\int_0^\infty m(y\|x)g(x)dx$ | integral estimated by direct age-adjusted death rate | 165 |
| $\equiv$ | identical to | 112 |
| $j$ | often an index, especially if $i$ is applied to represent $\sqrt{-1}$ | 109 |
| $j'$ | index prior to ranking | 220 |
| $(j)$ | rank | 220 |
| K | often a constant | 24 |
| K | fractional power denominator parameter | 218 |
| K | threshold asymmetry parameter | 313 |
| $K(\mu,\sigma)$ | truncation constant of $\{I_{[A,B]}(x)\}\{\phi([y-\mu]/\sigma)\}/\{\sigma K(\mu,\sigma)\}$ | 226 |
| $k$ | often a constant or an index | |
| $L_1$ | absolute value metric | 191 |
| $L_2$ | square-based metric | 191 |
| $\log(o_j/e_j)$ | log ratio of $j^{th}$ observed count to $j^{th}$ expected count | 175 |
| $\log[P/(1-P)]$ | logit model | 133 |
| $\lambda$ | Poisson parameter (this Greek letter is chosen instead of $\mu$ to underscore that $\lambda$ is *not* a location parameter) | 109 |
| MA | method of averages | 186 |
| MEDIAN | median as a curve property | 27 |
| MEDIAN | median as percentile | 222 |
| MEDIAN$(Y\|x,u,v)$ | median-based regression curve | 83 |
| ML | maximum likelihood | 191 |
| $m(1\|x)$ | odds of dying curve | 160 |
| $\max\|\hat{F}(x)-F(x)\|$ | Kolmogorov-Smirnov criterion | 58 |
| $\mu$ or $\mu$ | the usual symbol for a location parameter | 335 |
| $\hat{\mu}$ | location parameter estimator | 84 |
| $\mu(x)$ | the curve traced by the location parameter as $x$ changes | 93 |
| $\mu(x)$ | location parameter-based regression curve | 93 |

Appendix III. General Statistical and Mathematical Notation 361

| | | |
|---|---|---|
| $\mu(x_0)$ | the location parameter at the value $x_0$ | 93 |
| $\mu^{(x)}$ | $X$ variate's location parameter | 114 |
| $\mu + \sigma Z$ | general simulated variate based on standard variate $Z$ | 92 |
| $\mu \pm z\sigma$ | critical points | 303 |
| $\mu \pm z\sigma$ | common table margin that designates either or both, interval from $\mu - z\sigma$ to $\mu + z\sigma$ or the places where a curve symmetric about $\mu$ is evaluated | 303 |
| $\mu_0$ | null hypothesized value of location parameter $\mu$ | 281 |
| $\mu_A$ | alternative hypothesized value of location parameter $\mu$ | 289 |
| $\mu_A$ | true parameter value | 289 |
| $\mu_A \neq \mu_0$ | usual situation where null and alternative locations differ | 289 |
| $\mu_k$ | $\int_{-\infty}^{\infty}(x-\mu_1)^k f(x)dx$, $k^{th}$ moment about $\mu_1$ | 43 |
| $\mu_x$ | same as $\mu(x)$ | 93 |
| $\mu_x = \alpha + \beta x$ | simple linear regression line | 93 |
| $\mu_{x,z} = \alpha_z + \beta_z x$ | regression where parameters are related to variable $z$ | 113 |
| $\mu_{Y\|x}$ | $Y$'s density's parameter $\mu$ as a function of variable $x$ | 82 |
| $\mu. = (\sum_{j=1}^{m}\mu_j)/c$ | unweighted average of population means | 241 |
| $N(c_1, c_2, c_3)$ | trivarite joint normal density | 182 |
| $N(c_1, c_2\|c_3)$ | bivariate conditional normal density | 183 |
| $N(y; \mu_x, \sigma)$ | $Y$ is homoscedastic normal with a location parameter that is a function of the variable $x$ | 125 |
| $n$ | usual symbol for sample size | 36 |
| $n$ | degree of abridgement | 162 |
| $n(1)$ | sample size of $1^{st}$ group | 247 |
| $n_{low} = (n+3)/4$ | index of lower box hinge | 224 |
| $n_{high} = (3n+1)/4$ | index of upper box hinge | 224 |
| $_nD_x/_nP_x$ | observed death rate for ages $x$ to $x+n$ | 162 |
| $_nm_x$ | mortality rate for ages $x$ to $x+n$ | 162 |
| $_nP_x$ | population with ages $x$ to $x+n$ | 162 |
| $n^{-1}\sum_{j=1}^{n}x_j$ | sample mean | 110 |
| $(n-1)^{-1}\sum_{j=1}^{n}(x_j-\bar{x})^2$ | sample variance | 114 |
| $n^{-1}\sum_{j=1}^{n}(x_j-\bar{x})(y_j-\bar{y})$ | sample covariance | 115 |

# Appendix III. General Statistical and Mathematical Notation

| Symbol | Description | Page |
|---|---|---|
| $\nu$ | degrees of freedom | 249 |
| $o_j$ | $j^{th}$ observed number of cases (observed count) | 175 |
| $\mathcal{O}$ | ordered variate | 199 |
| $\omega_1, \omega_2 ... \omega_q$ | parameters that specify curve identity | 124 |
| $P$ | probablility as a parameter | 36 |
| $P^{(date)}$ | population at specific date | 331 |
| $P_s$ | population of the $s^{th}$ group | 72 |
| $p$-value | significance level (minimum $\alpha$ that suggests rejection) | 285 |
| $p$ | probability value | 97 |
| $p$ or $P$ | mixing parameter | 100 |
| pdf | probability density function (frequency curve) | 18 |
| $\hat{p}$ | proportion | 97 |
| $p$ | often the index of the last parameter | 127 |
| $p_k$ | usual symbol for $k^{th}$ mixing parameter | 98 |
| $\pi$ | 3.14159265....... | 100 |
| $\pi_k$ | occasional symbol for $k^{th}$ mixing parameter | 100 |
| $Q - Q$ | curve applied to check fit and for many other purposes | 85 |
| $q$ | alternative index of the last parameter | 125 |
| $\theta$ | generic parameter | 123 |
| $\theta_1, \theta_2 ... \theta_p$ | $p$ generic parameters that help describe a single density $f$ | 123 |
| $\theta_x = h(x; \omega_1, \omega_2 ... \omega_q)$ | the parameter sequence $\omega_1, \omega_2 ... \omega_q$ helps specify the function $h$ of variable $x$ that determines the value of the parameter $\theta$ | 123 |
| $\Theta = [\sqrt{\beta_1}(\beta_2 + 3)] / [2(5\beta_2 - 6\beta_1 - 9)]$ | archaic measure of skewness | 233 |
| Random | pseudo random number Mathematica | 95 |
| $r$ | usual symbol for Pearsonian correlation | 115 |
| $r$ | number of rows | 250 |
| rand | pseudo random number MATLAB & C$^{++}$ | 95 |
| rnd | pseudo random number Mathcad | 97 |
| $r_s = 1 - 6(\text{SDS})/(n[n^2 - 1])$ | rank correlation in terms of sum of squared rank differences | 115 |
| runif | pseudo random number $S^+$ | 95 |

# Appendix III. General Statistical and Mathematical Notation

| | | |
|---|---|---|
| $\rho$ | association or correlation parameter (or curve property) | 105 |
| $\rho_{12}$ | same as above for first and second variates | 105 |
| $\rho_{12\bullet3}$ | partial correlation of the $1^{st}$ and $2^{nd}$ given the $3^{rd}$ variate | 166 |
| $\rho_{12\bullet3}$ | $(\rho_{12} - \rho_{13}\rho_{23})/\{\sqrt{1-\rho_{13}^2}\sqrt{1-\rho_{23}^2}\}$ | 166 |
| $\rho_{12\bullet3}\sqrt{1-\rho_{13}^2}/\sqrt{1-\rho_{23}^2}$ | regression slope parameter's usual value (dual variable case) | 166 |
| $\rho\sigma^{(x)}\sigma^{(y)}$ | covariance | 115 |
| SDS | sum of squared rank differences | 115 |
| Sign$[x]$ | sign function | 98 |
| SS$_m$ | mean square for (or between) cells (or categories) | 240 |
| $S(x)$ | survival curve | 161 |
| $s$ | common symbol for the sample standard deviation | 269 |
| slash, $\vert$ | usual symbol for the word *given* | 82 |
| $s^{(n)}$ | standard deviation computed from size n sample | 281 |
| $s^{(x)}$ | sample standard deviation of variate X, when X = x | 110 |
| $s^2$ | sample variance | 114 |
| $s_m^2$ | mean square for cells | 242 |
| $s_p^2$ | pooled estimate of variance | 238 |
| $\sqrt{Var(X)}$ | population standard deviation | 44 |
| $\sum_{j=1}^{n} a_j X_{(j)}$ | linear estimator based on order statistics | 264 |
| $\sum_{j=1}^{c} p_j \phi(y;\mu_j,\sigma)$ | c-component normal mixture density | 99 |
| $\sum_{j=1}^{c} p_j \mu_j$ | often the c-component mixture population mean | 100 |
| $\sum_{i=1}^{m} (\bar{y}_i - \bar{y}.)^2$ | mean square for (or between) cells (or categories) | 240 |
| $\{\sum_{i=1}^{m} n(i)(\bar{y}_i - \bar{y}.)^2\}/(m-1)$ | mean square for cells | 240 |
| $\sigma$ | scale parameter | 74 |
| $\sigma^{(n)} = \tilde{\sigma}/\sqrt{n}$ | mean's standard deviation | 278 |
| $\sigma^{(x)}$ | scale parameter of model for X variate's marginal density | 114 |
| $\sigma(x)$ | scale parameter as a function of variable x | 87 |

# Appendix III. General Statistical and Mathematical Notation

| Symbol | Description | Page |
|---|---|---|
| $\sigma^{(y)}$ | scale parameter of model for Y variate's marginal density | 114 |
| $\sigma^2$ | the squared scale parameter that often equals the curve property Var | 87 |
| $\sigma^2_{col}$ | column variance | 247 |
| $\sigma^2_{row}$ | row variance | 247 |
| $\tilde{\sigma}$ | population standard deviation that may, or may not, be numerically equal to a scale parameter's value | 281 |
| $\tilde{\sigma}$ | the tilda, $\tilde{\ }$, indicates that $\tilde{\sigma}$ is not necessarily a scale parameter | 278 |
| $t$ | common symbol for time variate value or Student t statistic | 252 |
| $t_o$ | date of study completion | 258 |
| $\tau$ | threshold parameter | 149 |
| $\tau_1$ | lower valued of two threshold parameters | 312 |
| $\tau_u$ | higher (upper) valued of two threshold parameters | 312 |
| $\tau'$ | truncation parameter of an indicator function like $I^{[\tau',\infty)}(y)$ | 149 |
| $\tilde{\ }$ | tilda symbol | 278 |
| $U$ | variate that is uniformly distributed on the unit interval | 105 |
| $\mathcal{U}(x)$ | Cramer's nonstandard uniform density | 298 |
| $\text{Var}[h(X)]$ | variance of transformed variate $h(X)$ | 345 |
| $\text{Var}(X; u, x)$ | partial integral that defines the population variance | 273 |
| $\text{Var}([X+Y]/2)$ | variance of the composite variate $[X+Y]/2$ | 273 |
| $\text{Var}(\bar{x})$ | variance of $\bar{x}$ | 273 |
| $\text{Var}(Y\|x)$ | variance of variate Y given variable $x$ | 85 |
| var | older symbol for variance | 273 |
| $v(y)$ | variate transformation function | 339 |
| $X$ | common symbol for a variate | 27 |
| $X**2$ | common computer notation for X squared | 330 |
| $X_{([n+1]/2)}$ | median from n = odd sized sample | 261 |
| $[X_{(n/2)} + X_{(1+n/2)}]/2$ | median from n = even sized sample | 261 |
| $X_1 = \rho_1 X_{nui} + \sqrt{1-\rho_1^2} Z_2$ | key variate with nuisance component $X_{nui}$ | 144 |
| $X_j$ and $X_{(j)}$ | unordered and ordered variates | 220 |

# Appendix III. General Statistical and Mathematical Notation

| | | |
|---|---|---|
| $\{X_j\}; j = 1, ..., n$ | size $n$ sample | 220 |
| $X = \mu + \sigma\Phi^{-1}(Z)$ | general normal variate in terms of standard uniform variate | 51 |
| $0.5(X_{(1)} + X_{(n)})$ | sample midrange | 261 |
| $X_{(n_{\text{low}})}$ | lower box hinge | 225 |
| $X_{(n_{\text{high}})}$ | upper box hinge | 225 |
| $x$ | common symbol for a running variable or a variate's value | 27 |
| $x_1 \neq x_2$ | distinct points on $x$ axis | 81 |
| $x_0$ | fixed value of variable $x$ | 256 |
| $\bar{x}$ | common symbol sample mean | 114 |
| $\bar{x}^{(n)}$ | sample mean of a size $n$ sample | 276 |
| $(\bar{x} - \mu_0)/\sigma^{(n)}$ | sample mean test statistic ($z$ statistic) | 283 |
| $(x_1, \hat{y}_1)$ and $(x_2, \hat{y}_2)$ | points used by method of averages | 156 |
| $Y$ | common symbol for a response variate | 130 |
| $Y = \mu + \sigma\Phi^{-1}(Z)$ | standard Normal variate generation model | 74 |
| $\text{YMODE}_f(Y\|x)$ | mode-based regression curve | 85 |
| $Y'$ | transformed variate such as $\log(Y - \tau)$ | 149 |
| $Y_1, Y_2, ..., Y_u$ | sequence consisting of u distinct variates | 123 |
| $Y_{(j)} = \mu + \sigma Z_{(j)}$ | ordered observation in terms of ordered standard | 262 |
| $Y\|x \in N(y; \sigma\|x, \beta)$ | $Y$ given $x$ is homoscedastic normal with a location parameter that is associated with $x$ as parametrized by slope parameter $\beta$ | 112 |
| $y$ | common symbol for a variate's value | 130 |
| $\bar{y}_1 - \bar{y}_2$ | cell mean difference | 238 |
| $\bar{y}_\bullet$ | $(1/n)\sum_{i=1}^{m} n(i)\bar{y}_i$ weighted cell mean average | 240 |
| $\bar{y}_1, \bar{y}_2, ..., \bar{y}_m$ | sequence of m cell means | 240 |
| $Z$ | usual symbol for a standard variate (especially one that has density $g(z), \phi(z)$ or $I_{[0,1)}(z)$) | 19 |
| $Z \in f$ | the variate $Z$ is distributed with density $f$ | 105 |
| $\{Z_{(j)}\}; j = 1, .., n$ | ordered standard variates | 262 |
| $Z = \frac{\bar{x} - E(X)}{\sigma^{(n)}}$ | standardized sample mean test statistic | 281 |
| $z = f(y\|x)$ | conditional density evaluation function | 82 |
| $z_0$ | critical point | 293 |

# References

*The American Heritage Dictionary of the English Language,* Third Edition, Houghton Mifflin Company, Boston & New York, (1992).

Anderson, T.W. (1971), *An Introduction to Multivariate Statistical Analysis, Second Edition,* Wiley, New York.

Bartlett, J. (1948), *Familiar Quotations, Twelfth Edition,* Little, Brown & Company, Boston.

Bell, E.T. (1934), *The Search for Truth,* Reynal & Hitchcock, New York.

Bell, E.T. (1937), *Men of Mathematics,* Simon & Schuster, New York.

Bellman, R.E. (1961), *Adaptive Control Processes,* Princeton University Press, Princeton, New Jersey.

Benjamin, B. (1959), *Elements of Vital Statistics,* Quadrangle Books, Chicago.

Bennett, C.A., Franklin, N.L. (1954), *Statistical Analysis in Chemistry and the Chemical Industry,* Wiley, New York.

Bergamini, D. (1963), *Life Sciences Library-Mathematics,* Time Inc., New York.

Berkson, J. (1951), Why I prefer logits to probits, *Biometrics,* **7**:327- 339.

Blackwell, D. and Girshick M.A. (1954), *The Theory of Games and Statistical Decisions,* Wiley, New York.

Blom, G. (1962), Nearly best linear estimates of location and scale parameters" (In Sarhan, A. E. and Greenberg, B. G.), *Contributions to Order Statistics,* Wiley, New York.

Bowley, A.L. (1901), *Elements of Statistics*, Staples Press, New York.

Box, G.E.P., and Muller, M.E. (1958), A note on the generation of normal deviates, *Annals of Mathematical Statistics,* **28**:610-611.

Carroll, L. (1872), *Through the Looking-Glass,* Chapter 6.

Chambers, J.M., and Hastie, T.J. (1992), *Statistical Models in S.* Wadsworth & Brooks/Cole Advanced Books & Software, Pacific Grove, California.

Chambers, J.M.; Cleveland, W.S., Kleiner, B.; and Tukey, P.A. (1983), *Graphical Methods for Data Analysis,* Belmont, California.

Cleveland, W.S. (1985), *The Elements of Graphing Data,* Wadsworth, Pacific Grove, California.

Cochran, W.G. (1968), Errors of measurement in statistics, *Technometrics,* **10**:637-666.

Cohen, A.C. (1951), Estimating parameters of logarithmic-normal distributions by maximum likelihood, *Journal of the American Statistical Association,* **46**:206-212.

*Compton's Interactive Encyclopedia.* Compton's NewMedia & Merriam-Webster, Springfield, Massachusetts, (1993, 1994).

Cramer, H. (1945), *Mathematical Methods of Statistics,* Princeton University Press, Princeton, New Jersey.

Crow, E.L. and Shimizo. K. (1988), *Lognormal distributions,* Dekker, New York.

Cullen, M.J. (1975), *The Statistical Movement in Early Victorian Britain, The Foundations of Empirical Social Research,* Barnes & Noble Books, New York.

David, H.A. (1970), *Order Statistics,* Wiley, New York.

Devroye, L. and Gyorfi, L. (1985), *Nonparametric Density Estimation, the L1 View,* Wiley, New York.

Dickerson, P. and Johnson, P. (1999), International Programs Center, U.S. Bureau of the Census, Washington, D.C. 20233-8860, http://www.ac.wwu.edu/ stephan/Animation/pyramid.html.

Dixon, W.J. and Massey, F.J. (1983), *Introduction to Statistical Analysis, Fourth Edition*, McGraw Hill, New York.

Elderton, W.P. (1953), *Frequency Curves and Correlation*, Harren Press, Washington D. C.

*Encyclopedia Britannica, Vol. XXII, Poll to Reeves*, $11^{th}$ Edition, University Press, Cambridge (1911).

Finkel, A.M. (1990), *Confronting Uncertainty in Risk Management*, Center for Risk Management Resources for the Future, 1616 P Street, N.W. Washington, D.C. 20036

Fisher, R.A. (1920), A mathematical examination of the methods of determining the accuracy of an observation by the mean error, and by the mean square error, *Monthly Notices of the Royal Astronomical Society*, **53**:758-770.

Fisher, R.A. (1922), On the mathematical foundations of theoretical statistics, *Philosophical Transactions of the Royal Society of London, Series A*, **222**:309-368.

Fisher, R.A. (1936), The use of multiple measurements in taxonomic problems, *Annals of Eugenics*, **7**:179-188.

Fisz, M., (1963), *Probability Theory and Mathematical Statistics*, Wiley, New York.

Fox, M., (1956), Charts of the power of the F-test, *The Annals of Mathematical Statistics*, **27**:484-497.

Good, I.J. and Gaskins, R.A. (1980), Density estimation and bump-hunting by the penalized likelihood method exemplified by scattering and meteorite data, *Journal of the American Statistical Association*, **75**:42-73.

Greenwood, J. A., Hartley, H. O. (1962), *Guide to Tables in Mathematical Statistics*, Princeton University Press, Princeton, New Jersey.

Grizzle, J.E. (1967), Continuity correction in the ?2-test for 2x2 tables, *The American Statistician*, **21**:28-32.

Gross, A.J. (1975), Clark, V., *Survival Distributions: Reliability Applications in the Biomedical Sciences*, Wiley, New York.

Gumbel, E.J. (1961), Bivariate logistic distributions, *Journal of the American Statistical Association*, **56**:335-349.

Hald, A. (1952), *Statistical Theory with Engineering Applications*, Wiley, New York.

Heath, C.W. (1946), *What People are: A Study of Normal Young Men*, Harvard University Press, Cambridge.

Hoel, P.G. (1954), *Introduction to Mathematical Statistics*, Wiley, New York.

James, G. and James, R.C. (1959), *Mathematics Dictionary*, Van Nostrand, Princeton, New Jersey.

Jansson, B. (1966), *Random Number Generators*, Victor Pettersons Bokindustri Akiebolag, Stockholm.

Johnson, N.L. and Kotz, S. (1970), *Distributions in Statistics*, Houghton Mifflin, Boston.

Kapteyn, J.C. (1903). *Skew Frequency Curves in Biology and Statistics*, Astonomical Laboratory, Groningen: Hotsema Brothers, Inc.

Kendall, M.G. and Stuart, A. (1979), *The Advanced Theory of Statistics, Volume 2, Inference and Relationship*, Macmillan, New York.

Kendall, M.G., Buckland, W.R. (1960), *A Dictionary of Statistical Terms*, Hafner, New York.

Kotz, S. (1996), *Encyclopedia of Statistical Science*, Wiley, New York.

Kowalski, C.J. (1970), The performance of some rough tests for bivariate normality before and after coordinate transformations to normality, *Technometrices*, **12**:517-544.

Kowalski, C. and Tarter, M.E. (1969), Co-ordinate transformations to normality and the power of normal tests of independence, *Biometrika*, **56**:139-148.

Kronmal, R.A. and Tarter, M.E. (1968), The estimation of probability densities and cumulatives by Fourier series methods, *Journal of the American Statistical Association*, **63**:925-952.

Kurtz, A.K., and Edgerton, H.A.(1939), *Statistical Dictionary of Terms and Symbols*, Wiley & Chapman and Hall, New York and London.

Lake, F. (1901), *Roulette at Monte Carlo; 5,356 Actual Monte Carlo Numbers Classified & Arranged - Rules of the Game - 20 Systems of the Game*, self published.

Lexis, W.H.R.A. (1877), *Zur Theorie der Massenercheinigungen in der menschlicken Gesellschaft*, Freiburg.

Lipka, J. (1918), *Graphical and Mechanical Computation*, Wiley, New York.

Lloyd, E., H. (1952), Least-squares estimation of location and scale parameters using order statistics, *Biometrika*, **39**:88-95.

Mann, N. (1968), Point and interval estimation procedures for the two-parameter Weibull and Extreme-Value distributions, *Technometrics*, **10**:231-256.

Mardia, K.V., Kent, J.T. and Bibby, J.M. (1979), *Multivariate Analysis*, London, Academic Press.

Marriott, F.H.C. (1960), *A Dictionary of Statistical Terms*, $5^{th}$ Edition, Longman Scientific & Technical, New York.

*Mathematics,* Life Science Library (1963), Eds. Bergamini, D. et al., Time Inc., New York, 144-5.

Neyman, J. (1949), Theory of $\chi 2$ Tests, In *Proceedings of the Berkeley Symposium on Mathematical Statistics and Probability* (Ed. Neyman, J.), University of California Press, Berkeley: 239-273.

Pearson, K. (1894), Contributions to the mathematical theory of evolution, *Philosophical Transactions of the Royal Society*, **185A**:71-110.

Pearson, K. (1978), *The History of Statistics in the 17th and 18th Centuries — against the changing background of intellectual, scientific and religious thought* (Edited by E. S. Pearson), Griffen, London.

Pearson, E.S. and Hartley, H. O. (1956), *Biometrika Tables for Statisticians, Volume 1*, Cambridge University Press, Cambridge.

Plato (360 BC), *The Republic,* Translated by Benjamin Jowett , Library of the Future Series First Edition Ver. 3.01 Electronically Enhanced Text 1991, World Library, Inc.

Quandt, R.E. (1958), The estimation of the parameters of a linear regression system obeying two separate regimes, *Journal of the American Statistical Association,* **51**:873-880

Quandt, R.E. (1972), New approach to estimating switching regressions, *Journal of the American Statistical Association,* **67**:306-310.

Quandt, R.E. and Ramsey, J.B. (1978), Estimating mixtures of normal distributions and switching regressions. *Journal of the American Statistical Association,* **73**:730-751.

Rosenlicht, M. (1975), Differential extension fields of exponential type, *Pacific Journal of Mathematics,* **57**:289-300.

Sarhan, A.E. and Greenberg, B.G. (1962), *Contributions to Order Statistics,* Wiley, New York.

Sarndal, C.E. (1971), Studies in the History of Probability and Statistics. XXVII. - The hypothesis of elementary errors and the Scandinavian School in statistical theory, *Biometrika,* **58**:375-391.

Savage, I.R. (1968), *Statistics: Uncertainty and Behavior,* Houghton Mifflin Company, Boston & New York.

Savage, L.J. (1954), *The Foundations of Statistics.*

Scheffé H. (1959), *The Analysis of Variance,* Wiley, New York.

Scott, D.W. (1992), *Multivariate Density Estimation: Theory, Practice, and Visualization,* Wiley, New York.

Shryrock, H.S., Siegel, J.S., Bayo, F., Davidson, M., Demeny, P., Glick, P.C., Grabill, W.H., Grove, R.D., Israel, R.A., Jaffe, A.J., Kindermann, C.R., Larmon, E.A., Nam, C.B. (1973 ), *The Methods and Materials of Demography,* U.S. Department of Commerce, Washington D.C.

Spanier, J. and Oldham, K.B. (1987), *An Atlas of Functions,* Hemisphere Publishing Corporation, New York.

*SPSS for WINDOWS* — *,Release 8.0,* Chicago, (1998).

Stein, C. (1945), A two-sample test for a linear hypothesis whose power is independent of the variance, *Annals of Mathematical Statistics,* **16**:243-58.

*SYSTAT for WINDOWS* — *,Release 5.05,* Chicago, (1994).

Tapia, R.A. and Thompson, J.R. (1978), *Nonparametric Probability Density Estimation,* Johns Hopkins University Press, Baltimore.

Tarter, M.E. (1998), Fourier Series Curve Estimation and Antismoothing, *Encyclopedia of Statistical Sciences, Update Volume 2* (Kotz, S. ed.), Wiley, New York, 263-269.

Tarter, M.E. and Hill, M.D. (1997), On thresholds and environmental curve tensiometers, *Ecological and Environmental Statistics*, **4**:283-299.

Tarter, M.E. and Lock, M.D. (1991), Model-free curve estimation: mutuality and disparity of approaches, *Journal of Official Statistics*, **17**:58-73.

Tarter, M.E. and Lock, M.D. (1993), *Model-Free Curve Estimation*, Chapman and Hall, New York.

Tarter, M.E. and Silvers, A. (1975), Implementation and Applications of Bivariate Gaussian Mixture Decomposition, *Journal of the American Statistical Association*, **70**:47-45.

Tarter, M.E. Cooper, R.C. and Freeman, W. (1983), A graphical analysis of the interrelationship among waterborne asbestos, digestive system cancer and population density, *Environmental Health Perspectives*, **53**:79-89.

Tarter, M.E., Freeman, W. and Polissar, L. (1990), Applications of modular nonparametric subsurvival estimation, *Journal of the American Statistical Association*, **85**:29-37.

Tarter, M.E., Lock, M.D., Ray, R.M., (1995), Conditional switching: a new variety of regression with many potential environmental applications, *Environmental Health Perspectives*, **103**:2-10.

Taylor, A.E., (1955), *Advanced Calculus*, Ginn and Company, New York.

Thorndike, E.L. (1914), *Mental Work and Fatigue and Individual Differences and Their Causes*, Teachers College, Columbia University, New York.

Tocher, K.D. (1963), *The Art of Simulation*, Van Nostrand, Princeton, New Jersey.

Todhunter, I. (1865), *A History of the Mathematical Theory of Probability*, (republished in 1948 by Chelsea, New York).

Tukey, J. (1958), The log of a Cauchy density is not convex, *The Annals Math. Statist.*, **29**:591.

Tukey, J. (1965), The technical tools of statistics, *The American Statistician*, **19**:23-36.

Tukey, J (1977), *Exploratory Data Analysis*, Addison-Wesley, Reading, Massachusetts.

Wald, A. (1939), *Statistical Decision Functions*, Wiley, New York.

Whitworth, W.A. (1959), *Choice and Chance*, Hafner, New York.

Wilks, S.S. (1962), *Mathematical Statistics*, Wiley, New York.

# Index

acceptance, 282, 297, 305
Achilles, 155
additive linear model, 129, 132, 243
adjustment procedures, 142, 159, 162, 167, 168
age adjustment, 142, 147, 175, 249
age distribution, 153, 161, 166, 170
age-adjusted death rate, direct, 160, 161, 167, 168
age-adjusted death rate, indirect, 168, 170, 171, 173
algebraic manipulation, 192, 242
algorithms, 76, 91, 156, 270, 286
alpha level, 284, 285, 287
alphabet, 6, 125, 200, 243, 329, 330
alternative hypothesis, 56, 289
analysis of covariance, 104, 127, 128, 138, 145, 200, 244, 249, 278, 282, 319
analysis of variance, 90, 104, 105, 127, 128, 138, 145–147, 200, 219, 237, 238, 243, 245–249, 278, 282, 285, 288, 295, 309, 319
Anderson, T.W., 166
ANOVA designs, 146
ANOVA, see analysis of variance
ants, 126
approximately unbiased estimates, 319
arithmetic averages, also see mean, 156, 313
Asgard, 337
association parameter, 105–109, 113, 115, 116, 118, 120

assumption,
  bivariate normality, 119
  homoscedasticity, 87, 94, 123, 156, 238
asymmetry, 42, 52, 135, 138, 315, 317, 319, 320
asymmetry, truncation induction, 315
asymptotic efficiency, 303
atlas of functions, 7
attribute-valued variate, 75, 189, 196, 197, 199, 218, 230, 244, 249, 250
autopsy, 322
average
  arithmetic, 58
  weighted, 100, 241, 275
averaged kernels, 62
averages, method of, 155–157, 185

baby booms, 153
balanced design, 247, 248
balancing point, 18, 317, 319
bands, confidence, 80, 217, 228, 229, 288, 289
bands, prediction, 217, 228, 229, 279
bathtub function, 38, 153–155, 339
Bayesian statistics, 342–344
Belgian, 150
bell-shaped curve, 13, 38, 152, 337
Bellman, R.E., 123
benchmark, 50
Benjamin, 226
Berkson, J., 93

375

Bernoulli trials, 66
Bernoulli variation, 65, 66
best linear estimation (BLU), 255, 256, 261–271, 274, 278, 282, 291, 306–309
bias, 123, 317–320
Bibby, J.M., 151
Bills of Mortality, 38
binomial, also see density, binomial 36–38, 58, 59, 66
binomial index of dispersion, 58–60
bins, 78, 348
biology, 37, 334
bivariate Normality, 69, 119
Blom, G., 267, 277
borders, 42, 189, 213, 220, 316, 340, 349
boundaries, 1, 37, 46, 196, 228
Bowley, A.L., 329
box plot, 80, 190, 217, 223–225, 227, 229, 230, 232
box plot hinge, 10, 232
box sizes, 226
Box, G.E.P., 106
box-and-whisker plot, 222–224, 226–228, 230
Buckland, W.R., 10, 11, 42, 47, 145, 157, 222, 227, 245, 259, 312, 319, 320, 347

cancer, 175
cardinal number, 331
cards, 22, 312, 334
Carrol, L., 5
cases, expected, 175
casewise data, 196
Cassandra, 295
catenary, equation for, 142
catenary, history of, 18
Cauchy model, 54, 65, 68, 195, 196, 281, 319, 321, 337, 339, 341
Cauchy standard deviation, 54
Cauchy variate, 19, 20, 66, 74, 108, 110, 111, 118, 195, 196, 212, 222, 223, 235, 239, 243, 251, 252, 257, 259, 260, 281, 296, 300, 301, 312, 313, 315, 316, 319, 320, 322, 333, 337, 341
causation, 250
ceil function, 98
cells, in ANOVA, 238, 239, 241, 242, 247
censoring, general, 225, 256, 257
censoring, Type I, 226
censoring, Type II, 226
Central Limit Theorem (CLT), 68, 69, 100, 238, 300, 301, 311, 317–319, 326, 327
central tendency, 15
chain, 18
Chambers, J.M., 135, 224
chi-square ($X^2$) distribution, 56, 58, 60, 61, 63, 189, 250
Chui, C.K., 209
Cleveland, W.S., 224
Cochran's Theorem, 74, 248
Cochran, W.G., 246
coefficient, 143, 144, 340, 341
Cohen, A.C., 37
collection, 17, 312
combinations, 23, 150, 151, 312, 340, 345
comet, 142
commercial programs, 190
complete factorial, 244, 257, 258
composite function, 218
composite parameter, 238, 239, 242, 293, 294
composite variates, 35, 252, 273–275, 279, 280, 283, 307
computationally intensive, 196
computer, 9, 38, 146, 150, 330, 331, 334
condition, homoscedasticity, 94, 123, 156, 238
condition density, see density, conditional
confidence bands, 80, 217, 228, 229, 288, 289
confidence intervals, 288

confounding, 141, 145, 146, 282
constants, 37, 149, 157, 318, 323, 325, 326, 330, 344
contagious distribution, 343
contagious models, 101, 344
contamination, 145, 322
contingency tables, 60, 164, 249, 250
contour shape, 213
control, 64, 126, 127, 146, 177, 248, 249, 278, 293, 309, 320
Cooper, R.C., 176
correlation
    partial, 127, 177, 247–249
    Pearson, 329
    rank, 115
    tetrachoric, 115, 116
counts, 70, 78, 250
covariance (Cov), 232, 262, 267–269
Cox models, 258
Cramer, H., 10, 277, 278, 290, 300, 301
critical point, 294–298, 304, 307
critical region, 228, 288, 294, 296–298, 304, 305
Cullen, M.J., 37
curse of dimensionality, 152, 153
curve argument, 46, 149, 154, 315, 319, 335, 337–339, 345
curve fitting, 77, 157, 256
curve, general features 9–23, 25–31, 33–38, 43–45, 142, 149–157, 312–315, 317–319, 320–325, 330–343, 345, 346
curve properites
    kurtosis, 42, 45, 46, 50–54, 60, 86, 205, 208, 219, 232, 235
    skewness, 42, 45, 50, 136, 205, 208, 219, 232, 233, 235
    symmetry, 50, 54, 313
curve shapes
    bathtub, 153–155, 339
    bell-shaped, 13, 38, 152, 340
    heavy tailed, 235, 313, 337
    Laplace pointed, 51
    tapering over tails, 315
    truncated, 51
curve tails, 52–54, 155, 311, 313
curve types, 11, 14
    definition of, 6, 7, 13
    density, 12, 41, 149, 155, 323, 325, 335, 336, 339, 343
    dual regression, 136
    Laplace, 337, 341
    limiting, 229
    loglogistic, 339
    marginal, 350
    normal, translated, 335
    polynomial, 340, 341
    posterior, 344
    prior, 17
    probit, 91
    standard, 85, 204, 262, 265, 268, 270, 332, 335
    switching regression, 84
    translated, 334, 335
    truncated, 149, 311
curves prior, 344
curves, degenerate, 27, 251, 304
curves, demographic, 315

data, 9, 12, 14, 16, 17, 21, 23, 34, 35, 39, 142, 143, 146, 148, 151, 156, 157, 312–314, 319, 320, 331, 341, 343, 344
data analysis, 12, 14, 23, 34, 151, 319, 341
data availability, 34, 191
data files, 198
data generation models, 143, 146, 331
data type, 31, 36, 37
data type
    attribute, 9, 13, 14, 16, 42, 44, 146, 148–151, 153, 157, 199, 316, 317, 325, 337, 340
    continuous, 18, 22, 23, 28, 43, 44, 199, 217, 312, 320, 340
    discrete, 43, 44, 199
    ordered, 199
    pathological, 151

David, H.A., 225
de Moivre's normal, 47
decision criterion, 191
default setting, 259, 331
degenerate curves, 251
degree of freedom, 19, 68, 70, 195
demographic curves, 315
demography, 153, 154, 313
density, *see also* curves, desity, 12, 28, 31, 44, 156, 314, 315, 325
   binomial, 36–38
   Cauchy, 43, 50, 51, 53, 54, 65, 68, 70, 71, 118, 195, 196, 235, 251, 252, 312, 313, 315, 316, 319, 320, 333, 337, 341
   joint, 63, 69, 111, 116, 117, 124, 126, 127, 163, 164, 169, 198, 199, 213, 222, 255, 349
   Laplace, 50, 51, 53, 337, 341
   logistic, 50–53, 108, 133, 222
   loglog Cauchy, 195
   non-central-t, 68
   normal, 19, 33, 41, 43, 46–54, 71, 101, 107, 108, 111, 112, 118, 124, 127, 135, 154, 193, 205, 215, 237, 270, 300, 314, 315, 318, 319, 333–336, 341
   Poisson, 55, 56, 61
   Student-t, 195, 251, 252
   Weibull, 338
dependent variates, 215, 267
depressions, 153
Descartes, 1, 12, 18, 330, 344
descriptive statistics, 34
deviations, 26–28, 30–37, 155, 156, 319, 320, 333–335
Devroye, L., 190
dice, 38, 312
dimensionality, 123, 132, 149, 152, 153, 220, 270
discontinuity, 48
discriminant analysis, 234
disease occurrences, 65
distribution
   limiting, 318

   general, 10, 11, 13, 14, 16, 19–23, 25–27, 30, 31, 33–38, 143, 148, 149, 155, 311, 312, 314, 316, 318, 319, 322, 326, 331, 335, 338, 340, 342, 343, 345
   uniform, 91, 101, 165, 221, 325, 348
distributions allied to the rectangular, 300
Dixon, W.J., 41, 42, 219, 307, 319, 351
drug use, 70

e raised to the z power, 345
economics, 37, 151, 334
Edgerton, H.A., 227
Edgeworth series, 193
education, 17
efficiency, discussion of, 295
Einstein and Mercury's perihelion, 259
Elderton, W.P., 207
elementary functions, *see* function, elementary
enumerable population, 313
equations, linear, 352
error of the first kind, 298
error of the second kind, 298
estimation phase, 192
estimator
   approximately unbiased, 319
   inconsistent, 318, 319
   linear, 268, 275
   of a parameter, 279, 351
   unbiased, 261
expected value, 315, 317, 345
experiments, 35, 145
exponent notation, 330
exponential density, 47
expression, 146, 150, 311, 312, 315, 334, 335, 341
extreme values, 225

factorial, complete, 244
factors, 36, 145
family, 340, 341

family of curves, 340, 341
fences, 227
Fermat, 36–38, 330, 344
Finkel, A.M., 344
Fisher, R.A., 55, 191–193, 204, 217, 234, 242, 251, 255, 271, 330
Fisz, M., 155, 255
fitted model, 18
flexibility, 340
floor function, 98
Fourier series, 58, 62, 192, 193, 208–211
Fourier series, unweighted, 192
fractional power transformation, 218
frail data-with-model pairings, 295
frailty, 151
France, 331
Freeman, W., 176, 308
freqentist methodology, 9
frequency, relative, 153
frequentist, 9, 312
frequentist approach, 312
frequentist methodology, 312
frequentist power, 293
function
    bathtub, 38, 152–155, 339
    ceil, 98
    elementary, 46, 92, 106, 117, 133, 191, 193, 196, 203, 209, 210, 246, 257, 262, 280, 314, 318
    floor, 98
    indicator, 46, 149
    Heaviside, 95, 96
    link, 135
    log, 149, 345
    sign, 95
    tails of, 155

Galton's lognormal, 38, 50
gambling strategy, 312
gambling tables, 110, 349
games of chance, 36, 38
Gaskins, R.A., 191
Gauss, 6, 47, 50, 138, 142, 155–157, 185, 336

Gauss-Markoff Theorem, 74, 155, 156, 255, 262, 263, 265
Girshick, M.A., 9
Good, I.J., 191
goodness of fit, 55–61, 63, 70–72, 191, 205
Gram-Charlier expansion, 340
graphics, 9, 38, 330
Greek letters, 12, 13, 35, 150, 337
Grizzle, J.E., 189
Gross, A.J., 161
Gumbel, E.J., 151
Gyorfi, L., 191

Hald, A., 136
half open intervals, 22, 28
Hastie, T.J., 135
Heath, C.W., 226
heavy tailed curve, 235, 313, 337
heteroscedasticity, 94
Hill, M.D., 234, 309, 342
hinge, 10
histogram, 14, 16, 17, 20, 21, 39, 58, 62, 63, 78, 204, 205, 208, 209, 280
history of statistics, 10, 18, 37, 332
HIV, 70, 258
Hoel, P.G., 60
homoscedasticity, 87, 94, 123, 156, 238
Humpty Dumpty, 5
hypotheses, null and alternative, 56, 57, 60, 238, 242, 245, 282
hypothesis test, 55, 60, 190, 228, 238, 283, 284, 287, 294

ideals, 19, 145, 151
identically distributed, 16, 65, 91, 100, 112, 126, 259, 276, 300, 352, 353
iid standard form variates, 300, 301, 312, 318, 319
imperfections, 145
implicit notation, 334
imputation, 260
incomplete measurement, 197, 258

inconsistent estimator, 318, 319
independence, 116, 347, 348, 354
independent standard normal variates, 38, 143, 144, 145
indicator function, 45, 46, 53, 73, 96, 149
indirect age-adjustment, 175, 249
individuals at risk, 44
infinite parameter space, 192
integral, partial, 336, 346
integral, probability, 47
integrated square error, 191, 310
interaction, 244–246
interdependence, 145, 152
interquartile range, 320
invariance, 243, 245, 246
inverse Normal scores, 221
inverse relationships, 128
Iris setosa; versicolor, 234
ISE, 57–60, 71
I-V drug use, 70

James, G., 47
James, R.C., 47
Jansson, B., 97
Johnson, 341
Johnson, N.L., 7, 47, 50
joint density, 63, 69, 111, 116, 117, 124, 126, 127, 163, 164, 169, 198, 199, 213, 222, 255, 349
joint Normality, 69

Kapteyn, J.C., 50
Kendall, M.G., 10, 42, 47, 145, 157, 181, 222, 227, 245, 259, 312, 318–320, 344, 347
Kent, J.T., 151
kernels, 58
key variates or variables, 18, 144, 145
Kleiner, B., 224
Kolmogoroff-Smirnoff, 58, 63–65
Kotz, S., 7, 10, 47, 50
Kowalski, C.J., 185, 222

kurtosis, 42, 45, 46, 50–54, 60, 86, 205, 208, 219, 232, 235
Kurtz, A.K., 227

Laplace, 336, 337, 339, 341
Laplace distributed, 48
Laplace transform, 47
Laplace variate, 48
larvae, 126
latus rectum, 12, 18
law of large numbers, 259
law of parsimony, 2
LD status, 163
least squares, 156, 157, 262, 263
Leibnitz's Calculus, 14
levels, 81
Lexis, W.H.R.A., 36
life tables, 38, 39, 142, 312, 324
light bulbs, 197, 257
likelihood, 57, 191, 192, 204, 205, 217, 234, 255
likelihood ratio (LR), 57
likelihood, penalized, 191–193
likelihood, unpenalized, 192
limited range of precision, 197
limiting curve, 229
limiting distribution, 318
linear equations, 352
linear estimators, 256, 260, 261, 264, 268
link functions, 135
Lipka, J., 157
lizards, 316
Lloyd, E.H., 264
location parameter, 10, 11, 15, 16, 23, 31, 41, 62, 69, 70, 74, 84–86, 91, 92, 94, 99–102, 105, 108, 113, 115, 118, 123–126, 128, 136, 139, 142, 143, 146–149 203–206, 213–215, 235, 237, 239, 240, 242, 243, 255, 259–261, 265, 267, 277–279, 284, 286, 290, 291, 293, 295, 298, 302, 305, 307, 308, 311, 313–315, 317–319, 321, 322,

*Index* 381

326, 330, 332, 334, 335, 337, 342, 344, 351
location parameter average, 239
Lock, M.D., 57, 63, 86, 121, 195, 229, 301
log function, 149, 345
log-extreme-value distribution, 338
logCauchy, 339
logistic model, 53, 93, 107, 133, 134, 159, 197, 222
logit function, 48, 134
lognormal model, 50, 339
Loki, 337
lower case letters, 345
lurking variates and variables, 37, 53–55, 72, 74, 138, 235, 342

Mann, N., 338
Mardia, K.V., 151
margin of the table, 284
Marriott, F.H.C., 10, 222, 225, 227
Massey, F.J., 41, 42, 219, 307, 319, 351
MATHCAD, 346
Mathematica, 346
mathematical induction, 278
mathematical statistics, 13, 22, 23, 38, 142, 156, 337, 345, 346
MATLAB, 346
mean square error, 156, 265
mean square for cells, 241
mean, population, 313, 318
measure of central tendency, 15
measurement range, 313
measurement, incomplete, 197, 258
method of averages, 155–157, 185
method of least squares, 156, 157, 262, 263
method of moments, 157
metric, dual valued, 282, 283
metrics, 9
middle age, 152
migrant field workers, 99
migration, 153
MISE, 62, 63, 191

mixing parameter, 72, 99–101, 146, 149, 208, 211, 212, 235, 321, 342, 344
mixture data, generation, 97
mixture models, 102
mixture of normals, 74
mode, 48, 84, 138, 274, 333–335, 340
mode-based regression, 138
model, 12, 13, 15–18, 19 20, 142–146, 148–152, 154, 315–319, 321, 330, 333, 334, 336–343
  Cauchy, 54, 65, 68, 195, 196, 257, 281, 314, 319, 321, 337, 339, 341
  Cauchy, folded, 212
  contagious, 344
  data generation, 143, 146, 331
  definition of, 76
  explicit, 76, 77
  exponential, 47
  fitted, 18
  general, 9, 31, 35–37
  Laplace, 73, 336, 337
  linear additive, 129, 132
  logCauchy, 297, 339
  logit, 132–136
  mixture, 74
  Poisson, 56, 59
  polynomial, 340, 341
  truncated, 311
  untruncated, 319
  Weibull, 6, 340
model-free, 76
moments
  higher moments, 41–43, 45, 50, 53, 205, 208
  lower, 157
  method of , 157
  ordinary, 10, 11, 41, 42, 45, 46, 51, 52, 208
Monte-Carlo methods, 91
MSE, 49, 265, 270
Muller, M.E., 106
multiple comparisons, 265
multivariate analysis, 145, 150–152, 154, 155

neonatal mortality, 152
Neyman, J., 55–61, 64, 71
noncentrality parameter, 237, 247, 298, 299, 309
nonparametric, 62, 63, 76, 84, 86
normal equation, 227, 264
normal range, 227
normal variate, standard, 19, 73, 105, 106, 112, 221, 249, 250, 252, 284, 291, 293, 304
normal, truncated, 51
normality, bivariate, 69, 119
normality, joint, 69
normalized score, 222, 246
notation implicit, 334
notation, implicit, 334
nuisance contamination, 145
nuisance parameters, 285
nuisance variate, 144, 145
null hypothesis, 56, 60, 238, 242, 245, 248, 252, 282, 286, 287, 289, 291, 294, 298, 306–308
null, or trial, guess, 283
number
  cardinal, 331
  of observations, 37
  of subsample observations, 232
  of trials, 329
  ordinal, 331
  prime, 330

observational studies, 35
Ockham's razor, 2
Oldham, K.B., 7
ordered indexing, 331
ordinal number, 331
ordinate-axis, 78, 83, 156, 176, 321
outliers, 320
outside value, 225, 227

painted image, 150
parabola, 7, 12
parameter
  association, 177, 179, 181, 182, 184, 204, 215, 222
  broad-brush, 113

composite, 239, 242, 252, 293, 294
compsite, 238
location, 10, 11, 15, 16, 23, 31, 41, 62, 69, 70, 74, 84–86, 91, 92, 94, 99–102, 105, 108, 113, 115, 118, 123–126, 128, 136, 139, 142, 143, 146–149 203–206, 213–215, 235, 237, 239, 240, 242, 243, 255, 259–261, 265, 267, 277–279, 284, 286, 290, 291, 293, 295, 298, 302, 305, 307, 308, 311, 313–315, 317–319, 321, 322, 326, 330, 332, 334, 335, 337, 342, 344, 351
mixing, 72, 149, 342, 344
noncentrality parameter, 237, 247, 298, 299, 309
nuisance, 148, 222, 251, 327
regression slope, 157
scale, 15–17, 20, 25, 28, 36, 37, 69, 70, 91, 94, 102, 105, 108, 112, 115, 117, 118, 120, 121, 124, 125, 128, 136, 137, 139, 146–149, 203, 205–207, 212–215, 217, 218, 222, 235, 238, 239, 245, 251, 255, 261, 265–270, 277, 278, 280, 281, 283–285, 289, 291, 295, 302, 306, 308, 319, 321, 330, 332, 336, 339, 342, 344
shape-determining, 218
shift, 23
threshold, 149, 339
truncation, 53
partial correlation, 177, 183–185, 247–249
partial integral, 336, 346
Pascal, 36–38
pathological data, 65, 151
pdf, see curves, density, 12, 146, 155, 323, 325, 335, 336, 339, 343
pdf, see curves, density and,, 12
peakedness, 42
Pearson family, 208, 209, 340, 341
Pearson, E.S., 295
Pearson, K., 55, 57–62, 64, 71, 72,

115, 117, 182, 183, 205, 207, 210, 212, 214, 250, 329
Pearsonian sample correlation, 329
percentile, 222
perihelion, 258
perinatal mortality, 152
permutations, 83
personalistic approach, 110
pilot studies, 288
Plato, 142, 151
Poisson density, 55, 56, 61
Poisson density, truncated, 56
Poisson index of dispersion, 58
Poisson variates, 56, 139, 218
polynomial, *see* curves, polynomial
polynomial model, 340, 341
polynomial, degree of, 351
pooled estimate of variance, 238, 242
population
    characteristics 12, 15–18, 21, 35, 37, 38, 153, 155, 313, 312, 315, 319, 320, 322, 324, 331, 333, 334, 342
    enumerable, 313
    mean, 313, 318
    pyramid, 153, 315
    standard deviation, 34–36, 44, 45, 319, 333, 334
possible outcomes, 141
power, 5, 7, 38, 45–47, 77, 104, 192, 206, 207, 209, 218, 222, 270, 282, 286, 293–295, 297–299, 302, 303, 305–308, 332, 345
power, definition of, 293
power, frequentist, 293
prediction bands, 217, 228, 229, 279
prime symbol, 330
probability, 10–13, 22, 36–38, 312, 344
probability integral, 47
probit curve, 135
progrssion, AIDS, 223
properties, 12, 15, 17–21, 23, 25, 26, 28, 31, 34–36, 43, 45, 142, 150, 154–157, 334, 336, 345

proportion, 36, 142, 150, 344
proportion variate, 36, 66, 68
prospective studies, 104
pull-down menus, 190
pyramid, population, 153, 315

Quandt, R.E., 85
quantiles, 217, 222, 223, 226, 227, 232
quartile, 54, 222, 224, 225

Ramsey, J.B., 85
random numbers, 90
randomly collected data, 34
randomness, 347
range, interquartile, 320
range, normal, 227
range, of measurement, 313
ranges, 26, 37, 54, 225–227, 313, 320
ranges, unweighted, 227
rank correlation, *see* correlation, rank
rare events, 55
rates, 68, 159, 160, 162–164, 166, 167, 169–172, 175, 310, 312
ratio, 42, 44, 46, 52, 66, 68–71, 118, 134, 161, 172, 173, 175, 177, 242, 252, 253, 302, 341
ratios, 341
Ray, R.M., 121, 229
reciprocal, 86, 331, 333, 337
reciprocal operation, 4
rectangular density, 215, 300, 301
regression curves, nonparametric, 155
regression model, 142, 143, 145, 152
regression
    error-based, 143, 145
    logistic, 134, 135
    median-based, 229
    nonparametric, 155
    nonparametric, 157
    slope A or $\alpha$, 126
    slope B or $\beta$, 126
    switching, 84, 85
rejection, 282, 287, 294, 306, 307
relative frequency, 153
reliability, 313

replicates, 219
replicator, 4
representational complexity, 191
representational scope, 336
representative number, 12
residual, 69, 93, 190, 256, 319
response variate, 144, 145, 147, 156
retrospective studies, 104
risers, 22, 23, 67, 98, 99, 220
robustness, 9, 151, 256, 270, 286, 289, 295, 302, 303, 306, 308, 341
Rosenlicht, M., 106
roughness penalty, 191
roulette, 38, 312
roulette wheel, 38, 348
rounding, 98

sample, 14, 15, 21, 34, 35, 312, 313, 317, 319, 320, 344
sample cumulative distribution function (scdf), 220, 221, 277
sample space, 349, 350
sample standard deviation, 319
sampling distributions, 276, 279, 280, 283, 285, 289, 290, 296, 300, 302, 306–308
Sarhan, A.E., 267, 268, 278, 308
Sarndal, C.E., 36
Savage, L.J., 9
scale parameter, 15–17, 20, 25, 28, 36, 37, 69, 70, 91, 94, 102, 105, 108, 112, 115, 117, 118, 120, 121, 124, 125, 128, 136, 137, 139, 146–149, 203, 205–207, 212–215, 217, 218, 222, 235, 238, 239, 245, 251, 255, 261, 265–270, 277, 278, 280, 281, 283–285, 289, 291, 295, 302, 306, 308, 319, 321, 330, 332, 336, 339, 342, 344
Scheffe, H., 295
Scott, D.W., 123
Second World War, 191
series, Fourier, 58, 62, 192, 193, 208–2112
series, unweighted Fourier, 192

seventeenth century, 36
shape-determining parameter, 218
Sherlock Holmes, 259
shift alternative, 23
Shryrock, H.S., 154
Siegel, J.S., 154
sign function, *see* function sign
significance, 246, 259, 282, 289, 291
significance test, two-tailed, 288
Silvers, A., 99, 213
simulation, 35, 334
simultaneous assessment, 265
skewness, 42, 45, 50, 136, 205, 208, 219, 232, 233, 235
slash symbol, *see also* 17, 151
smoothness, 315
social science, 38, 151, 338
social sciences, 334
spacings, 232
Spanier, J., 7
special function, 47
specification phase, 192
spread, discussion of, 23
spurious association, *see also* lurking variables 162
squared deviation, 67, 206
standard curve, 85, 217, 262, 265, 268, 270, 332, 335
standard deviation, 20, 41, 44, 45, 49, 51, 54, 70, 94, 108, 148, 206, 207, 226, 227, 232–234, 251, 264, 278, 280, 281, 283, 286, 289, 300, 302, 303, 306, 333, 319
standard error of the sample mean, 62, 303
standard normal variate, 49, 68, 69
standarization, based on quartiles, 320
Star Trek, 4
starting position, 192
statistic, 220, 227, 228, 238–240, 242, 243, 245, 247–252, 256, 259–268, 270, 273, 277, 279, 280, 282, 284–286, 288, 290, 291, 293–298, 300, 302, 303, 305–308

# Index

Stein, C., 288
step function, 26
Stuart, A., 181, 318
Student-t density, 19, 68, 195, 252
studies
    observational, 35
    prospective, 104
    retrospective, 104
subgroup medians, 229
subjects, 247
subsamples, 219, 224, 232, 238, 240, 241
subscript, 12, 125, 199–201, 219, 222, 237, 240, 244, 260, 266, 277, 279, 280, 312, 314, 330, 331, 344, 351, 352
support region, 26, 149, 194, 209, 214, 218, 235, 314, 315, 319, 322, 323, 345
survival, 38, 63, 75, 80, 103, 155, 163, 197, 257, 258, 280, 302, 315, 322, 323, 338
survival curve, 80, 161, 280, 302, 322
survival time, 38, 75, 103, 257, 258, 315, 322
switching regression, *see* regression
symbol, 12, 17, 21, 145, 148, 323, 330, 331, 334, 335, 337
symbolic representation, 17, 148, 150, 151, 332, 345
symmetry, 33, 50, 54, 153, 313, 315, 318, 320, 326, 334, 335
SYSTAT, 19, 225

t, non-central, 68, 70
table availability, 283
table margin, 284
tables, 39
Tapia, R.A., 191
Tarter, M.E., 57, 63, 86, 99, 121, 149, 176, 195, 209, 213, 222, 229, 234, 301, 308, 309
Tchebychev's Inequality, 20
test construction process, 284
test statistic, 228, 284, 285, 288, 291, 294, 296–298, 305, 307, 308

test yardstick, 298, 305
the history of statistics, 38
third absolute value moment, 51
Thompson, J.R., 191
Thorndike, E.L., 329
threshold parameter, 149, 339
Tocher, K.D., 91
Todhunter, I., 38
transformations
    general, 193, 196, 213, 218, 219, 221, 226, 245, 247, 339, 341
    fractional power, 218
    log, 194, 195, 213
    log(-log), 194
translated normal, 334, 335
translated, see curves, translated, 334, 335
trigonometry, 46
trimodal, 195
trivariate data, 177
Trollope's novels, 10
trucation, asymmetric, 315
truncated density, 51, 53
truncated normal, 51, 54, 316–318, 321
truncated Poisson density, 56
truncation, 141, 142, 149, 313–320
truncation, asymmetric, 51, 315–318, 321
Tukey, J., 196, 224
Tukey, P.A., 195, 224
two-factor additive linear, 243
two-stage procedure, 288
two-tailed significance test, 288
type I error, 298
type II error, 298, 299

unbiased estimator, 261, 264, 265, 267, 318, 319
unbiased linear estimators, 261, 264
uncertainty
    about Cauchyness, 66
    about sampling error, 273, 278, 279
    and measurement error, 273

and variablity, 25
and variate with parameter linkage, 36
unconfounded, 166, 246, 248
uniformly distributed, 51, 91, 165, 221, 325
uniqueness, 150
unmixed Poisson, 55
unpenalized likelihood, 192
unweighted Fourier series, 192
unweighted sample ranges, 227
upper case letters, 345

Van der Waerden's test, 221
variables, lurking, 53, 72, 342
variate, 16–18, 22, 23, 35, 36, 38, 144–146, 148, 149, 151, 153, 155–157, 311, 312, 318–322, 334, 335, 337–339, 341–345
    attributed valued 75, 189, 196, 197, 199, 218, 230, 244, 249, 250
    composite, 35, 252, 273–275, 279, 280, 283, 307
    continuous, 199

discrete, 9, 109, 110, 132, 139, 167, 196, 197, 199, 201, 217, 218, 221
nuisance, 124, 127, 138, 139, 143–145
ordered, 196, 197, 199–201, 217, 218
proportion, 66, 68
response, 144, 145, 147, 156
uniformly distributed, 325
Viète, 330, 344
vital statistics, 153, 155

Wald, A., 9
wars, 18, 153
wavelets, 62, 208
Weibull, 338–340
weighted sum, 146, 147, 220, 261, 264, 277, 290, 322
well-defined quantities, 27, 331, 333, 339, 340
whisker endpoints, 224, 227, 228
whiskers, 223–225
Whitworth, W.A., 329